The Design of Mammals

A Scaling Approach

Despite an astonishing 100 million-fold range in adult body mass from bumblebee bat to blue whale, all mammals are formed of the same kinds of molecules, cells, tissues and organs and to the same overall body plan. A scaling approach investigates the principles of mammal design by examining the ways in which mammals of diverse size and taxonomy are quantitatively comparable.

This book presents an extensive re-analysis of scaling data collected over a quarter of a century, including many sources that are rarely or never cited. The result is an unparalleled contribution to understanding scaling in mammals, addressing a uniquely extensive range of mammal attributes and using substantially larger and more rigorously screened samples than in any prior works. An invaluable resource for all those interested in the "design" of mammals, this is an ideal text for postgraduates and researchers in a range of fields from comparative physiology to ecology.

John William Prothero served on the faculty of the Department of Biological Structure at the University of Washington from 1965 to 1999. During this time, he taught histology for fifteen years and subsequently functional neuroanatomy for nearly twenty years. He has a long-term interest in many aspects of scaling.

The Design of Mammals

A Scaling Approach

JOHN WILLIAM PROTHERO
University of Washington, Seattle

CAMBRIDGE
UNIVERSITY PRESS

University Printing House, Cambridge CB2 8BS, United Kingdom

One Liberty Plaza, 20th Floor, New York, NY 10006, USA

477 Williamstown Road, Port Melbourne, VIC 3207, Australia

314-321, 3rd Floor, Plot 3, Splendor Forum, Jasola District Centre, New Delhi - 110025, India

79 Anson Road, #06-04/06, Singapore 079906

Cambridge University Press is part of the University of Cambridge.

It furthers the University's mission by disseminating knowledge in the pursuit of
education, learning and research at the highest international levels of excellence.

www.cambridge.org
Information on this title: www.cambridge.org/9781108828864

First published 2015
First paperback edition 2020

A catalogue record for this publication is available from the British Library

Library of Congress Cataloging in Publication data
Prothero, John William.
The design of mammals : a scaling approach / John William Prothero,
University of Washington, Seattle.
      pages   cm
Includes bibliographical references and index.
ISBN 978-1-107-11047-2 (Hardback : alk. paper)   1. Mammals – Morphology.
2. Morphology (Animals)   3. Mammals – Anatomy.   I. Title.
QL739.P76 2015
599–dc23   2015013888

ISBN 978-1-107-11047-2 Hardback
ISBN 978-1-108-82886-4 Paperback

Additional resources for this publication at www.cambridge.org/9781108828864

For Joyce

I met a traveller from an antique land
Who said: "Two vast and trunkless legs of stone
Stand in the desert. Near them, on the sand,
Half sunk, a shattered visage lies, whose frown,
And wrinkled lip, and sneer of cold command,
Tell that its sculptor well those passions read
Which yet survive, stamped on these lifeless things,
The hand that mocked them and the heart that fed:
And on the pedestal these words appear:
'My name is Ozymandias, king of kings:
Look on my works, ye Mighty, and despair!'
Nothing beside remains. Round the decay
Of that colossal wreck, boundless and bare
The lone and level sands stretch far away."

Ozymandias
Shelley, 1818

Contents

Acknowledgments

Cambridge University Press: I am deeply grateful to the staff at Cambridge University Press, and especially to Ilaria Tassistro and Megan Waddington, for their unfailing and cheerful support during the editing and production of this book. I am also very grateful to Lindsay Nightingale, a freelance copy-editor who worked tirelessly on this book.

Computer assistance: This scaling study could not have been completed but for the fact that we live in an age of computer hardware and software. I thank Timothy Cahill and Grant Eckberg for setting up and maintaining (over many years) my computer system and its local area network. I also thank Paul Sampson (Seattle) and then-graduate student Benjamin Ely, for computational assistance (Chapter 23), and Steve Fedberg for writing a program for the calculation of slopes by what I term the single median method (Chapter 23). In addition, Tom Fisher and Mark McNair played a key role in setting up the computerized mammal taxonomy and body weight database used throughout this study (Chapter 4).

Editing: I cannot say how much I am indebted to my friend and colleague, Klaus Jürgens, for his tireless editing and assistance through several successive drafts of this composition. Klaus suggested literally hundreds of concrete changes, nearly all of which I incorporated. The work is much improved as a result of his sustained efforts. Those shortcomings that remain in the work are of course my responsibility.

Library searches: This study depended critically on a very extensive library search. Many librarians gave me a helping hand. I particularly wish to thank Nancy Blase (University of Washington) and Melynda Okulitch (Salt Spring Island) who located studies I sorely needed but would never have located but for their expertise and persistence.

Mammals in the wild: I am indebted to the National Science Foundation (US) for partially funding an excursion that allowed me to study first-hand (mostly large) mammals in their natural habitats in national parks in Kenya, India, and Nepal. This memorable experience provided a crucial "grounding" in the scaling of mammals that I, at least, could not have obtained in any other way.

Provision of "raw" data: Many workers responded warmly to my requests for unpublished raw data. These include: William Calder, John Eisenberg, Heiko Hörnicke, Colin Jones, Klaus Jürgens, Robert Kenney, Robert Martin, James Mead (Smithsonian), Fritz Meijler, Margaret Mullender, Daniel Promislow, and especially Tom Shepard (Seattle).

Scaling collaborators: I much enjoyed and substantially benefited from the opportunity of conducting varied scaling studies with my friends and colleagues Jonathan Gallant (Seattle), Klaus Jürgens (Hannover and Seattle), and John Sundsten (Seattle).

Scaling colleagues: I have been privileged to meet and discuss scaling with many fellow workers, including: Heinz Bartels (Hannover, Germany), George Bartholomew, William Calder, Terrence Dawson (Sydney, Australia), Robert Dedrick, Roderick Dewar (Sydney, Australia), Peter Hochachka (Vancouver, Canada), Klaus Jürgens (Hannover, Germany and Seattle), James Kenagy (Seattle), Geoffrey Maloiy (Nairobi, Kenya), John Maynard Smith (Brighton, England), Bastian Meeuse (Seattle), Fritz Meijler, Manfred Pietschmann (Hannover, Germany), George Sacher (Seattle), Victor Scheffer (Seattle), Knut Schmidt-Nielsen, and David Western (Nairobi, Kenya). Heinz Bartels and Terrence Dawson kindly provided me with a desk and access to extensive scaling data I would not have otherwise seen, during separate visits to their laboratories extending from 4 to 8 weeks. I am appreciative of their generous hospitality.

Those helping hands: On my journey through life I have been helped professionally by many individuals beyond those already mentioned, some of whom have long since passed on. I am especially appreciative of two former high school teachers, Reginald Bocking and Garnet Trevithick, who, perhaps unknowingly, kindled my interest in mathematics and physics, as well as to the following individuals: John Blackwell (applied mathematics, physics), Alan Burton (biophysics), Helen Scouloudi and Michael Rossmann (molecular structure), Arnold Tamarin (*in vivo* microscopy) and Edward Roosen-Runge (histology). They enriched my life in more ways than they knew.

Family and friends: Last, but certainly not least, I thank my (now deceased) always-supportive mother, Isobel, and my two younger brothers, George (teacher and nursery gardener) and Frank (teacher and historian). This rather lengthy study would not have come to fruition but for the direct and sometimes indirect encouragement and assistance of my remarkably understanding and caring wife, Joyce, and our three very exceptional children, Jeffrey, Jacqueline, and Jerrold. I am also pleased to call to mind my life-long friendship with Bruce Hill (Perth, Australia), a geologist. All of the above individuals contributed in diverse and sometimes subtle ways to the work you now hold in your hands.

Part I

Background

1 Introduction

*There are two... schools of biologists, one which emphasizes the average and general
and the other the individual and particular.*

Brody 1945

My interest in *scaling* in mammals was first sparked when, as a fledgling graduate
student adapting fitfully to a human-centered and largely qualitative medical school
environment, I read a paper by Günther and Guerra kindly passed on to me by my thesis
advisor, Alan Burton. This paper showed (as was already well known to others) that
there are quantitative regularities in the construction of mammals of varied sizes not
apparent from studies limited to human anatomy and physiology. Little did I then
realize that scaling would subsequently come to occupy such a large and beguiling part
of my adult vocation.

The term scaling is concerned with how the properties and proportions of a given
system vary (or fail to vary) with system size or with time. Discussions of so-called
spatial and *temporal* scaling are common to many disciplines. It should be said that
scaling is notable for its breadth of applications more than its depth. Scaling studies, in
the first instance, tend to be descriptive rather than explanatory. The term "system" is
open to different interpretations in fields as diverse as architecture, astronomy, climate
change, ecology, economics, engineering, landscape design, paleontology, physics, and
physiology. An important assumption in scaling studies, usually implicit rather than
explicit, is that two or more examples of a given system differing in size (or time
interval) are qualitatively similar with respect to the properties under study. That is to
say, variation in these properties is strictly quantitative. It is generally accepted that
regular scaling with size in any system always breaks down at sufficiently small or large
sizes because of *scale effects* (see Chapter 3).

Scaling in biology has been defined as the study of body *size* and its consequences. It
hardly needs saying that body size affects virtually every aspect of mammalian life.
Adult mammals vary in size by a factor of 100 million (from bumblebee bat to blue
whale; *Craseonycteris thonglongyai* to *Balaenoptera musculus*). Apart from fish, no
other vertebrate group, living or extinct, spans a comparable size range (Chapter 2).
Notwithstanding this remarkable 100-million-fold size range, all mammals broadly
share the same cell types, the same tissues and organs, and the same overall body plan.
Nevertheless, no one seriously believes that scale-up by a factor of a hundred million in
a complex system can be achieved without a diverse suite of internal and external

changes, both structural and functional. These adjustments to increased body size are brought into being by natural selection.

Hence, scaling in mammals may be defined as the quantitative study of those adjustments (or self-adaptations) as well as non-adjustments made in response to increased (or decreased) body size, mediated by natural selection (rather than size per se). Scaling studies stand to reveal the many respects in which adult mammals generally are similar as well as (in perhaps fewer but still important respects) dissimilar.

It follows that a major aim of mammalian scaling studies is to determine, and ultimately to understand, the nature of these – frequently very regular – internal and external changes made as a function of changing body size in mammals generally. These studies may be developmental (ontogenetic) in a single species, adult interspecific (across multiple species) or, as is often the case in this study, both. We will see that these internal changes involve, for example, at increasing body sizes, a reallocation of some biological resources away from energetically expensive organs and tissues (e.g., brain, kidney, liver) to less expensive ones (e.g., bone and fat). Thus, scaling in mammals is concerned in part with what may be termed "bio-economics."

The long-term goal of scaling studies in mammals is to describe the adjustments and non-adjustments to changes in body size, to identify the underlying mechanisms of adjustment, to estimate when, in the course of evolution, the adjustments were made and to determine why, insofar as that may prove possible, certain adjustments were made but not others. Such explorations stand to appreciably deepen our understanding of mammalian "design."

Mammalian scaling studies usually involve two variables, here denoted by x and y. In the first instance, we aim to answer empirically the question of how y scales (varies) with x. Commonly, the x-variable is a measure of body or organ size. The y-variable may be either structural or functional. Examples of *structural* y-variables include the number of components, body length, body surface area, and organ volume. These y-variables have dimensions of zero (number), one (length), two (area), and three (volume), respectively. Instances of *functional* y-variables include blood pressure, body temperature, and heart rate. Generally speaking, functional y-variables are dimensionally more disparate than structural variables.

In practice we plot the logarithm of a y-variable against the logarithm of an x-variable. The answer to the initial question of how y varies with x is given by the *slope* (b) of the straight line expressing y as a function of x. In the expression for a power function, given by $y = ax^b$, the constant a is numerically equal to the antilog of the intercept (i.e., where $x = 1$) and the exponent b to the slope of the straight line, assuming that x and y are given in log–log coordinates. The slope may be viewed (see Chapter 23) as an average – or perhaps better as a median – and as a generalization (or simplification). Scaling studies inherently involve quantitative generalizations (see above quote from Brody).

This work on scaling (see Contents) is divided into five Parts. Proceeding in reverse order, Part V (Chapter 25) takes a broader view of scaling, by examining the potential application of scaling concepts to currently pressing global problems – climate change, soil erosion, and water shortages. Methodological aspects of scaling studies in mammals are discussed in Part IV (Chapters 23, 24). Methodology is important, because numbers are no better than the methods used to acquire them.

A concise summary of many – but not all – of the results derived in Part II is provided in Part III (Chapters 19 to 22). The possible influences of dimensions and invariance on empirical slopes are discussed. Part II itself (Chapters 6 to 18), the core of this work, brings together the results of an extensive empirical analysis of scaling in mammals, involving approximately 100 different y-variables. The organization of the work is intended to facilitate finding those results in which the reader may be most interested.

Part II is organized mainly on a *systemic* basis. Recall that for more than a century mammals have been analyzed mainly in terms of subsystems. This is reflected, for example, in the specialities of cardiovascular, respiratory, and urinary anatomy and physiology. Whole body functions such as body temperature and energy metabolism are discussed in Chapter 17. Part II closes (Chapter 18) with a discussion of lethal limits. The way in which lethal limits scale may be useful for understanding how exacting are homeostatic constraints when placed under external stress.

Part I (Chapters 1 to 5) presents background material intended to facilitate an understanding of the (partly novel) presentations given in Parts II and III. Keep in mind that the present study does not attempt to make either ecological or phylogenetic inferences. On the contrary, the aim here is to describe, and eventually understand, structural and functional scaling in mammals *generally*, independent, so far as possible, of ecological and phylogenetic considerations. For example, water molecules, which make up about two-thirds of an adult mammal, are all the same in size and shape. Increase in adult body size across species entails an increase in the *number* of water molecules, all identical to one another. This well-ordered increase, like numerous others, appears to be essentially independent, on average, of either ecology or phylogeny.

Adult mammals vary both *taxonomically* and in *body size*. It is helpful to have a quantitative measure of diversity that tells us how nearly a given dataset comes to representing mammals generally or whether one dataset is more representative of mammals than another. We first consider *taxonomic* variation. Let us denote the number of species in a given sample by N_{sp} and the number of orders by N_{ord}. We take taxonomic variation into account by comparing the number of species and orders present in a dataset with the number of species and orders of mammals generally (see Chapter 2).

A measure of *size variation* is provided by pWR, calculated as the logarithm of the ratio of the largest body mass in a sample to the smallest. We know (see above) that pWR for mammals generally is about eight (pWR = $\log(10^8)$ = 8). In Part I, Chapter 2, the values of N_{sp}, N_{ord}, and pWR are combined to provide a single *diversity index* (DI); this index is used throughout the work. These four parameters (N_{ord}, N_{sp}, pWR, DI) are collectively referred to as evaluative criteria.

It will be found that *structural dimensions* (for number, length, area, and volume or mass) are of considerable relevance to scaling in adult mammals. Much of Part II is concerned with how these dimensions scale. *Invariance* is also of particular importance in scaling studies. Invariance simply means that a given property of a system remains constant (invariant) under a specific transformation. In a broad sense invariance speaks to the issue of "constant design." We may define two types of biological invariance: *absolute* and *relative* (Chapter 3). Absolute invariance means that component attributes such as number, diameter, area, and volume or mass show little if any dependence on adult body size. In Part II we encounter absolute invariance for the mean size and shape of most types of small and large molecules (e.g., glucose, hemoglobin, water) and for mean size and shape of some cell types (e.g., red blood cells). Mean body temperature and mean blood pressure and other homeostatic variables may exhibit absolute or near-absolute invariance. It will be argued that absolute invariance, both structural and functional, is relatively common in mammalian scaling.

It is also useful in considerations of mammalian *structural* scaling to define *relative invariance* (Chapter 3). This term means that a structural parameter (e.g., mass, volume or component number) increases in direct proportion to body size. For example, blood volume, the mass of the lungs, and the mass of skeletal muscle in adult mammals scale in reasonable accord, on average, with relative invariance. The same is likely to be true of the total number of water molecules or red blood cells in the adult mammal. Absolute and relative invariance are of considerable importance in mammalian scaling studies (see Chapter 3). Observe that some forms of absolute invariance are physically deter-mined, while other forms are *elective* (subject to natural selection). For example, the size of water molecules is physically determined, whereas the size of hemoglobin molecules is probably elective. In all probability, all forms of relative invariance are elective.

The scaling parameter of most interest is the *slope* (*b*) of the best-fit line expressing *y* as a function of *x*, both in log–log coordinates. If we are told that a given best-fit line has a slope of 2/3, we want to know the implications of that number for the scaling of the *y*-variable. Strictly speaking, it is the meaning of numbers – not the numbers themselves – that is of scientific importance. Perhaps the simplest approach to understanding the implication of a slope is to compute the *factor* by which the *y*-variable changes for a given *factor* of change in the *x*-variable (see Chapter 3). If, for example, the *x*-variable doubles, what is the factor by which the *y*-variable changes?

Commonly we have two or more independent datasets specifying how a given *y*-variable scales with the *x*-variable. If, as is usually the case, two specific datasets give rise to best-fit lines with somewhat different slopes, we want to know the scaling implications of this *difference* in slope. We are interested in the scaling implications of a difference in slope for scaling in mammals generally. That is, we elect to assess the impact of a difference in slope on the *hypothesis* that the best-fit lines apply over the *whole* mammalian size range (pWR = 8) (see above). We provide three different assessments of the discrepancy between the end-points of two lines with different slopes assumed to have a common intercept (Chapter 3). The first assessment employs *percent difference*; the second is expressed in terms of *number of doublings*; the third is based on the *factor* by which one end-point is larger than the other. These three assessments are inter-related.

Authors who provide scaling data (i.e., measures of some *y*-variable with correspond-ing measures of body size) sometimes state that the body size measurements were made on *adult* mammals. Often this consideration is ignored. Further, when animals are said to be adult it is uncommon to find a statement of the *criterion* used to make this judgement. A reasonable assumption is that adulthood is regularly equated with sexual maturity. We know that for numerous species, including our own, sexual maturity is reached before physical maturity. The danger for scaling studies is that a systematic bias in body mass may be introduced if small species are truly adults and large species are not, or vice versa.

In order to reduce the magnitude of this problem, a computer-based body weight (or mass) table was constructed (largely completed by 1995). Frequently, several estimates of body weight (or mass) from different works were available for a given species. In those cases, representative body weights were selected that were closest to the mean of the several estimates. The database constructed in this way is here referred to, for convenience, as the *standardized body weight table*, denoted by *SBT* (see Chapter 4 and Appendix B). The standardized body weight table contains body weight (or mass) data for some 1,700 species of mammals.

The great majority of datasets analyzed in Part II were *screened* using the standardized body weight table and the criteria given in Appendix B. This means that when a dataset from a particular author was screened the resultant dataset was usually somewhat smaller than the one originally reported by the author. On the other hand, my aggregated datasets, as a rule drawing on screened data from multiple authors, are usually larger than those previously reported as of the time of my analysis.

When one derives a *best-fit line*, usually by least squares (LSQ), for a given scaling dataset, we need *evaluative* criteria by which to judge how well a best-fit line describes the underlying dataset. I have used three different criteria: *mean percent deviation* (MPD), the *coefficient of determination* (r^2), and the *99% confidence limits* (CL) (in most cases) on the slope (see Chapter 5). The MPD provides a *non-dimensional* measure of the scatter around the best-fit line. The coefficient of determination tells how much of the variation in the y-variable is attributable to variation in the x-variable. The 99% CL provide a measure of the confidence we can place in the slope. Thus the great majority of best-fit lines reported in Part II are characterized by a total of seven *evaluative* parameters (N_{ord}, N_{sp}, MPD, pWR, DI, r^2, and the 99% CL on the slope). For a brief review of these evaluative criteria and other matters see Chapters 3–5.

The above seven evaluative criteria do not bear on the question of how *sensitive* a slope derived by LSQ is to sample size and to the range in body size. To examine this question I carried out the following experiments. First, I looked at 24 datasets (18 structural and 6 functional) based on 100 (or more) entries (species). For each of these 24 datasets (not including the separate case of body length – see Chapter 6) I constructed an *end-sample*, consisting of just those records for the ten smallest and the ten largest species by body mass (i.e., 20 records in all). Then a *mid*-sample was derived by deleting from the full dataset (FDS) the ten smallest and ten largest species. Thus, a mid-sample comprised 80 or more records. A LSQ analysis was carried out for each end-sample, mid-sample, and FDS. For the possibly unexpected results of these experiments see Chapters 6 and 21.

The analytical work underlying this study proceeded in two overlapping phases. The first phase, spread over several decades, was *bibliographic*. All the relevant references were entered into a computer-based bibliographic program. Beyond the many papers widely cited in the mammalian scaling literature, many other useful papers (especially in the biomedical literature) were found that have rarely been cited. Bibliographic research on a given topic (say the circulatory system) was curtailed when diminishing returns set in. The aggregate bibliographic database contains more than 10,000 references. Many, but not all, of these references are directly pertinent to scaling in mammals.

The second phase of the work consisted of *numerical analysis*. Raw scaling data on a given topic, normally from multiple different sources, were entered into a spreadsheet program. In nearly all cases the "raw" data were screened for body weight or mass and by other criteria (see Appendix B). After screening, a best-fit line was computed using LSQ. Once the above two phases had been completed for a given topic, further work was generally not undertaken. As a consequence, the coverage of some topics is more up-to-date than for others.

Note that each dataset consists of some number of records (or rows in a spreadsheet). For most datasets the number of records is the same as the number of species. But two datasets – here termed "instances" – for the same *y*-variable (e.g., body length) will usually differ in the number of records and in the spectrum of species. Nonetheless, two datasets for the same *y*-variable will commonly contain some records for the same species. Thus not all the records in two (or more) datasets will be unique. Part II is based on the analysis of roughly 16,000 records in all, for nearly 100 different *y*-variables. Part III is based on roughly 7,500 records (all unique) for 72 *y*-variables (52 structural and 20 functional), all in tabular form.

This is the first extensive scaling study devoted almost exclusively to mammals. It is the first to assemble a large number of databases and to *analyze* them uniformly, using LSQ analysis and seven different *evaluative criteria* (see above). Likewise, it is the first to take a systemic approach to scaling for eight different subsystems, running from the circulatory to the urinary (see Contents). It is the first study to compare and contrast physical and biological scaling in some detail. Also new is the extensive comparison between *ontogenetic* (chiefly human) scaling and *adult interspecific* scaling.

In addition, this work is apparently the first to compare adult humans with other adult mammals across 28 structural variables (see Chapter 22). For an extended discussion of body composition (elements to organs) see Chapter 8. Lethal limits are reviewed at greater length than previously in Chapter 18. Finally, there is a wider discussion of scaling methodology (see Chapters 3, 23, 24). In Chapter 23 it is shown that methods of analysis other than least squares may give similar values for the slope.

Whether the reader coming to this book is experienced in scaling studies, or a relative novice, it may prove useful to read through the chapters of Part I first before proceeding to other parts of the work. The primary purpose of Part I is to provide essential information needed to quickly and easily absorb the tabular and figurative material presented in the later chapters. (For the reader's convenience, this information is itemized in Chapter 5.) The subsequent chapters (6 through 25) are largely independent of one another. They may be read in any order. The price of this independence is a certain amount of repetition, designed to reduce (but not eliminate) the need for cross-checking across chapters. Chapters 6 through 24 each provide a summary statement at the end of the chapter.

With three exceptions (Chapters 6, 7, 18), each chapter in Part II also provides a summary figure showing the distribution of slopes encountered in the given system expressed in terms of rank order. Those distributions that are especially distinctive are compared in Chapter 20. In the long run, the pattern of slopes for each system is likely to prove as interesting and important as the narrower study of individual slopes. A partial interpretation of these distributions of slopes is given throughout the text in terms of absolute and relative invariance as well as in terms of those dimensions associated with number, diameter, area, and volume (e.g., see Chapter 12, Figure 12.3). A significant challenge for future work related to scaling in mammals will be to discover explanations for these rather varied distributions of slope.

An engaged reader who embarks on this excursion through the varied terrain of scaling in mammals should emerge with a detailed picture of a selection of the manifold adjustments mammals make to increased body size. Just as important as structural and functional *variance* are the many ways in which mammals exhibit *invariance*, on average, despite enormous changes in body size. Like newly-weds, variance and invariance go hand in hand.

2 The mammals

Mammals are the most diversified of all creatures living on earth today.

Boitani and Bartoli 1982

The time scale

Our universe is 13.8 billion years old; planet earth coalesced 4.5 billion years ago. Fossilized stromatolites, the products of colonial cyanobacteria, are dated to 3.4 billion years ago. (Contemporary stromatolites may be seen in Australia and elsewhere.) They are the earliest widely accepted evidence of life on earth. Eukaryotes (nucleated cells) first appeared 1.5 billion years ago. Vertebrate fossils recently found in China date to 450 million years ago (mya). The earliest mammal-like reptiles, the cynodonts (Order Therapsida, Class Synapsida), surfaced in the fossil record of the Late Permian and early Triassic periods (245 mya), slightly after the dinosaurs (in geological terms). Until near the end of the Cretaceous period (65 mya) mammals were shrew- to rat-sized creatures, slinking about in the dark to evade the dominating dinosaurs. Within ten million years – a wink in the timespan of the universe – following the dinosaur extinction a significant radiation of the mammals took place, giving rise to all the contemporary taxonomic orders, with some individual species of large size [1].

The primary purpose of this chapter is to introduce some qualitative but mainly quantitative measures of mammalian diversity, both size- and taxonomy-related. These measures involve a somewhat different way of looking at mammals than has been customary in scaling studies.

Mammals as a divergent vertebrate

Structural features believed to be unique to mammals are recapped in Table 2.1; of the ten structures listed, seven are considered to be universal. In principle any one of these structures might be regarded as *the* defining characteristic of mammals (see above quote from Boitani and Bartoli). But mammary glands are the most important of these structures from an evolutionary standpoint; lactation has profound implications for mammalian social behavior [2].

Table 2.1 Structural features considered to be unique to mammals

Feature	Comments	Feature	Comments
Anucleated RBC [a]	Universal	Fur (hair)	Sparse in most cetaceans
Articulated lower jaw	Universal	Mammary glands	Universal
Auditory ossicles (3)	Universal	Seven cervical vertebra	Universal except manatee and sloths
Diaphragmatic muscle	Universal	Sweat glands	Most mammals
Differentiated teeth	Universal	Urinary bladder	Universal

[a] RBC = red blood cells.

Table 2.2 Body size ranges in adults of the major vertebrate groups

Group	Minimal body mass (g)	Maximal body mass (g)	pWR	Comments
Column	a	b	c	d
Vertebrates	0.001	150,000,000	11.0	Stout infantfish to blue whale
Dinosaurs	1,000	80,000,000	4.9	Saltopus to brachiosaurus
Fish	0.001	21,500,000	10.0	Stout infantfish to whale shark
Reptiles	0.117	1,097,000	7.0	Least gecko to saltwater crocodile
Birds	2	127,100	4.8	Hummingbird to ostrich
Land mammals	1.5	10,886,000	6.9	Shrew to elephant
Mammals	1.5	150,000,000	8.0	Bumblebee bat to blue whale

Body size range in adult vertebrates

Publication of size range should be mandatory.

Calder 1987

Body size influences virtually every aspect of animal life. It is fundamental to mammalian scaling studies (see above quote from Calder). Often in scaling studies a measure is needed of the size range for some sample of mammals. Of special significance is the size range for mammals generally. In order to compare size ranges quantitatively the parameter pWR may be used. It is defined as:

$$pWR = \log(BW_{max}/BW_{min}) \tag{2.1}$$

Thus pWR is the logarithm of the ratio of the largest body weight (BW_{max}) or mass to the smallest (BW_{min}) found in a sample of (normally adult) mammals [3]. Being a ratio of like quantities, pWR is independent of units. The size range of mammals in relation to other vertebrates is shown in Table 2.2. An adult blue whale is about twice as large as the largest known dinosaur fossil (brachiosaurus). The smallest vertebrate is the stout infantfish; adults weigh 1 mg. Among these vertebrate groups, excepting only fish

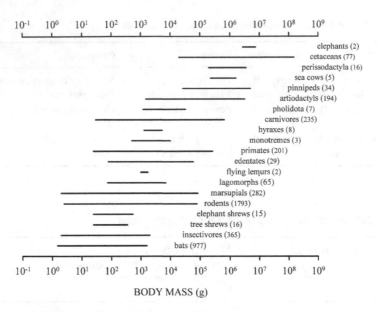

Figure 2.1 Size range of adult mammals by taxonomic order.

(pWR = 10.0), the mammals have the greatest body size range (pWR = 8.0). Indeed, the most remarkable feature of mammals, from a scaling standpoint, is their immense size range (100 million-fold in mass).

Body size range in adult mammals by taxonomic order

Scaling studies aim to establish how each of many variables, such as heart size or heart rate, varies with, or is independent of body size in mammals generally. At present most empirical studies are made with small samples, representing limited size ranges and restricted taxonomic variation. Extrapolations from small samples to all mammals may be questionable (but see Chapters 6, 21). Quantitative criteria are needed by which to judge how nearly a sample represents mammals generally. The first question to be asked is how the range in body size varies from one taxonomic order to another. For an answer, see Figure 2.1. The data used to construct Figure 2.1 are summarized in Table 2.3.

The number of species per order in Figure 2.1 is given in parentheses [4]. Most of the body size values in Table 2.3 are taken from a standardized body weight (or mass) table (see Chapter 4). The eutherian orders are arranged in accord with increasing mean body size, on a logarithmic scale. Note the substantial variation in the size range spanned by individual orders. Three orders (elephants, flying lemurs, and hyraxes) have relatively narrow size ranges (pWR < 1.3). Four orders (carnivores, marsupials, primates, and rodents) span at least half the mammalian size range (pWR ≥ 4.0). The widest range, perhaps surprisingly, is for marsupials (pWR = 4.6); the next widest range is for carnivores (pWR = 4.3).

Table 2.3 Range in adult body sizes by taxonomic order

Order	Minimal body mass		Maximal body mass	
	Species	Mass (g)	Species	Mass (g)
Chiroptera	*Craseonycteris thonglongyai*	1.5	*Pteropus giganteus*	1,600
Insectivora	*Suncus etruscus*	2	*Echinosorex gymnurus*	2,000
Scandentia	*Ptilocercus lowi*	25	*Urogale everetti*	350
Macroscelidea	*Elephantulus rozeti*	25	*Rhynchocyon chrysopygus*	540
Rodentia	*Mus minutoides*	2.5	*Hydrochoerus hydrochaeris*	79,000
Marsupialia	*Ningaui timealeyi*	2.0	*Macropus rufus*	85,000
Lagomorpha	*Ochotona pusilla*	75	*Lepus arcticus*	6,800
Dermoptera	*Cynocephalus volans*	1,000	*Cynocephalus variegatus*	1,750
Edentata	*Chlamyphorus truncatus*	80	*Priodontes maximus*	60,000
Primates	*Microcebus myoxinus*	25	*Gorilla gorilla*	275,000
Monotremata	*Ornithorhynchus anatinus*	500	*Zaglossus bruijini*	10,000
Hyracoidea	*Heterohyrax brucei*	1,300	*Heterohyrax capensis*	5,400
Carnivora	*Mustela nivalis*	30	*Thalarctos maritimus*	650,000
Pholidota	*Manis tetradactyla*	1,200	*Manis gigantea*	33,000
Artiodactyla	*Neotragus pygmaeus*	1,500	*Hippopotamus amphibius*	3,200,000
Pinnipedia	*Arctocephalus galapagoensis*	27,000	*Mirounga leonina*	5,000,000
Sirenia	*Dugon dugon*	230,000	*Trichechus manatus*	1,600,000
Perissodactyla	*Equus ferus*	200,000	*Ceratotherium simum*	3,600,000
Cetacea	*Pontoporia blainvillei*	20,000	*Balaenoptera musculus*	150,000,000
Proboscidea	*Elephas maximus*	2,730,000	*Loxodonta africana*	7,500,000

The mean value of pWR for the 20 orders represented in Figure 2.1 is 2.2 (the order Tubulidentata, with one species, is omitted). On average, for a given order, the largest mammal is 160 times larger than the smallest ($10^{2.2} \approx 160$). The six orders present in the bottom rows of Figure 2.1 comprise about 80% of all mammalian species [4]. This finding implies that proportionately fewer ecological niches are available at larger body sizes [5]. We will find in Chapter 7 that within a given species the largest animals are typically about twice as large as the smallest.

The two smallest adult species of mammals are the white-toothed pygmy shrew (*Suncus etruscus*) and the bumblebee bat (*Craseonycteris thonglongyai*). Adults of each species may weigh as little as 1.5–2 g. At the other extreme is the blue whale (*Balaenoptera musculus*). An adult female blue whale, 23.7 m in length, weighed piecemeal, registered 136.4 tonnes [6].

One hundred and fifty tonnes is considered to be a representative value of body weight for blue whales (1 tonne = one metric ton). Larger blue whales have been reported: a figure of 170.5 tonnes was estimated for a blue whale 29.7 m long [7]. My colleague, the late Victor Scheffer, put the upper limit in body mass for a blue whale at a value (here corrected) of 195 tonnes [8]. For comparison, note that the world's tallest tree and the world's largest "living species" by volume is the Pacific Coast redwood (*Sequoia sempervirens*), at about 116 m high; but only about 1% of a tree consists of living matter. It turns out that a blue whale has about 12 times more volume of living

Table 2.4 Verbal definition of body size categories for adult mammals

Body size category	Mass range (g)	Examples
Small	1–100	Many bats, mouse, dusky antechinus
Medium	100–10,000	Dusky antechinus, weasel, coypu
Large	10,000–1,000,000	Human, lechwe, giraffe
Very large	1,000,000–100,000,000	Giraffe, elephant, blue whale

matter than a mature redwood tree. For convenience I have used a value of 8 for pWR in all mammals (i.e., $\text{pWR} = \log(150 \times 10^6/1.5) = 8.0$) throughout this work.

To put this size range in perspective, recall that 1.5 g corresponds to one-tenth of a (US) tablespoon of water. On the other hand, 150 tonnes is about the mass of an empty Boeing 777. A major challenge to students of mammalian scaling is to understand the adjustments making this enormous size range possible.

A verbal classification of mammalian size ranges

In addition to the above quantitative measure of the range (pWR) in body sizes seen in a given sample, it is useful to have a verbal classification; this is provided in Table 2.4.

Each category in Table 2.4 represents a 100-fold range in body size. The great majority of mammals fall into the small and medium categories. Most wild species for which adult data are available fall entirely within one of the four categories; a few species, such as the dusky antechinus (*Antechinus swainsoni*) and the giraffe (*Giraffa camelopardis*), straddle two categories.

Towards a measure of taxonomic diversity

Taxonomic considerations are important in scaling studies. A structural or functional scaling study that applies to only one order, such as artiodactyls or primates, may be an unreliable guide to scaling in mammals at large. Variations among species from different taxonomic orders are typically more striking than variations within an order. For example, humans (primates) and the giant anteater (edentata) show much greater differences across a range of parameters (body temperature, dentition, diet, lumbar vertebra) than do humans and chimpanzees. To know how far we have come, and especially how far we have yet to go (in addressing variation in mammals generally), we need a quantitative index of diversity.

To be useful in mammalian scaling studies, a diversity index should take into account the range in body size and taxonomy. That is, variation in both body size and taxonomy may confound the results of scaling studies based on small taxonomic diversity or restricted size ranges, if the aim – as here – is ultimately to characterize scaling in mammals generally rather than in particular groups (but see Chapter 21).

Throughout this work I have relied on Corbet and Hill [4] for taxonomy; they identify 21 different orders of mammals, comprising 4,327 species. Their book is structured in a manner that is exceptionally convenient for scaling studies. More recent works [9] on mammalian taxonomy give a higher number of species (see below).

A diversity index (DI)

The size range in any given sample may be normalized to the size range of mammals generally by dividing pWR by 8. The ratio (pWR/8.0) tells us how nearly a sample represents the size range for all mammals. If, for example, a given sample of bats spans the size range for all bats, then pWR is about 3 and pWR/8.0 is 0.375. Thus bats represent 3/8 of the size range of all mammals, on a logarithmic scale. To assess taxonomic variation, it is helpful to denote the number of orders and species present in a sample by N_{ord} and N_{sp}, respectively. Then the ratio ($N_{ord}/21$) (see above) tells us how nearly a sample represents all 21 orders of mammals [4]. It remains to take account of the number of species (N_{sp}) in a sample. There are roughly 1,000 species of bats, almost a quarter of all mammalian species [4]. Given scaling data for some variable (say heart rate) in all bats, it would be risky to infer that these data are reflective of the scaling of heart rate in mammals generally. The total number of species, by itself, can be a misleading estimate of diversity.

What is needed is a measure that assigns less influence to species number. This may be achieved by employing $\log(N_{sp})$ rather than N_{sp}. To normalize this measure to the total number of mammalian species, we form the ratio ($\log(N_{sp})/\log(4{,}327)$), where 4,327 is the total number of mammalian species [4]. A *diversity index* (DI) in percent may be defined by the relation:

$$\text{DI}\,(\%) = 100 \cdot (N_{ord}/21) \cdot (\log(N_{sp})/\log(4,327)) \cdot (\text{pWR}/8.0) \qquad (2.2)$$

If we used a more recent figure [9] of 5,416 for the total number of mammalian species the diversity index would be 3% smaller than given by Equation (2.2).

How the diversity index (DI) varies with three parameters

The way the diversity index changes as we progressively increase pWR, N_{ord}, and N_{sp} is shown in Table 2.5.

The most efficient way to increase the diversity index is by increasing pWR, N_{ord}, and N_{sp}, in that order. Individual laboratories commonly have experience with one or a few species of mammals. Achieving a high DI usually requires combining data from different laboratories. For the largest DI in the present study (76%) see Chapter 6. A diversity index of more than 50% could be achieved with as few as 73 different species if carefully chosen to span the whole mammalian body size range (pWR = 8.0) and 21 taxonomic orders. See Table 2.5, last column, bottom row. Nonetheless, achieving a diversity index of 50% or more represents a very considerable challenge.

Table 2.5 Diversity index as a function of pWR, N_{ord}, and N_{sp} for adult mammals

pWR	N_{ord}	N_{sp}	DI (%)	pWR	N_{ord}	N_{sp}	DI (%)
1	1	3	0.1	4	4	16	3.2
1	2	3	0.2	5	5	36	6.4
2	2	4	0.4	6	8	43	12.8
3	2	6	0.8	7	13	52	25.6
3	3	12	1.6	8	21	73	51.2

Summary

Mammals first appeared in the geological record about 245 million years ago. Ten specific structural traits, of which seven are considered to be universal, serve to differentiate mammals from other vertebrates. Of these ten unique structures, the mammary glands are the most important from an evolutionary standpoint. Mammals span eight orders of magnitude in body size; excepting fish, this is a greater range than for any other vertebrate group, including the extinct dinosaurs. The logarithm of the ratio of maximum to minimum body weight or mass in a sample, denoted by pWR, is a useful measure of size variation for mammalian scaling studies [3]. On average, each mammalian taxonomic order spans 2.2 orders of magnitude in body size from the smallest to the largest species (a factor of 160; $10^{2.2} \approx 160$). In the present sample marsupials span the greatest size range (pWR = 4.6).

Taxonomic diversity is defined in terms of the number of orders (N_{ord}) and the number of species (N_{sp}) in a sample. An overall index of diversity (DI) in percent is then defined in terms of pWR, the number of taxonomic orders represented (N_{ord}), and the *logarithm* of the number of species (N_{sp}). This index provides a compact single measure of how closely any given sample of mammals comes to representing mammals generally. These several indices (pWR, N_{ord}, N_{sp}, DI), along with others (see Chapter 5), are used throughout this work. The above indices may easily be amended so as to apply to scaling in particular groups of mammals (or to other classes) as desired.

References

1. Prothero, D.R. (2006). *After the Dinosaurs: The Age of Mammals*. Bloomington, IN: Indiana University Press.
2. Pond, C.M. (1977). The significance of lactation in the evolution of mammals. *Evolution*, **31**:177–199.
3. Prothero, J. (1986). Methodological aspects of scaling in biology. *Journal of Theoretical Biology*, **118**:259–286.
4. Corbet, G.B. and Hill, J.E. (1991). *A World List of Mammalian Species*, 3rd edn. Oxford: Oxford University Press.
5. Colvinaux, P. (1978). *Why Big Fierce Animals are Rare: An Ecologist's Perspective*. Princeton, NJ: Princeton University Press.

6. Winston, W.C. (1950). The largest whale ever weighed. *Natural History*, **59**:392–398.

7. Laurie, A.H. (1933). Some aspects of respiration in blue and fin whales. *Discovery Reports*, **7**:365–406.

8. Scheffer, V.B. (1974). The largest whale. *Defenders of Wildlife International*, **49**:272–274. Note: Scheffer based his estimate on an empirical relationship between body length and body weight. However, the figure of 215 tonnes, derived from this empirical relation and an estimated body length of 100 feet, is mistaken: the correct figure would be 195 tonnes.

9. Wilson, D.E. and Reeder, D.M. (eds.) (2005). *Mammal Species of the World: A Taxonomic and Geographic Reference*. Baltimore, MD: Johns Hopkins University Press.

3 The nature of scaling

It is the scale that makes the phenomenon.

Eliade 1958

This chapter provides a brief introduction to scaling generally, to scaling in biology, and more especially to scaling in mammals. Scaling is concerned with how the properties of a system change or resist change with variation in system size or over time (see above quote by Eliade). In its contemporary aspect, scaling studies arose in mathematics and physics. A classic example of scaling generally is the way in which human institutions tend to become more bureaucratic (and sometimes autocratic) with increased size. In harmony with this conception, mammalian scaling studies focus chiefly on how and possibly why selected properties change, or do not change, given variation in body size. Scaling studies are inherently quantitative: numbers are of the essence. Much of the present work is concerned not just with numbers, but also with what the various numbers derived from scaling studies entail in a biological context. We will see that scale affects many aspects of mammalian "design."

We begin with time scales and then turn briefly to physical scaling and a critique thereof. There follows a discussion of certain aspects of scaling, including scale effects, computation of best-fit lines, and the numerical properties of power functions. Some implications of differences in slope are discussed. (When a power function of the form $y = ax^b$ is plotted on log–log paper, a straight line is obtained. The slope (b) of this straight line is numerically the same as the exponent b. For this reason the terms "exponent" and "slope" are often here used interchangeably.) Three measures for judging how well a best-fit line describes a given dataset are introduced. It is shown that nature is not at liberty to choose structural scaling exponents greater than 1 arbitrarily. For a given intercept and size range there is a strict upper limit to slopes greater than 1 among which nature may "choose." Recognition of this fact could be a modest step towards the long-term goal of understanding scaling in mammals.

Emphasis is put on two types of *invariance*: absolute and relative. These two forms of constancy are likely to prove to be of fundamental significance to scaling in mammals. Two methods are put forward to test for non-linearity (on a log–log plot) in scaling datasets. The usual criteria for scaling at constant shape are stated and partially amended. Possible graphical relations between adult interspecific scaling and ontogeny (development) are considered. A prior "repeating units" model is briefly reviewed.

Finally, a tentative classification of possible scaling reference slopes is given. Such reference slopes may provide a useful guide to the interpretation of certain empirical slopes.

The antiquity of scaling

The scaling perspective is much older than one might suppose. Cave paintings, as found in France and Spain, some dating back at least 30 millennia, show drawings sufficiently detailed and accurate to allow individual species (e.g., bison) to be identified. Artists from that distant time had in mind a simple form of geometric scaling where animals were drawn in correct proportions but at a smaller size. Some of this seductive art work I had the good fortune to see first-hand at Altamira, where the first such paleolithic cave paintings – then very controversial – were discovered in 1868. These paintings are said to be 18,000 years old. Likewise, the fact that the base lengths (of the four sides) of the great pyramid of Khufu in Egypt deviate from the mean length by less than two-hundredths of a percent might imply that those master builders working 4,500 years ago (with little more than long cords, experienced eyes, carpenter's squares, the stars above, and water below) based their construction on precise scale-models. No overall working plans for Khufu have yet been found.

Scaling in applied physics and engineering

...we can only get from our dimensional formulae what we put into them.

Ellis 1966

Dimensional analysis is a means of processing information, not providing it.

Massey 1971

Scaling has been and remains of considerable practical importance in applied physics and engineering. The modern period in physical scaling studies is exemplified by the work of Froude (1810–1879). He was one of the first to show clearly the practical importance of using mathematical physics to convert measurements made on a scale-model into precise predictions of the behavior of a full-size prototype [1]. His work, along with that of others, ultimately had a major bearing on the design of aircraft, harbours, ships, and many other large structures. The long-term outcome has been a great increase in confidence and a substantial reduction in the cost of building large structures [2].

Mathematical extrapolation from physical measurements made on a scale-model to the behavior of a full-scale prototype is valid only when one or more quantitative constraints – termed *similarity criteria* – are enforced. For example, in certain hydro-dynamic problems a *non-dimensional* quantity known as the Reynolds number (the ratio of inertial to viscous forces) must be the same in the model as in the prototype. That is, experiments are designed so that the Reynolds number is the same for the scale-model as for the prototype. Given this invariance, measurements of, say, drag made on a scale-model can be used to predict the drag for a prototype. The Reynolds number is one example (of hundreds) of non-dimensional quantitative similarity criteria in wide use in applied physics and engineering. The mathematical process of deriving non-dimensional quantitative similarity criteria from physical laws is known as *dimensional analysis* (DA) [2, 3].

A number of extended attempts have been made to apply DA directly to scaling problems in mammals [4, 5]. For a more recent discussion of this topic see Lin [6]. In the following, I give a few reasons for thinking this line of attack may fail for the kind of scaling problems considered here [3].

Critique of physical scaling methodology as applied to biology

To an innocent bystander, dimensional analysis (DA) can give the impression of being an *a priori* method. However, this is not the case. Judgement and experience enter, for example, into the initial selection of relevant variables for the analysis. Omitting a relevant variable is likely to give the wrong answer, and DA usually gives no warning that such a variable has been omitted. It is well known that a scale-model and a prototype can be quantitatively similar in only one or at most a few respects, owing to constraints imposed by physical realities (such as gravity). A point often overlooked is that DA presupposes that the geometry of the system under study is essentially simple [2, 7]. Dimensional analysis is an analytical process, but not a biologically explanatory one (see above quotes by Ellis and Massey). Explanation in biology usually traces back to natural selection. Faults in natural selection are not uncommon; in the human they include ectopic births and the inability to synthesize vitamin C. These and other defects defy strictly physical explanations. The products of natural selection need not be strictly beneficial.

Scaling studies in mammals differ from physical scaling in several key respects. First, no contemporary species of small mammal is a phylogenetic ancestor of larger species of mammals. All mammalian species have diverged from an extinct common ancestor. Hence, no existing species of small mammal is likely to serve as a plausible scale-model for all larger mammals. Unlike physical scaling studies, which are usually concerned with one scale-model and one prototype, scaling in mammals potentially concerns thousands of "prototypes" spanning a wide range of body sizes and exhibiting great diversity. Second, few physical similarity criteria (e.g., see Reynolds number above) are likely to have useful application to scaling in all adult mammals (see Box 3.1). Lastly, DA cannot tell us how the *number* of biological components, such as molecules (e.g., water), macromolecules (e.g., hemoglobin) or organs (e.g., kidneys), should scale with increasing adult body size. These and other shortcomings severely limit the useful applications of DA to scaling in mammals at large [3].

Finally, we recognize that physical scaling studies have motivations orthogonal to those of biological scaling. In applied physics and engineering the primary purpose of scaling studies is to produce workable artefacts that are cost-effective, efficient, and safe. Biologists aim to understand how and why organisms change, or defy change, with

Box 3.1 Physics and biology

Consider, as an example, the equilibrium of a teeter-totter or see-saw (see panel a of the figure here). The teeter-totter is balanced at the fulcrum (F). Two weights, W_1 and W_2, are shown at distances L_1 and L_2 from the fulcrum, respectively. The physical condition for the teeter-totter to be in equilibrium (horizontal) is that the respective torques ($L_1 W_1$ and $L_2 W_2$) be equal:

$$L_1 \cdot W_1 = L_2 \cdot W_2$$

a.

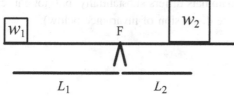

b.

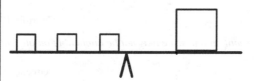

Box 3.1 Figure Equilibrium condition for a teeter-totter.

    If this were a biological problem, a biologist would of course be interested in the physical condition for equilibrium. But a biologist would also be quick to note that the equilibrium condition can be satisfied in an indefinitely large number of different ways. Thus we could have one large weight (say of a mother) at one end of a teeter-totter matched by many smaller weights (of a litter) distributed over the length of the other side (panel b.). Physics imposes a condition that must be satisfied for equilibrium; but physics does not dictate how that condition may be met.

    As another example, consider the maintenance of constant body temperature in endotherms. Physics states that on average, heat loss must match heat generated. But physics does not determine the level of body temperature. Mean body temperature has been set by the process of natural selection. Birds generally have higher mean body temperatures than do mammals. Endotherms can reduce heat loss by increasing insulation, or increase heat loss by sweating or panting. These tactics are consistent with, but not required by physics.

    Such examples support the inference that physics imposes important constraints on biological systems, but crucially physics does not determine how nature satisfies those constraints. For this reason, it is unlikely that the methods of scaling employed in

---

Box 3.1 *(cont.)*

applied physics and engineering can be used to predict many empirical scaling relations for mammals generally. These scaling relations represent nature's adjustments, made by trial-and-error, to increased body size. It is plausible to assume that in many instances, other adjustments would have worked about as well or better.

---

altered body size, in the ways they do. Much of biology is an exercise in "reverse engineering." We are given functioning systems, and our aim is understand how they are constructed and how they work. The methodology – largely empirical – one employs in scaling studies of mammals differs substantially, but not entirely, from that employed in physical studies (see discussion of invariance below).

Scale effects

Usually, but not always, events that take place at quite different scales have a negligible influence upon each other.

Courtois 1985

...concepts and conclusions relevant on one scale may not be applicable on all scales. This is a key principle in physics...

Greene 2004

In practice, physical systems scale uniformly only over some limited size range. Outside this range, regular scaling breaks down (see above quotes from Courtois and Greene). These breakdowns are termed *scale effects*. The importance of scale effects in building construction is well known. Vitruvius (*c.* 75–15 BC), a Roman engineer and architect, acknowledged the existence of scale effects [8] when he said "...that which may be effected on a small or moderately large scale cannot be executed beyond certain limits of size." The most widely quoted statement in biological scaling studies might be that of Galileo [9] (1564–1642): "From what has already been said, you can plainly see the impossibility of increasing the size of structures [at constant shape] to vast dimensions either in art or in nature...". Scientific consideration of scale effects and limits to scaling in mammals has so far been restricted largely to the skeletal system [10].

Consider modeling a ship in a water tank. If a scale-model of a ship is too small then its hydrodynamic behavior will be dominated by surface tension, a factor that is negligible in a large-scale prototype. This principle has wide applications. Our universe is divisible into largely autonomous domains. For example, chemistry is pursued with little or no regard for nuclear physics. Newtonian physics is valid for masses much smaller than the sun and for velocities much less than that of light. For larger masses and higher velocities, one must appeal to relativity theory. If everything in the universe were strongly coupled to everything else, as some New Age devotees seem to imply, the universe would likely be incomprehensible. Scale effects apparently underlie, to a significant degree, our ability to understand both our universe and ourselves (see also Chapter 25).

Scaling in biology

Living organisms, whether unicellular or multicellular, are apparently most similar at the lowest levels of resolution. That is, the types of molecules and macromolecules present are broadly similar across animal species and classes (e.g., see Chapter 8). Hence, the greatest scope for the application of scaling studies to biology will possibly prove to lie at these lower levels of resolution.

Scaling in mammals

The fact that adult mammals span a 100-million-fold size range is astonishing (Chapters 1, 2). No set of human artefacts scaling at essentially constant design rivals this size range. Understanding the scale-up of mammals is directly relevant to understanding the "design" of mammals. The purpose of the following discussion is to review briefly a dozen or so different aspects of mammalian scale-up from a quantitative standpoint. The issues reviewed below will prove to be germane at many places throughout the remainder of this work.

Most scaling studies in biology rely on data taken from the literature. In the majority of cases the published data were not collected for reasons having anything to do with scaling. Usually some taxonomic data are provided, often merely in the form of common names. Body weights (x) or masses are often – but not always – given; data for some y-variables (e.g., heart size or heart rate) are presented. Scaling studies proceed by collecting as much data for any given y-variable as is practicable. The next step is to construct a best-fit line describing how a y-variable varies with an x-variable (usually body weight or mass).

Computation of a best-fit line

In the empirical scaling studies discussed in this work (see Part II), a variable (y) is usually plotted, figuratively speaking, on log–log paper against some measure of body size (x). In most cases, the data are distributed linearly, to a first approximation, on this log–log plot, implying that y is related to x via a power function (see below and Chapters 23, 24). Using software, generally, a straight line is fitted to the data, normally by the method of least squares (LSQ). The computer output provides, at a minimum, the slope (b) of the best-fit line, the standard error (s.e.) of the slope (from which confidence limits are partly derived), and a coefficient of determination (r^2) (see Chapter 23). It is an empirical observation that structural and functional y-variables extending over several decades or more in body size have slopes falling between the extremes of –0.3 and +1.1 (see Chapters 19, 20). This seemingly narrow range of slopes is nonetheless consistent with substantial changes in the details of mammalian construction over wide size ranges (see Chapter 2).

Table 3.1 The factor increase in *y* for different factor increases in *x* and different exponents

	Exponents (*b*) of *x*							
	−1/3	0	1/4	1/3	2/3	3/4	1	10/9
Factor (*f*) increase in *x*	Factor increase in *y*							
1.5	0.87	1	1.11	1.14	1.31	1.36	1.5	1.57
2	0.79	1	1.19	1.26	1.59	1.68	2	2.16
5	0.58	1	1.50	1.71	2.92	3.34	5	5.98
10	0.46	1	1.78	2.15	4.64	5.62	10	12.92

See relation (3.2).

Power functions

The factor increase in *y* for a given factor increase in *x*

Perhaps the simplest way to obtain an intuition as to how mathematical power functions work is through numerical examples. A power function can be represented by the mathematical relation:

$$y \propto x^b \ (x > 0) \tag{3.1}$$

If *x* increases by a factor *f*, then *y* increases by the factor f^b:

$$y \propto f^b \tag{3.2}$$

Some results of evaluating relation (3.2) numerically are shown in Table 3.1. For example, if *x* doubles (*f* = 2) and the exponent (*b*) of *x* is 2/3, then *y* increases by the factor 1.59 ($2^{2/3}$ = 1.59). (See Table 3.1, row two for an exponent of 2/3.) Note that when the exponent *b* is negative (e.g., −1/3) the factor by which *y* changes is always less than 1. As the exponent *b* of *x* approaches 0, the factor of increase in the *y*-variable progressively approaches 1.

Scaling implications of pairwise differences in slope

...what appear to be small differences in exponents [may] *represent sizeable magnitudes when expressed arithmetically.*

Schmidt-Nielsen 1984

It is sometimes said in the scaling literature that a given empirical slope (say 0.97 or 1.03) is not significantly different from 1. In the absence of a relevant criterion, the statement is virtually meaningless. Many workers, myself included, are occasionally guilty of this practice. Still, one ought to bear in mind that astronomers call an asteroid approaching within a quarter of a million miles of the earth a "close" encounter; Andromeda, our nearest spiral galaxy, although 2.5 million light-years away, is nonetheless described by astronomers as "nearby." On the other hand, a micrometer is a

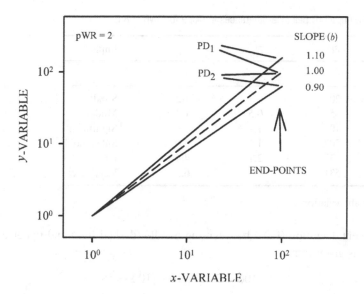

Figure 3.1 The impact of differences in slope at two end-points.

small distance on a human scale (about 10^{-6}:1) but a large distance on the atomic scale (about 10^4:1). Words such as "far" and "near" have, by themselves, no certain meaning.

Each branch of science scales the phenomena under study to a convenient size by adopting units appropriate to the circumstances at hand. Scaling in adult mammals involves distances varying from 10^{-10} m (diameter of a hydrogen atom) to about 30 m (length of an adult blue whale) (i.e., 11 orders of magnitude). In the following, three different measures are given on which a judgement of the biological significance or non-significance of a difference in scaling slopes may be based. These measures are denoted by *percent deviation* (PD), *number of doublings* (N_d), and *factor* (*f*) of increase.

Say we are given two slopes. What are the biological implications if one slope is 1.0 and the other is 0.9? Are the implications the same if one slope is 0.8 and the other is 0.7? Questions such as these are seldom raised in mammalian scaling studies. (But see above quote by Schmidt-Nielsen.) An informal derivation of the formulae presented below is given in Chapter 24.

Percent deviation between two end-points defined by a power function

A first approach to understanding the implications of a difference in slope (Δb) is illustrated in Figure 3.1 (in effect a log–log plot). Three lines are assumed to intersect at the point (1, 1). The dashed line is taken as the reference slope for the lower and upper lines. The right-hand end-point of the dashed line lies at (100, 100). The upper solid line with slope 1.1 ends at (100, 158.5) ($100^{1.1} = 158.5$). The lower solid line ends at (100, 63.1) ($100^{0.9} = 63.1$) (where $y = x^b$). Observe that in this example the three given straight lines span two orders of magnitude in the *x*-variable (i.e., from 1 to 100).

Table 3.2 The impact of pairwise differences in slope for pWR = 8.0

Δb	PD_y	N_d	f	Impact
a	b	c	d	e
0.01	20	0.3	1.2	Small
0.02	45	0.5	1.4	Modest
0.04	109	1.1	2.1	Significant
0.06	202	1.6	3.0	Substantial
0.08	337	2.1	4.4	Large
0.10	531	2.7	6.3	Very large

See text for abbreviations.

The percent deviation (PD_1) between the middle (dashed line) and upper end-point (solid line) is given by:

$$PD_1 = 100 \cdot (158.5 - 100)/100 = 58.5\% \tag{3.3}$$

That is, the end-points of two lines sharing a common intercept (1, 1), differing in slope by 0.1 and projected over two orders of magnitude in the x-variable differ from one another in the y-variable by 58.5%. The general formula for this relationship is given by:

$$PD_y = 100 \cdot \left(10^{(\text{pWR} \cdot \Delta b)} - 1\right)\% \tag{3.4}$$

where Δb is the absolute difference $\left(\left|\Delta b\right|\right)$ between two slopes and pWR is the logarithm of the mass range (see Chapter 2; pWR is equal to the logarithm of the ratio of the maximum body weight (or mass) to the minimum body weight (or mass) in a sample). For the numerical data just cited, pWR = 2.0 and Δb = 0.1. Substituting these values into relation (3.4), we obtain $PD_1 = PD_y = 58.5\%$. The same result applies to the lower solid line ($PD_2 = 100 \cdot (100 - 63.1)/63.1 = 58.5$). The values of percent difference in y (PD_y) for various values of Δb (column a) are given in Table 3.2, column b for pWR = 8.0 (see Chapter 2). We assume, as a rule of thumb, that a PD_y of 100% or greater will be of biological significance for most y-variables.

For example, say the difference in slopes (Δb) is 0.04 (column a, third row), and pWR = 8.0. Then PD_y is 109% (Table 3.2, column b, row three); i.e., $PD_y = 100 \cdot (10^{0.32} - 1) = 100 \cdot (2.09 - 1) = 109\%$, and by the above rule of thumb is thus biologically significant (see column e, row three).

Number of doublings and factor increase for a power function

A second way to characterize a difference in end-points (see Figure 3.1) for a given difference in slope is in terms of the number of doublings (N_d). If one y-value for an end-point is twice another, then there is a difference between the two end-points of one doubling. A general expression (see Chapter 24) for N_d is given by:

$$N_d = (\Delta b) \cdot \text{pWR}/\log(2) \tag{3.5}$$

Considering mammals generally, where pWR = 8.0, we find:

$$N_d = 26.57 \cdot \Delta b \approx 27 \cdot \Delta b \tag{3.6}$$

Relation (3.6) (i.e., $N_d \approx 27 \cdot \Delta b$) is easy to remember. Values of N_d for various values of Δb, assuming pWR = 8.0, are given in Table 3.2, column c. With a pocket calculator, N_d is easily converted to the factor (f) or multiplier by which one end-point exceeds another. Factor f is given by:

$$f = 2^{N_d} \tag{3.7}$$

If, for example, $N_d = 2.657$, then from relation (3.7) we find that $f = 6.31$ ($2^{2.657} = 6.31$). It is easily shown that percent deviation (PD_y) is related to f, the multiplicative factor, by the expression:

$$\text{PD}_y = 100 \cdot (f - 1)\% \tag{3.8}$$

For $f = 6.31$ we find from relation (3.8) that $\text{PD}_y = 531\%$. (See Table 3.2, column b, bottom row.) The numerical implications of a given difference in slope are expressed in Table 3.2 in terms of *percent difference* (column b), the *number of doublings* (column c), or a *multiplicative factor* (column d). Subjective and in part arbitrary expressions of the biological impact of differences in slope are given in Table 3.2, column e, for pWR = 8.0. As another example, consider the assertion sometimes made that the difference ($\Delta b = 0.083$) between slopes of two-thirds and three-quarters is not biologically significant with regard to (say energy) scaling in mammals. From relations (3.6–3.8), one finds that the end-points for these two slopes (assuming a common origin as in Figure 3.1, and pWR = 8.0) differ by 2.2 doublings, a factor of 4.6 ($2^{2.2} = 4.6$), or 360%. Compare this result with that for $\Delta b = 0.08$ shown in Table 3.2, second row from the bottom. Small to medium differences in slope can have a large impact over substantial size ranges (say pWR > 4.0); however, the reader can easily show that even relatively large differences in slope (e.g., $\Delta b > 0.15$) have a small impact ($f < 1.05$) over small size ranges (say pWR < 1.0). It should be stressed that differences in end-points judged to be biologically significant by these criteria (as when extrapolated to pWR = 8.0) might be found not to differ significantly statistically speaking; statistical significance is a separate question (see "On the invariance...," Chapter 23, for a related discussion).

Evaluative criteria

In Chapter 2 we described four *evaluative* measures (N_{ord}, N_{sp}, pWR, and DI). These measures are used to determine how representative a sample is of mammals generally. Here it is useful to introduce another three evaluative measures; these are designed to tell us how well a best-fit line (usually determined by least squares) describes an underlying dataset. The first such measure is called the mean percent deviation (MPD).

It provides a (non-dimensional) estimate of the scatter around a best-fit line. The greater the scatter, the less useful is the best-fit line for prediction of y-values from given x-values. The second measure is called the coefficient of determination (r^2). It tells us the fraction of the variation in a y-variable that is attributable to variation in the x-variable. No causation is implied. The third measure is given by the 99% confidence limit (CL) on the slope. This measure tells us how much confidence we can place in a given slope. These three measures (MPD, r^2, 99% CL) are reviewed in Chapters 5 and 23.

Inherent limits on structural slopes greater than 1

Natural selection is not free to scale the size of any one *structural* component (e.g., organ or tissue) of mammals (or other organisms) at arbitrary slopes greater than one ($b > 1.0$) over substantial size ranges. Here we invoke a self-evident premise, namely that no mammal consists entirely of one component, such as muscle or water. If a scaling exponent is greater than 1, it has a limiting y-value dependent on the values of the intercept and the size range. A striking example follows: assume that some tissue (y), such as *skeletal muscle*, comprises half (50%) of an idealized mammal with an adult body volume (BV) of 1 cm^3 (empirically, skeletal muscle comprises, on average, 42.6% of adult body volume or mass (see Chapter 12)). Further, suppose that skeletal muscle volume scales as the 1.5 power of body volume (i.e., $y = 0.5 \cdot BV^{1.5}$) (intercept is 0.5 and exponent is 1.5). This exponent would obtain if natural selection "aimed" to scale *total muscle strength* in direct proportion to body volume (BV). To see this, suppose that muscle diameter (d) and length (l) each scale as the square root of body volume (i.e., $d \propto BV^{0.5}$ and $l \propto BV^{0.5}$). In that case, total muscle cross-sectional area, and by implication total muscle strength, scale as the first power ($b = 1$) of body volume (i.e., cross-sectional area of skeletal muscle $\propto d^2 \propto BV^{0.5} \cdot BV^{0.5} \propto BV$).

A straightforward question relates to the body volume at which natural selection, thus conceived (i.e., scaling of skeletal muscle (y) subject to $y = 0.5 \cdot BV^{1.5}$) produces an adult mammal that is entirely skeletal muscle. The answer is 4 cm^3 ($0.5 \cdot 4^{1.5} = 4$)! That is, a four-fold increase in body volume converts a 1 cm^3 animal that is 50% skeletal muscle into an animal of 4 cm^3 in volume that is 100% skeletal muscle. This outcome shows quite strikingly that empirical exponents as high as 1.5 are mathematically impossible for organ and tissue scaling in mammals generally, where the intercept is appreciable (say $a > 0.1$) and pWR > 4.0 (see below). It is apparent that the exponent and intercept are inversely related. A larger intercept implies a smaller limiting exponent (i.e., b_{max}) (for exponents > 1) [11].

Here we are interested in the inverse question. Given a size range (pWR) and an intercept (a), what is the limiting value of the structural slope b_{max}? For convenience, we assume that the smallest mammal in any given dataset has an idealized body volume of 1 cm^3 or a mass of 1 gram. Let the body size range be denoted by pWR (Chapter 2). We now inquire as to the relation between any given intercept (a) and the exponent (b_{max}) at which one tissue or organ comes, by extrapolation, to comprise the whole animal. It can be shown (see Chapter 24) that the answer to this query is given by:

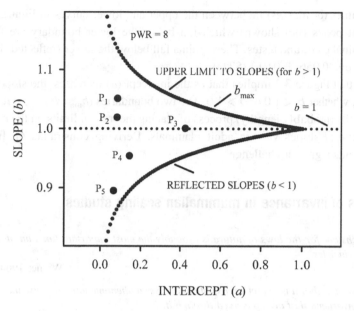

Figure. 3.2 An upper and tentative lower limit for organ and tissue slopes greater than 1.

$$b_{max} = 1 - (1/pWR) \cdot \log(a) , 0 < a < 1 \tag{3.9}$$

where b_{max} is the maximum exponent ($b > 1$) for any given intercept (a) and any given value of pWR (Chapter 2). The manner in which b_{max} varies with intercept (for pWR = 8.0) is shown in Figure 3.2. The reader can easily confirm that relation (3.9) works correctly for the above numerical example (pWR = log(4/1) = 0.6; log(a) = log(0.5) = –0.3; hence b_{max} = 1 – (–0.3/0.6) = 1.5). For the case where a = 0.1 and pWR = 8.0 we find that b_{max} = 1.125, far from a speculative slope of 1.5 (see above and Figure 3.2).

The lower curve in Figure 3.2 ($b < 1$) is simply a reflection of the upper curve. This reflected curve ($b_{reflected}$) is calculated from the relation:

$$b_{reflected} = 2 - b_{max} \tag{3.10}$$

In adult mammals, some 50% of body mass is collectively accounted for by skeletal muscle, lung mass, and blood mass (or volume) (see Chapters 12, 15, and 9, respectively). Only a few organs and tissues (chiefly bone and fat) may scale faster ($b > 1$) than body mass. Given these two compartments ($b \approx 1.0$, $b > 1.0$) there must be a third (aggregate) compartment, the components of which necessarily scale with slopes less than one ($b < 1$) [12]. The lower curve in Figure 3.2 might be considered as demarcating approximately the lower boundary for compensatory slopes. The five points (P_1, P_2, P_3, P_4, and P_5) represent fat, skeleton, skeletal muscle, skin, and liver (see Table 19.5, rows 30, 27, 24, 18, and 13, respectively).

It can be shown (by drawing on the data presented in Chapter 19, Table 19.5) that about 85% of paired structural intercepts and slopes in a sizeable sample of organ and tissues volumes (or masses) lie between the upper and lower curves in Figure 3.2. The four deviant points (not shown) which fall below the lower boundary are for brain, eyeballs, spinal cord, and testes. These points fall below the lower (reflected) boundary on average by -0.056 ± 0.012 (STD).

Observe that Figure 3.2 implies that as the intercept (a) increases, the slope tends to approach 1, whether $b < 1.0$ or $b > 1.0$. The two boundaries (b_{max}, $b_{reflected}$) represent a first step in the (probably lengthy) process of placing theoretical limits on the range over which *structural* slopes vary in adult mammals. Deriving constraints on functional slopes presents a greater challenge.

Two types of invariance in mammalian scaling studies

To be a touchstone for the laws of nature is probably the most important function of invariance principles.

Wigner 1964

It is readily seen. . . that it is in fact impossible to analyse a phenomenon otherwise than in terms of invariants that are conserved through it.

Monod 1972

Invariance (constancy) is of fundamental importance to contemporary physics and science generally (see above quotes by Wigner and Monod). It will be argued here that invariance (both absolute and relative) is also important in mammalian scaling (see Chapters 19 and 24). It was mentioned above that in applied physics and engineering, deducing the behavior of a full-size prototype from the behavior of a model (e.g., in a water tank) requires that one or a few quantitative criteria (e.g., the Reynolds number) be independent of size. Such independence is here termed absolute invariance (Schmidt-Nielsen [13] speaks of "non-scalable" variables and constants). See also the discussion by Niklas [14]. Absolute invariance is common in the physical world and plays an essential role in contemporary physics. A good example is the constancy of the velocity of light (in free space), independent of the velocity of the observer. The velocity of light provides an example of a physical dimensional constant (i.e., distance/time).

Absolute invariance also plays a central role in scaling in mammals. There are two types of absolute invariance: physical and biological. All adult mammals are composed of about two-thirds water. The water molecules in mammals are, for practical purposes, all of the same size and shape. Their physical and chemical properties are a given. These are important facts drawn from physics. At the same time, water molecules account on average for 62.4% of body mass or volume in adult mammals (Chapter 8), an important fact of biology.

Moreover, many quantities, such as mean blood pressure, diameter of the hemoglobin molecule (and diameters of thousands of other proteins), mean red blood cell diameter, surface or volume, and mean body temperature are all, on average, likely to be independent or nearly independent of adult body size. The respective y-variables exhibiting absolute invariance can be regarded as dimensional constants. Many epithelial cells might show size invariance, although quantitative evidence is lacking. From a

Table 3.3 Two forms of invariance pertinent to adult mammalian scaling studies

Form of invariance	Exponent (b) of body mass	Examples
Absolute	0.0	Size and shape of atoms, molecules, most macromolecules. RBC diameter, capillary diameter, mean body temperature.
Relative	1.0	Skeletal muscle and lung mass, blood volume. Total number of water or Hb molecules, or RBC.

RBC = red blood cells; Hb = hemoglobin.

quantitative standpoint, absolute invariance is the more common of the two types (absolute, relative) of scaling invariance found in mammals (see below). What is implied by this statement is that the slope (b) for the line relating each of many y-variables to body size is zero, or is sensibly close to zero ($b \approx 0$). It is true that some proteins, such as proteoglycans found in cartilage, are polydisperse (variable in size) and might (or might not) vary with body size, but this is exceptional. Most proteins studied thus far are monodisperse (uniform in size); their diameters and shapes are likely to be independent of body size.

The second type of constancy is termed "relative invariance." This implies that a given y-variable is directly or nearly directly proportional to an x-variable ($b \approx 1$). Note that when organ and body size are measured in the same units relative invariance implies that their ratio is non-dimensional. Alternatively, the number of components (such as hemoglobin or water molecules) may also scale directly or nearly so with body size ($b \approx 1$) (see Table 3.3). Recall that pure numbers are also non-dimensional.

Relative invariance is also of considerable quantitative significance in the scaling of adult mammals. For example, total body water scales nearly in accord with relative invariance ($b = 1.002 \pm 0.011$) (99% CL) (Chapter 8). Furthermore, it was found, in a small sample ($N_{sp} = 11$, pWR = 4.6) that bulk protein in adult mammals scales as the 0.985 ± 0.046 (99% CL) power of body size (Chapter 8). Assigning a slope of 1 (rather than 0.985) to the best-fit line suggests that the mass of bulk protein accounts for 16.5% of body mass in adult mammals. We see then that body water (see above) and bulk protein jointly account for about 80% of body mass in adult mammals (62.4 + 16.5 = 78.9).

Furthermore, the evidence is that blood volume (7.7%, Chapter 9), lung mass (1%, Chapter 15), and skeletal muscle (42.6%, Chapter 12) each scale nearly in accord with relative invariance, and together account for some 51% of body mass or volume (i.e., 7.7 + 1 + 42.6 = 51.3). See Table 3.3. Even allowing for double counting (e.g., estimates of muscle mass surely include some blood), it is likely that relative invariance of organs and tissues accounts for at least half of mammalian adult body size. In the examples just given, relative invariance has the same implication as constant volume-fraction or constant mass-fraction.

Note that both functional and structural y-variables can exhibit absolute invariance, but, as far as is known, no functional y-variables exhibit relative invariance in mammals generally. It is also true that absolute and relative invariances are the only examples of scaling in mammals that exhibit, in principle, essentially the same (linear) behavior on both linear–linear and log–log plots.

Keep in mind that absolute and relative invariance are often closely related. For example, water and many macromolecules (e.g., actin, myosin, and hemoglobin) each exhibit essentially constant size and shape at all body sizes (absolute invariance). These are among the most common molecules in the adult mammalian body. The *number* of water and hemoglobin molecules in the adult mammal evidently scales very nearly as the first power ($b = 1.0$) of body mass (i.e., relative invariance). Thus, relative invariance is a wider concept than constant volume-fraction. If the number of red blood cells per unit volume and blood volume both exhibit relative invariance, as the empirical evidence suggests, then their ratio (i.e., red blood cell concentration) must exhibit absolute invariance. It is clear that the primary mechanisms by which mammals increase body size involve both absolute and relative invariance. The issue of organ or tissue scaling at constant shape, another form of absolute invariance, is discussed briefly below.

We alluded above to the importance of invariant similarity criteria (such as the Reynolds number) in physical scaling studies. It was pointed out that blood volume, skeletal muscle, and lung mass each scale nearly in accord with relative invariance (Table 3.3). These tissues and organs account for about 50% of adult body mass. These forms of invariance might represent the most important respects in which mammals are quantitatively similar (see above discussion of scaling in physics).

Testing for non-linearity

In LSQ analyses of mammalian scaling data, it is usually implicitly assumed that the data are linearly distributed (usually in log–log coordinates). In substantial datasets, where the number of species (N_{sp}) is large (say $N_{sp} > 300$), and the size range is considerable (say pWR > 4.0) the scatter in the data could obscure any non-linearity present in the data in a log–log plot. One way to approach this problem is to "bin" the data. This is accomplished as follows. First, one rank-orders the data by increasing values of the x-variable (usually body mass). Then one computes the logarithms of the x- and y-variables. Next, one calculates the size range, as given by pWR. Finally, one computes bin size by dividing pWR by some integer, say in the range of 10 to 50.

This calculation defines 10 to 50 bins, each spanning (nearly) the same fraction of pWR (since body masses are usually not uniformly distributed in logarithmic coordinates, it is not feasible to divide body masses into bins each spanning exactly the same width). Lastly, one computes the mean values of the logarithms of the x- and y-variables for each bin. A plot of these logarithmic averages stands to reveal whether there is any obvious curvature present. Examples of this procedure are given in Chapter 17.

A second approach to revealing non-linearity is to partition a given dataset into two or more subsets spanning different size ranges (e.g., see a plot of brain size in Chapter 13, Figure 13.2). Both of these tests for non-linearity are reviewed in Chapter 24.

Criteria for scaling at constant shape

Body and organ shapes are of considerable interest from a scaling standpoint. Natural selection has likely molded these shapes to make them function more efficiently. In scaling studies, we mainly want to know whether shape is invariant. In reviewing the criteria for scaling at constant shape, it is useful to distinguish between two levels of analysis: whole body and body parts (organs and tissues). We consider first constant shape at the whole body level. It is well known that a series of objects of varying size but constant shape exhibit lengths and surface areas that scale as the one-third and two-thirds powers of body volume, respectively. If then lengths and areas scale on a log–log plot as the one-third and two-thirds power of body volume (or mass, assuming constant density), one infers constant shape.

However, this inference is open to doubt, as it is easily shown that either the lengths or the surface areas of objects of varying size and *dissimilar* shape may nevertheless scale nearly in the same way as for truly constant shape (i.e., with slopes near to 1/3 or 2/3 for lengths and surface areas, respectively) [15]. See Chapter 24. It is true, however, that if *both* lengths and surface areas scale nearly as prescribed (slopes of 1/3 and 2/3, respectively), then the given objects are more likely to share nearly the same shape.

We next consider the scaling of objects (organs and tissues) *embedded* inside the body. We assume that the way an organ or tissues scales with body volume (BV) can be defined in terms of three mutually perpendicular lengths (or diameters), denoted by (l_1, l_2, l_3), sharing a common origin, and expressed in terms of power functions of body volume. We write:

$$l_1 \propto BV^{b_1}; \quad l_2 \propto BV^{b_2}; \quad l_3 \propto BV^{b_3} \tag{3.11}$$

where the exponents (slopes) are b_1, b_2, and b_3. The volume of the organ (OV) in question (expressed in terms of body volume BV) is given by:

$$OV \propto l_1 \cdot l_2 \cdot l_3 \propto BV^{(b_1+b_2+b_3)} \tag{3.12}$$

Now we examine the condition for scaling at constant volume-fraction (OV $\propto$ BV^1), where the exponent is one. Thus, relation (3.12) implies that:

$$b_1 + b_2 + b_3 = 1 \tag{3.13}$$

Note that relation (3.13) does not require scaling at constant shape. That is, relative invariance with respect to organ or tissue volume or mass need not imply constant shape. (It is not uncommon in science that implicit and unstated assumptions prove, in particular applications, to be mistaken.)

If we wish to enforce constant shape, independently of constant volume-fraction, we write:

$$b_1 = b_2 = b_3 \tag{3.14}$$

Relation (3.14) means that each organ length (or diameter) scales as the same power of body volume (as is required for scaling at constant shape). Relation (3.14) does not

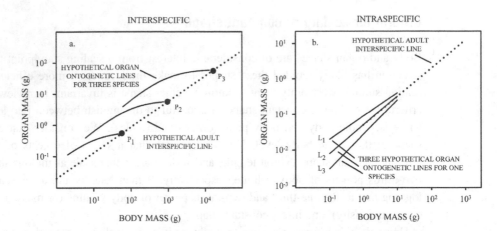

Figure 3.3 Organ scaling in mammals during ontogeny.

require that these exponents be 1/3, merely that they be equal (see Chapter 24). Again, scaling at constant shape does not require scaling at constant volume-fraction (3.13). If we wish to enforce scaling at *both* constant shape and constant volume-fraction then we write:

$$b_1 = b_2 = b_3 = 1/3 \qquad (3.15)$$

Relation (3.15) is derived by combining relations (3.13) and (3.14). It can be shown (see Chapter 24) that invariant shape is a special case of absolute invariance, where ratios of side-lengths (e.g., l_1/l_2) are held constant (see relation (3.11).

The usual criterion for scaling at constant organ shape states that a slope or exponent (say for organ length or diameter) must take the value of 1/3 when referred to body volume (or mass at constant density). This criterion is correct when organ volume scales as, or very nearly as, the first power of body mass ($b = 1.0$), but might be misleading otherwise. We will see in Chapter 19 (Table 19.5) that organ and tissue volumes reviewed here scale with body size with slopes ranging between 0.686 (eyeball) and 1.051 (fat). For the eyeball, the requirement for scaling at constant shape is that eye diameter scale as the 0.23 (i.e., 0.686/3) power of body volume or mass (far from a slope of 0.333). A more detailed derivation of the above relations for scaling at constant shape is given in Chapter 24.

Ontogeny

Most scaling studies in mammals in the past few decades have concentrated on how a y-variable such as mean body temperature or heart size varies or varies not, in adults of species with varying body sizes. Such analysis is usually based on log–log plots of a y-variable expressed as a function of adult body size (the x-variable). The immediate causal explanation of such adult interspecific trajectories is to be found not in the study of adults per se, but rather in the study of ontogeny. This conception is illustrated in Figure 3.3, panel a (adapted from an illustration of brain scaling in Holt *et al.* [16]).

The dotted line (Figure 3.3, panel a) represents a hypothetical adult interspecific trajectory for some given organ. The three solid lines (panel a) are meant to represent hypothetical scaling for the same organ in three different species of varying adult size. The three points P_1, P_2, and P_3 (panel a) have a double significance; first, they lie on (or more usually near) an adult interspecific line (dotted) and second, each lies at the end of an ontogenetic line (solid curve).

An understanding of ontogeny stands to throw a light on the (probably complex) mechanisms that lead to adult interspecific scaling of organs and tissues. Such understanding no doubt lies well in the future. However, there is no reason to expect that physics generally, or dimensional analysis in particular, will play a significant role in this long-term endeavor (see above).

The question of how ontogenetic scaling of organs and tissues relates to adult interspecific scaling has been largely ignored (see most chapters in Part II). The simplest possible realistic relationships are shown in Figure 3.3b. Here we envisage three hypothetical scaling trajectories for some given organ (solid lines) in a single species, relative to an adult interspecific line (dotted). The trajectories, on a log–log plot, here assumed to be linear, show an ontogenetic line approaching an adult interspecific organ line from above (L_1), collinearly (L_2), and from below (L_3). It will be seen in Part II that this simple classification works reasonably well on average (for mainly human ontogenetic trajectories).

A repeating units model

Very much of our anatomy and physiology is repetitive, or partly repetitive...

Crick 1966

The first pattern of organic order consists in the universal occurrence of standard parts or units.

Riedl 1978

One of the most important features underlying evolutionary increases in animal and plant size, complexity and diversity has been their modular construction from reiterated parts.

Carroll 2001

The fact that adult mammals, all with qualitatively similar body and organ plans, have body sizes dispersed over at least eight orders of magnitude (i.e., pWR = 8.0) suggests that their constituent organs and tissues are "designed" on inherently scalable principles. The widespread deployment in nature of assemblies, modules, standard parts, subunits, or entities here termed *repeating units* (RU), testifies to their fundamental importance [17]. Nevertheless, it is odd that the most works on human anatomy or on structural biology mention repeating units only in passing, or, more often, ignore them all together.

Modular construction is evident at every hierarchical level in biology (and indeed in the universe at large), from the molecular to the organ. Thus, at the macromolecular

Table 3.4 The parameters for an organ scaling model based on repeating units

Entity	Diameter (d)	Surface area (sa)	Volume (v)	Number (#)	Total surface area (totsa)	Total volume (totv)
a	b	c	d	e	f	g
RU	RU_d	RU_{sa}	RU_v	$RU_\#$	RU_{totsa}	RU_{totv}
Exponent [a]	1/9	2/9	3/9	6/9	8/9	9/9

RU = repeating unit; [a] Exponent of organ volume rather than body volume.

level we have polymers constructed of subunits, such as amino acids and bases. At the subcellular level, where component size tends to be fixed, and variation is mostly in number, we find centrioles, chromosomes, cilia, genes, microtubules, microvilli, mitochondria, nuclei, ribosomes, and sarcomeres, among others. Certain differentiated cells, such as blood, muscle and nerve cells, ova and sperm, can be regarded as repeating units.

At the level of tissues and tissue products we find acini, alveoli, capillaries, glomeruli, hair follicles, hepatic and pulmonary lobules, islets, nephrons, osteons, seminiferous tubules, teeth, villi, and so on. Finally, at the organ level we have many paired structures such as adrenal glands, cerebral and cerebellar hemispheres, ears, eyes, kidneys, lungs, mammary glands, nostrils, and vertebrae. The hierarchical organization of repeating units is perhaps best illustrated in skeletal muscle (Chapter 12, Table 12.1). It is difficult to avoid the conclusion that repeating units are a major factor accounting for scalability in mammals, and perhaps in multicellular organisms generally (see above quotes by Crick, Riedl, and Carroll). Often, repeating units are themselves composed of repeating units (e.g., myofibers contain sarcomeres and ribs contain osteons).

At least some types of repeating units (e.g., pulmonary alveoli and renal corpuscles) increase in both size and number with increasing organ size. In a prior study, I described how a repeating units model can apply to the scaling of kidneys and lungs in mammals and birds [17]. The essential exponents (slopes) derived from this model (expressed in terms of organ mass or volume) are summarized in Table 3.4.

In this model, repeating unit diameter (RU_d) (column b) scales as the one-ninth power of *organ volume* (bottom row). Assuming constant shape for repeating units, it follows that *repeating unit* surface area (RU_{sa}) and volume (RU_v) scale as the two-ninths and three-ninths powers of organ volume, respectively (columns c and d). Repeating unit number ($RU_\#$) (column e) scales as the two-thirds power of organ volume. Total repeating unit surface area (RU_{totsa}) and volume (RU_{totv}) (i.e., the surface area and volume of all repeating units) scale as the eight-ninths (2/9 + 6/9) and first (3/9 + 6/9) powers of organ volume (columns f and g), respectively. Keep in mind that lung mass scales as the first power of body mass (see Chapter 15).

It is likely that some other organs and tissues will require somewhat different scaling rules than those given in Table 3.4. Elsewhere, I have suggested that tree rings may be regarded, from a scaling standpoint, as a form of repeating units [18].

Table 3.5 A tentative classification of reference slopes for scaling studies in adult mammals

$b < 0$		$b = 0$	$b > 0$						$b = 1$	$b > 1$
a	b	c	d	e	f	g	h	i	j	k
−1/3	−1/4	0	1/9	1/4	1/3	2/3	3/4	8/9	1	10/9
LL	U	AI	RU	U	L	SA	U	RU	RI	UL

LL = lower limit; U = ubiquitous; AI = absolute invariance; RU = repeating unit; L = length; SA = surface area; RI = relative invariance; UL = upper limit.

Adjustments to physical constraints

...knowing the outer environment cannot teach us the actions born in, and proper to, the internal environment.

Bernard 1865

Any engineer will appreciate that a pump weighing half a ton cannot perform as many cycles per minute as a similar one weighing a pound.

Hill 1950

We have seen above that the primary mechanism by which mammals increase body size is to multiply the number of molecular, macromolecular, and in some cases microscopic components, each component generally being of fixed size and shape. Having achieved an increase in body size, mammals must then make internal adjustments in the name of a sustainable and efficient domestic economy. We tend to think of the adjustments made by natural selection as improving the ability of organisms to contend with an often demanding external environment. Nonetheless, adjustments to the internal environment (see above quote from Claude Bernard) might well be just as important a contribution to overall fitness as those to the external environment.

The major adjustments to increased body size could prove to lie chiefly in changes to functional rates rather than to the size of organs and tissues (see above quote from Hill). We will see in Chapter 9, for example, that maximal heart rate scales as the −0.146 ± 0.025 (99% CL) power of body size. This is far from the exponent of −1/3 suggested by arguments from physical similarity [19]. Generally, there is little reason to suppose that detailed adjustments in maximal functional rates (e.g., heart rate, respiration rate, oxygen consumption VO_2) will prove to be predictable from strictly physical considerations.

Towards a classification of reference slopes

Table 3.5 is intended to provide a schema (i.e., a set of reference slopes) for thinking about empirical slopes. No causality is implied between slopes and the given classification. The most that can be claimed is that each of the various reference slopes may be *correlated* with one or more empirical slopes. The slopes given in columns (a, k) are considered to be extremes. The most negative slope reported in this work is −0.265 for resting respiratory rate in 103 species of adult mammals (Chapter 15). A slope as negative as minus 1/3 in a large sample is possible but perhaps unlikely. Thus minus 1/3 is a plausible lower limit (LL) for scaling in adult mammals generally (column a).

No single function or structure in this study scales over a sizeable size range with a slope greater than 1.051 (i.e., that for fat; Chapter 8). It seems safe to assume (say for $a > 0.1$) that the upper limit (UL) to slope in a sizeable representative sample of adult mammals will not exceed 10/9 (column k), for pWR = 8.0. In both the above cases the difference between the relevant empirical slopes and the postulated extremes is at least 0.06 ($-0.265 - (-0.333) = 0.068$; $1.111 - 1.051 = 0.06$); that is, the assumed extremes (Table 3.5, columns a, k) might prove to be wider than is strictly necessary.

The slopes given in Table 3.5 (columns b, e, and h), all multiples of $\pm 1/4$, are considered to be ubiquitous (U). Ubiquity is believed to be associated with an underlying fractal geometry [20]. A slope of zero (column c), representing absolute invariance (AI) is a candidate for the most common scaling slope representing adult mammals generally.

The slopes of 1/9 (column d) and 8/9 (column i) are those for the diameter and total surface area derived from a scaling model of repeating units (see Table 3.4 above), assuming that organ volume scales as the first power of body volume. The slopes of 1/3 and 2/3 (columns f and g) correspond to the physical dimensions for length (L) and surface (or cross-sectional) area (SA), respectively. These slopes have also often been invoked (problematically) to characterize scaling at constant shape (see above). The penultimate class (column j) refers to relative invariance (RI) ($b = 1.0$). This is also likely to be a commonly occurring slope for adult mammals when we include the number of individual species of small molecules (e.g., water) and macromolecules (such as actin, myosin, and hemoglobin). See also Chapters 19 and 20.

Recall that only y-variables scaling in accord with absolute ($b = 0.0$) or relative ($b = 1.0$) invariance represent a naturally occurring form of scaling constancy. We assume that an empirical slope concurs with a reference slope if the two slopes lie within the 99% CL on the empirical slope (see Chapter 5, Figure 5.1). However, there are thousands of different types of proteins that doubtless exhibit absolute invariance with respect to size and shape. For this reason, scaling with absolute invariance is likely to be very common in adult mammals when viewed at, say, the molecular level.

Summary

A simple form of geometric scaling was practiced in prehistoric times. Modern scaling studies began in mathematics and physics, where they gave rise to a sophisticated methodology based in major part on dimensional analysis. It is argued that this methodology has at most limited application to scaling in mammals at large [3, 21]. We then turn to a brief discussion of scale effects. Insofar as known regular scaling in all systems breaks down at sufficiently large or small sizes. This is an important principle in physics.

The primary analytical focus in mammalian scaling studies is on the slopes (exponents) determined from log–log plots. It is reasonable to interpret apparent linearity on a log–log plot as evidence that a y-variable may be expressed as a power function of the x-variable (i.e., $y = ax^b$). A few remarks are made about the numerical properties of power functions. The scaling implications of differences in slope (Δb) are discussed. Given two line-segments sharing a common intercept (i.e., the lower end-point), the divergence between the upper end-points is expressed in terms of percent difference (PD), the number of doublings (N_d), and a multiplicative factor (f) (see Figure 3.1). It is shown that small differences in slope projected over large size ranges can produce large differences in the magnitude of the end-points. At the same time, large differences in slope may have only a minor effect over small size ranges.

Three additional evaluative criteria (see also Chapter 2) are introduced, bearing on the question of how well a best-fit line describes the underlying data. These include mean percent deviation (MPD), which measures the degree of scatter around a best-fit line; the coefficient of determination (r^2); and the 99% confidence limits on the slope. It is shown that there exist inherent limits on *structural* slopes greater than 1; these limits vary with intercept and size range. The argument is extended to include structural slopes less than one. Possibly this is the first attempt in mammalian scaling studies to impose upper and lower limits on structural slopes as a function of intercept and pWR.

Two forms of invariance are introduced: absolute ($b = 0$) and relative ($b = 1$). It is suggested that these forms of invariance play an important role in scaling in adult mammals. Moreover, these forms of invariance provide linkages to scaling in applied physics and engineering, where scaling theory is much more developed. In addition, two tests for non-linearity are introduced, one dependent on "binning" and one on partitioning datasets into non-overlapping subsets. Criteria for assessing whether animals or organs scale at constant shape are reviewed and partially amended.

The above material leads to a preliminary discussion of possible relations between ontogenetic scaling (mostly in the human) and adult interspecific scaling. Ontogeny (Chapter 22) stands to explain the mechanisms accounting for adult interspecific scaling. A scaling model for repeating units, a topic usually ignored in scaling studies in mammals, is introduced. The chapter ends with a tentative classification of reference slopes. These reference slopes reflect, in major part, invariance and dimensionality (lengths, surface areas). It is concluded that scaling in mammals is now, and will likely remain for some time a largely empirical undertaking (but for published examples of modeling on diverse topics, some concerning mammals, see Brown and West [22]). A synopsis of some of the above points is given in Chapter 5.

References

1. Froude, W. (1955). *The Papers of William Froude*. London: Institute of Naval Architects.
2. Langhaar, H.L. (1951). *Dimensional Analysis and Theory of Models*. New York: John Wiley & Sons.
3. Prothero, J. (2002). Perspectives on dimensional analysis in scaling studies. *Perspectives in Biology and Medicine*, **4**:175–189.

4. Günther, B. (1975). On theories of biological similarity. *Fortschritte der experimenttellen und theoretischen Biophysik*, **19**:1–111.
5. Stahl, W.R. (1963). The analysis of biological similarity. In: Lawrence, J.H. and Newman, J.W. (eds.) *Advances in Biological and Medical Physics*. New York/London: Academic Press, pp. 355–489.
6. Lin, H. (1982). Fundamentals of zoological scaling. *American Journal of Physics*, **50**:72–81.
7. Rosen, R. (1983). Role of similarity principles in data extrapolation. *American Journal of Physiology*, **244**:R591–R592.
8. Becker, H.A. (1976). *Dimensionless Parameters: Theory and Methodology*. New York: John Wiley & Sons (see Vitruvius quote on p. 22).
9. Galilei, G. (1954). *Dialogues Concerning Two New Sciences*. New York: Dover Publications. Republication of 1914 translation; original work published 1638.
10. Hokkanen, J.E.I. (1986). The size of the largest land animal. *Journal of Theoretical Biology*, **118**:491–499.
11. Prothero, J. (1995). Bone and fat as a function of body weight in adult mammals. *Comparative Biochemistry and Physiology*, **111A**:633–639.
12. Adolph, E.F. (1949). Quantitative relations in the physiological constitutions of mammals. *Science*, **109**:579–585.
13. Schmidt-Nielsen, K. (1984). *Scaling: Why is Animal Size so Important?* Cambridge: Cambridge University Press.
14. Niklas, K.J. (1994). *Plant Allometry: The Scaling of Form and Process*. Chicago, IL: University of Chicago Press.
15. Prothero, J. (1986). Methodological aspects of scaling in biology. *Journal of Theoretical Biology*, **18**:259–286.
16. Holt, A.B., Cheek, D.B., Mellitus, E.D. and Hill, D.G. (1975). Brain size and the relation of the primate to the nonprimate. In: Cheek, D.B. (ed.) *Fetal and Postnatal Cellular Growth: Hormones and Nutrition*. New York: John Wiley & Sons, pp. 23–44.
17. Prothero, J. (1996). Scaling of organ subunits in adult mammals and birds: a model. *Comparative Biochemistry and Physiology*, **113A**:97–106.
18. Prothero, J. (1999). Scaling of tree height and trunk diameter as a function of ring number. *Trees*, **14**:43–48.
19. Hill, A.V. (1950). The dimensions of animals and their muscular dynamics. *Science Progress*, **38**:209–230.
20. Savage, V.M., Gillooly, J.F., Woodruff, W.H. *et al.* (2004). The predominance of quarter-power scaling in biology. *Functional Ecology*, **18**:257–282.
21. Bridgman, P.W. (1931). *Dimensional Analysis*. New Haven, CT: Yale University Press, p. 53.
22. Brown, J.H. and West, G.B. (eds.) (2000). *Scaling in Biology*. Oxford: Oxford University Press.

4 Towards a standardized body weight table

Estimates of body weight are hard to come by... they are invariably accumulated from disparate sources, based on only a few specimens and often biased by inadequate sampling of geographic variation. In any case, they are only species averages over highly variable taxa.

Gould 1975

Appeal to a standardized computer-based body weight table is useful and perhaps even essential to a cautious approach to scaling studies in adult mammals generally. When a worker reports the measurement of some scaling y-variable, say blood volume, it is customary to give a figure for body size, usually in terms of body weight or mass (the x-variable). Sometimes it will be explicitly stated that the animals in question are adults. Only rarely are criteria given by which adulthood was determined. Doubtless the term adult often connotes sexual maturity. However, sexual maturity can be reached before a plateau in body size is reached. This concern may be significant, as developing mammals often display different values of any given scaling y-variable, expressed as a function of body size, from those exhibited by fully grown adults (e.g., Chapter 22).

A second problem, especially with domesticated species, is that body mass may be skewed upwards by high fat levels. Body sizes above or below the normal adult range for a given species could confound the results of an adult interspecific scaling study. Thus the primary purpose of a standardized body weight table (SBT) is to provide guidelines as to when published body sizes may be accepted or should be put aside (for details see Appendix B). By adopting this protocol, one plainly stands a better chance of computing slopes that are reflective of normal adult mammals.

Many field guides to mammal identification give measurements of average body length (nose to tip of tail) and tail length. No doubt body length is the easiest parameter relating to body size to measure in the field, especially for very large mammals, such as elephants and whales. And body length is of use for partially characterizing body shape (Chapter 6). The chief problem is that there are few self-consistent ways of partitioning length in mammals into meaningful components. This is more obvious for diffuse tissues (e.g., connective tissue; it is not useful to speak of a length of blood – except perhaps when measuring hematocrit (Chapter 9)).

The ideal measure of body size for scaling purposes would not be length, weight, or mass, but *volume*. The reason is that organisms can be considered, for scaling purposes, predominantly as chemical "machines," for which surface area and volume are likely to be functionally the more significant measures of size. Body volume can be exhaustively partitioned into component organ systems, organs and tissues, each of which is potentially of intrinsic meaning.

Body weight or mass is the commonly used measure of body size in mammalian scaling studies [1]. Still, body mass, per se, is strictly speaking called for only in a restricted domain of mammalian scaling studies; i.e., mainly problems involving inertia, as in locomotion. There are other difficulties inherent in any measure of body size: in a given species it may vary with age, degree of exercise, diet, gender, latitude, or season, as well as with stage of pregnancy or lactation in females. Sample sizes are often not given, and when given are frequently small (<10) [2] (see above quote from Gould).

Biologists do not usually weigh mammals using an analytical equal-arm balance (or equivalent), by which mass is empirically determined. Typically, some form of *spring balance* seems to be employed, which gives a measure of body *weight* (i.e., a force), not body mass. By convention, the units of weight (a force) and mass (a quantity of matter) are chosen so as to give numerically the same results, at sea-level and at mid-latitudes. It seems likely that it is the weight, not mass, of mammals that is commonly measured in the field or laboratory; but this is unlikely to make much difference from the biological standpoint. In the present work, body size (usually in g) is used as a measure of body, organ, or tissue size. This choice was motivated by the need to expedite the process of transcribing a large heterogeneous body of empirical data (more than 16,000 records in total) into computer-readable form and then back-checking these data against the original sources (where body size has usually been given in g) to correct for possible errors in transcription.

Construction of a prototype computerized and standardized body weight table (SBT) was initiated in the early 1980s and continued intermittently up to the mid-1990s. The table was constructed for use in ongoing in-house scaling studies. All of the data were taken from *standard* sources (e.g., Nowak and Paradiso [3] and MacDonald [4]). Often multiple values of body size from different sources were available for a given species. Those values nearer the mean were employed, rather than the extremes. A summary of the current contents of the SBT is provided in Table 4.1. As before, the number of species present in any given order is denoted by N_{sp}. Altogether, size-data for 1,410 different species in the "general" category (gender not specified) are stored in the database (Table 4.1, column b, bottom row). Observe that none of the information presented below exists explicitly in the SBT database.

The number of species represented in the SBT for each taxonomic order is listed in Table 4.1, column b, along with the number of extant species (column c), based on Corbet and Hill [5]. The number of species in the SBT is expressed as a percent of extant species in column d (mean = 60% ± 26 (STD)). Altogether 33% (100·(1,410/ 4,327) = 33) of extant mammals are represented by some body size data in the SBT (see also Tables 4.2 and 4.3). The three largest orders (rodents, bats, and insectivores) are represented by 24% or fewer extant species (Table 4.1). A more recent taxonomic compilation (Wilson and Reeder [6]) increases the number of extant species from 4,327 to 5,416.

Table 4.1 Contents of the standardized body weight table by taxonomic order and species number (N_{sp})

	Taxonomic order	N_{sp} represented in the SBT	N_{sp} extant	SBT as % of extant
	a	b	c	d
1	Artiodactyla	155	194	80
2	Carnivora	183	235	78
3	Cetacea	63	77	82
4	Chiroptera	239	977	24
5	Dermoptera	2	2	100
6	Edentata	18	29	62
7	Hyracoidea	4	8	50
8	Insectivora	72	365	20
9	Lagomorpha	23	65	35
10	Macroscelidea	7	15	47
11	Marsupialia	120	282	43
12	Monotremata	3	3	100
13	Perissodactyla	13	16	81
14	Pholidota	4	7	57
15	Pinnipedia	22	34	65
16	Primates	85	201	42
17	Proboscidea	2	2	100
18	Rodentia	386	1,793	22
19	Scandentia	4	16	25
20	Sirenia	4	5	80
21	Tubulidentata	1	1	100
Sum	-	**1,410**	**4,327**	-

A summary of the SBT contents for three classes (general, male, female) is given in Table 4.2. The class "general" (columns a, b, c) refers to those cases where body weight is given without gender; it might imply male, female, or both. For each class, the number of species is specified for which minimum (min), maximum (max), and mean body weights or masses occur in Table 4.2.

In most instances it is likely that the minimum (min) and maximum (max) body weights in the general class span the body weights for both males (columns d, e, and f) and females (columns g, h, and i). Columns c, f, and i represent mean body weights. The majority of body weights in the database fall into the "general" class, with only about a third as many in the male and female categories.

Table 4.3 provides a summary of the contents of the SBT from the perspective of four different *evaluative* measures, as described in Chapters 2 and 3. The first two rows in Table 4.3 are simply a restatement of the information in Table 4.2. Row three gives values of the size range (pWR) and row four the values for the diversity index (DI) for each column (a to i). It is worth noting that for the general class (min, max), the size range (pWR) reaches 7.5 or 7.6 and the diversity index reaches 79/80. These results show that high values of pWR (approaching eight) and diversity (at least 80%) are achievable in practice with some current databases (see also Chapter 6).

Table 4.2 Number of species for which body weight or mass is specified in the SBT by order and for the three classes: general, male, and female

Taxonomic order	General			Male			Female		
	Min	Max	Mean	Min	Max	Mean	Min	Max	Mean
Column	a	b	c	d	e	f	g	h	i
Artiodactyla	144	144	11	42	42	20	35	35	22
Carnivora	140	140	44	37	38	21	34	34	18
Cetacea	32	32	31	5	8	18	5	6	17
Chiroptera	181	182	100	60	60	68	52	52	68
Dermoptera	2	2	-	-	-	-	-	-	-
Edentata	13	13	5	-	-	-	-	-	-
Hyracoidea	3	3	1	-	-	-	-	-	-
Insectivora	64	64	19	15	15	13	14	14	12
Lagomorpha	20	20	3	10	10	8	9	9	10
Macroscelidea	3	3	4	-	-	-	-	-	1
Marsupialia	112	112	31	97	97	91	88	87	86
Monotremata	3	3	-	3	3	1	2	2	2
Perissodactyla	7	7	6	1	4	1	-	-	1
Pholidota	3	3	1	-	-	-	-	-	-
Pinnipedia	19	19	3	12	12	17	9	9	21
Primates	60	60	26	14	14	42	12	12	38
Proboscidea	1	1	1	-	-	-	-	-	-
Rodentia	317	317	96	105	105	100	102	102	105
Scandentia	2	2	2	-	-	1	-	-	-
Sirenia	2	2	2	-	-	-	-	-	-
Tubulidentata	1	1	-	-	-	-	-	-	-
Total	**1,129**	**1,130**	**386**	**401**	**408**	**401**	**362**	**362**	**401**

Min = minimum; max = maximum.

Table 4.3 Number of orders (N_{ord}), number of species (N_{sp}), size range (pWR), and diversity index (DI) for the SBT, as a function of the three classes: general, male, or female

Measure	General			Male			Female		
	Min	Max	Mean	Min	Max	Mean	Min	Max	Mean
Column	a	b	c	d	e	f	g	h	i
N_{ord}	21	21	18	12	12	13	11	11	13
N_{sp}	1129	1130	386	401	408	401	362	362	401
pWR	7.6	7.5	7.7	7.4	7.4	7.0	7.2	7.5	7.0
DI (%)	80	79	59	38	38	39	33	35	39

See Table 4.2.

Summary

A standardized computer-based body weight table (SBT) was constructed, as time permitted, for the most part over a period of 15 years or so. The primary aim was to provide guidelines (see Appendix B) for the screening of data prior to the calculation of best-fit LSQ lines. This database was found in practice to provide useful data for the great majority of the species encountered in this study. Using newer data from contemporary mammalian studies, it would be easy to extend the SBT by the inclusion of body size data for at least several hundred additional species. Before leaving this chapter, note that the SBT was also employed to study body size distributions in adult mammals (see Chapter 7).

References

1. Wang, Z-M., Pierson, R.N. Jr. and Heymsfield, S.B. (1992). The five-level model: a new approach to organizing body-composition research. *American Journal of Clinical Nutrition, Comparative Biochemistry and Physiology*, **56**:19–28.
2. Gould, S.J. (1975). On the scaling of tooth size in mammals. *American Journal of Zoology*, **15**:351–362.
3. Nowak, R.M. and Paradiso, J.L. (1983). *Walker's Mammals of the World*, 4th edn. Baltimore, MD: Johns Hopkins University Press.
4. MacDonald, D. (1984). *The Encyclopedia of Mammals*. New York: Facts on File.
5. Corbet, G.B. and Hill, J.E. (1991). *A World List of Mammalian Species*, 3rd edn. Oxford: Oxford University Press.
6. Wilson, D.E. and Reeder, D.M. (eds.) (2005). *Mammals of the World: A Taxonomic and Geographic Reference*. Baltimore, MD: Johns Hopkins University Press.

5 A reader's guide

Part II presents the results of scaling analyses for nearly 100 y-variables. The LSQ data (chiefly slopes and intercepts) for 72 of these y-variables are summarized in Part III. The purpose of this chapter is to review the (partly novel) presentation of data as seen in Part II. It is meant to provide a one-stop summary of needed information. Some information not presented in prior chapters is included here. Altogether two dozen topics are addressed briefly. See also Chapters 1–4, 23, and 24.

Number of orders (N_{ord})

The number of *taxonomic orders* represented in any particular dataset is denoted by N_{ord}. This number can, in principle, vary between 1 and 21 [1], but on average for structural y-variables in this study it is about 10. See Chapter 2.

Number of species (N_{sp})

The number of *species* in a dataset is designated by N_{sp}. In the majority of instances in Part II, the number of species is the same as the number of points appearing in a plot. On average, in this work, the number of species for any stated structural y-variable is about 100. The ideal value of 4,327 extant species (Table 5.3, column b, bottom row) is derived from Corbet and Hill [1]. A more recent source furnished a value of 5,416 species [2]. The number of species on which scaling measurements have been made is typically a small fraction of either total. See Chapter 2.

The logarithm of the body mass ratio (pWR)

Let the maximum and minimum body weights or masses in a dataset be represented by BW_{max} and BW_{min}, respectively. Then a useful measure of the range in body size, denoted by pWR, is defined by:

$$pWR = \log(BW_{max}/BW_{min}) \tag{5.1}$$

where pWR varies between 0 and 8 in adult mammals. On average, in Part II, pWR for structural variables is about 6 (i.e., six orders of magnitude in body size). See Chapter 2.

Diversity index (DI)

It was suggested in Chapter 2 that the number of taxonomic orders (N_{ord}) and species (N_{sp}) be combined with pWR into a single *diversity index*, denoted by DI, and given by:

$$\text{DI}\,(\%) = 100 \cdot (N_{\text{ord}}/21) \cdot \left(\log(N_{\text{sp}})/\log(4,327)\right) \cdot (\text{pWR}/8.0) \tag{5.2}$$

By virtue of this calculation, we have an objective assessment of how *representative* a specific dataset is of mammals generally. It may also be used to determine whether one dataset for a particular y-variable is preferable to another (as being more representative of mammals at large). In principle, DI could vary between 0 and 100%. A typical value of DI for structural variables in Part II is around 20%. The ratio of two diversity indices is independent of the maximum number of taxonomic orders (21) or species (4,327). Moreover, the definition of DI (relation 5.2) is easily amended to allow for changes in the estimated total number of orders or species.

Mean percent deviation (MPD)

We need measures of how well any particular best-fit line characterizes an individual dataset; one such measure relates to the amount of *scatter* present in the data. If the scatter is large, a best-fit LSQ line will be a poor predictor of the value of a y-variable at any stated body size. The scatter around a best-fit line may be assessed by the *mean percent deviation* (MPD) (Chapter 23). We first provide an expression for the absolute percent deviation (PD) between an observed y-value (y_{obs}) of any specific y-variable and the value of the y-variable (y_{pred}) predicted by a best-fit line. That is:

$$\text{PD} = 100 \cdot |(y_{\text{obs}} - y_{\text{pred}})|/y_{\text{obs}} \tag{5.3}$$

In relation (5.3) the vertical bars represent absolute value; thus PD is always a positive number. PD has an advantage over other measures of dispersion – such as standard deviation or standard error of estimate – in that it is *non-dimensional*, thereby perhaps legitimizing comparisons made across qualitatively (dimensionally) different variables. In order to compute a *mean* PD (denoted by MPD), one averages over all computed values of PD (the *i*th member of the set is denoted by PD_i). The MPD is then calculated as the average of the several PD_i from the relation

$$\text{MPD}\,(\%) = (1/n)\sum \text{PD}_i \tag{5.4}$$

where the summation ($\sum$) is over the number (n) of points in the sample ($i = 1, \ldots, n$). In the present work, "n" is generally the same as the number of species (N_{sp}). MPD is typically between 30% and 40% for structural variables discussed in Part II. The way in which MPD is calculated is illustrated in Table 5.1.

Table 5.1 Calculation of MPD from a set of percent deviations (PD)

x_{obs}	y_{obs}	y_{pred}	PD
a	b	c	d
1	3	2.5	16.7
5	4	5.5	37.5
7	8	7	12.5
MPD	-	-	22.2

Table 5.2 Objective criteria, together with a subjective evaluation, for assessing mammalian scaling datasets and the best-fit LSQ lines derived from them

N_{ord}	N_{sp}	MPD	pWR	DI (%)	r^2	99% CL	Subjective confidence level
a	b	c	d	e	f	g	h
≥ 15	$>1,000$	<10	>6	>60	>0.99	≤ 0.02	Very high
>10	≥ 100	>10	>4	≥ 40	>0.97	≤ 0.04	High
>5	>10	>30	>2	>10	>0.95	≤ 0.06	Medium
≤ 5	≤ 10	>50	≤ 2	≤ 10	<0.95	>0.06	Low

N_{ord} = number of orders; N_{sp} = number of species; MPD = mean percent deviation; pWR = logarithm of weight or mass ratio; DI = diversity index; r^2 = coefficient of determination; CL = confidence limits.

Columns a and b provide three values for an x-variable (X_{obs}) and a y-variable (Y_{obs}), respectively. Column c offers hypothetical y-values predicted by a best-fit LSQ line. The percent deviations between columns b and c are given in column d (e.g., $100 \cdot |3 - 2.5| / 3 = 16.7$). The mean percent deviation (MPD) of the three percent deviations is 22.2 (Table 5.2, column d, bottom row).

Coefficient of determination (r^2)

The degree to which variation in the y-values is accounted for by variation in the x-values for any specific dataset is given by the coefficient of determination (i.e., the correlation coefficient (r) squared). See Chapter 23. Values of the coefficient of determination may vary between 0 and 1. The ideal value is 1 (for those lines with slopes not equal to zero). For structural y-variables in Part II, the coefficient of determination averages about 0.975 (see Table 5.2, column f).

Slopes, confidence limits, and confidence intervals

The primary parameter of interest in scaling studies is the slope of a best-fit (LSQ) line calculated from the x- and y-values, normally after transforming to logarithmic coordinates. In adult mammals generally, functional and structural slopes jointly vary between $-1/3$ and $10/9$. I have usually calculated 99% *confidence limits* on the slopes; these are more rigorous than the 95% confidence limits often employed. The common practice of not reporting confidence limits on slopes (in particular) is unfortunate, as the stated values of the slopes are then difficult to interpret. Often we are interested in the *confidence interval*. If we are given a slope of 0.95 and confidence limits of ±0.05 then the confidence interval for the slope is 0.90 to 1.00.

Evaluative parameters

Seven *evaluative* parameters employed in this study were defined above. These comprise the number of orders (N_{ord}) (see Table 5.2, column a) in a sample, the number of species (N_{sp}) (column b), the mean percent deviation (MPD) (column c), the logarithm of the body weight or mass range (pWR) (column d), the diversity index (DI) (column e), the coefficient of determination (r^2) (column f), and the 99% confidence limits (CL) on the slope (column g) (see Chapters 2, 3, 23).

It is useful to define numerical classes for these parameters in terms of their *qualitative* implications. This was carried out after a careful analysis of the structural data displayed in Part II. The findings of this study, formulated partly with an eye to simplicity, are brought together in Table 5.2. Very few existing datasets contain data for more than a thousand species. If such datasets are taxonomically diverse ($N_{ord} > 15$) and span six or more orders of magnitude in body size (pWR), we are likely to find very high diversity indices (DI $> 60\%$), narrow 99% confidence limits ($\leq\pm0.02$), and high coefficients of determination ($r^2 > 0.99$). We can reasonably have *very high* confidence that such datasets are representative of *mammals generally* (see Table 5.2, column h, row one).

On the other hand, if the N_{ord} is less than five and the number of species (N_{sp}) is less than ten (and so on), we will have *low* confidence that the stated slope is a useful measure of the (usually unknown) slope that might apply to mammals generally (see Table 5.2, column h, row four). Such considerations entered into the construction of Table 5.2. Inevitably, any attempt along these lines has a degree of arbitrariness.

A typical table appearing in Part II

Table 5.3 illustrates the format of many of the tables seen in Part II. There are six parameters (columns a–f) plus the 99% confidence limits on the slope (column g), giving a total of seven evaluative parameters provided for every slope, insofar as possible. Idealized values for the several parameters are shown in Table 5.3, row four.

Data sources

Table 5.3, column h, shows how *data sources* are referenced. In Part II the sources here denoted by 1 and 2 are provided in the references for each chapter. Note that in the great

Table 5.3 The format of a typical table appearing in Part II

N_{ord}	N_{sp}	MPD	pWR	DI (%)	r^2	Slope $\pm$ 99% CL	Source
a	b	c	d	e	f	g	h
2	5	-	2.6	0.6	0.99	0.942 ± 0.007	1
4	33	17.4	3.1	3.1	0.99	0.936 ± 0.045	2
12	74	23	6	22	0.993	0.954 ± 0.024	From Lit.
21	4,327	0	8	100	1.00	$- \pm 0.000$	-

See Table 5.2. Lit. = literature. In row 2, columns d and e, the fact that pWR and DI are numerically the same (3.1) is a coincidence.

majority of cases the data have been "screened" (see Appendix B), so that the resultant datasets are usually somewhat smaller than those cited in the given papers. Consequently, the slopes (or exponents) reported here are often slightly different from those originally reported by the author(s). Row three, pertaining to the scaling of skin mass, is taken from Chapter 11, Table 11.2; it is based on an extensive review of the literature. The phrase "From Lit." refers to such literature searches, which, in total, contain too many references to make citing more than a few of them practicable. Note, as is typically the case in Part II, that the sample size (N_{sp}) in Table 5.3, column b, row three, is larger than in rows one and two.

Exceptions

The reader may have noticed that the MPD is not provided for the first row in Table 5.3, column c. This was not an oversight. It means that a particular author provided data from which the number of orders and species, and the size range, could be inferred. Hence, the diversity index could be calculated. However, the raw data were not provided. Thus the MPD could not be calculated. Moreover, the slope is that provided by the author, not as recalculated here after data screening. Sometimes it was possible to estimate the 99% percent confidence limits and sometimes not. For example, the author(s) may have used more than one point per species, but the total number of points may be uncertain. In some other cases, it might have been possible to estimate the 99% confidence limits, but I chose simply to report the confidence limits (usually 95%) stated by the author(s). On occasion, data for other evaluative parameters are not provided for similar reasons.

Significant differences in slope for a given *y*-variable

Often we want to know whether two slopes for some y-variable, often with different confidence limits, are significantly different, statistically speaking. This question may be answered rigorously using a t-test [3]. However, a useful rule of thumb, found to work empirically in a number of test cases in the present study, is that if the difference between two slopes is less than the larger of the two confidence intervals,

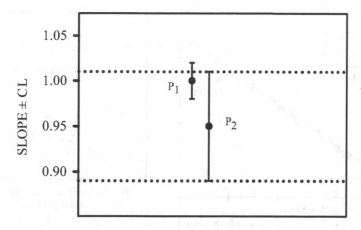

Figure 5.1 A rule of thumb for judging significant differences in slopes.

then the difference in slopes is unlikely to be significant (say at the 99% confidence limit). See Figure 5.1.

In Figure 5.1, the two points P_1 and P_2, representing slopes of (1.00 ± 0.02) and (0.95 ± 0.06) ($\pm$CL), respectively may be judged not significantly different, at some confidence level (to be specified). That is, the two slopes differ by less than the larger confidence limit, since P_1 has a value of 1.00 whereas the upper confidence limit for P_2 extends to 1.01 $(0.95 + 0.06 = 1.01)$.

A typical plot appearing in Part II

Figure 5.2 shows a plot typical of many of those found in the various chapters of Part II. There are a number of special things to be observed. Three panels are labeled a, b, and c. Panel a is a log–log adult interspecific plot. Note that the evaluative parameters ($N_{ord} = 12$, $N_{sp} = 74...$) in the upper left corner are identical to those listed in Table 5.3 (row three). The slope ($b = 0.954 \pm 0.009$) (99% CL) refers to the slope of the best-fit LSQ line. The slope is the same as that shown in Table 5.3 (row three). However, in accord with common practice, the uncertainty in the slope is here specified in terms of the *standard error* (s.e.) in Figure 5.2 (panels a, b), whereas in Table 5.3 the 99% confidence limits (0.024) on the slope are provided (rather than s.e.). These points refer to the general practice throughout Part II. Recall that given the standard error and the number of points (species) one is able to compute any desired confidence level. In Figure 5.2, panel b, the dotted line is equivalent to (i.e., has the same slope and intercept as) that shown in panel a. The three solid circles in panel b represent values for the y-variable exhibited in human ontogeny. Note that tick marks on the right and left vertical axes of panel a are meant to be "read through," as applying equally to the left vertical axis in panel b. This *read-through* convention is often employed in Part II.

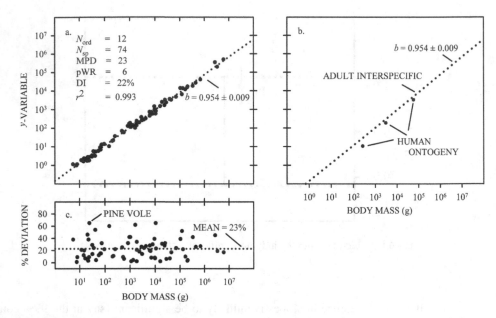

Figure 5.2 The detailed structure of a typical plot appearing in Part II.

A useful but also partly disadvantageous feature of log–log plots is the *compression* that takes place. By plotting the percent deviations on a linear scale (as in panel c) one sees at a glance the actual size of the deviations. The dotted line in panel c refers to the MPD (23%), as also stated in panel a (upper left) and Table 5.3 (column c, row three). Often one or more of the most deviant species are identified by common name (for brevity). For example, in panel c, the point for the pine vole (body mass = 24.3 g) differs from the best-fit LSQ line by 65%. This fact is not at all obvious from panel a. Note that tick marks on the horizontal axes of panels a and c are also meant to be "read through" (see above).

Ontogeny

Every point on an adult interspecific line represents the end-point of an ontogenetic trajectory for some particular species. In Part II, ontogenetic trajectories (usually for the human) are often shown in relation to adult interspecific lines. The ontogenetic trajectories are classified as approaching an adult interspecific line from above, co-linearly or from below. See Chapters 3 and 22.

Pairwise differences in slope

We consider two lines on a log–log plot, having different slopes and diverging from a common intercept (1, 1) (see Chapter 3, Figure 3.3). We aim to characterize the difference between the end-points (y_1, y_2). We do this by expressing the said difference in terms of percent deviation (PD_y), number of doublings (N_d), and a factor (f) by which y_1 is greater (or smaller) than y_2, for pWR = 8.0. These three terms are expressed by:

$$PD_y(\%) = 100 \cdot \left(10^{(pWR \cdot \Delta b)} - 1\right) \tag{5.5}$$

$$N_d = \Delta b \cdot pWR / \log(2) = 26.6 \cdot \Delta b \tag{5.6}$$

$$f = 2^{N_d} \tag{5.7}$$

where Δb is the absolute difference in slope between the two particular lines. See Chapter 3. These three measures (PD, N_d, and f) are meant to provide a quantitative basis for judging whether a stated difference in slope is likely to be biologically significant (e.g., when extrapolated to pWR = 8).

End-sample, mid-sample, and full dataset. A sensitivity analysis

Often one would like to know how *sensitive* the slope and confidence intervals of a best-fit line are to sample size and size range (pWR). An empirical approach to answering this question is as follows. First, we restrict our attention to those datasets representing 100 or more points (species). Such a dataset may be termed a full dataset (FDS). Each such FDS is rank-ordered by body size.

Next one derives a *subset* of the FDS by deleting specifically those records for the ten smallest and ten largest species, as measured by body size. This reduced dataset is called the mid-sample. It consists of 20 fewer species than does the FDS, and generally spans a

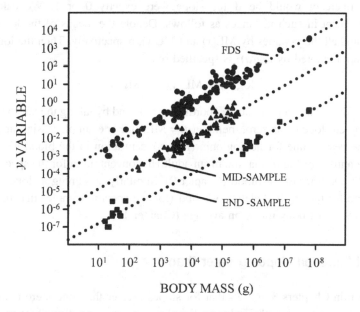

Figure 5.3 Definition of end- and mid-samples and FDS.

Table 5.4 Typical values for number of species (N_{sp}) and size range (pWR) for end- and mid-samples and FDS

Parameter	End-sample	Mid-sample	FDS
N_{sp}	20	83	103
pWR	6.6	3.3	6.6

smaller weight range than does the FDS. Lastly, we construct a second subset of the FDS, termed the end-sample, consisting of the ten smallest and the ten largest species. The end-sample always has the same value of pWR as does the FDS, but has at least 80 fewer species ($100 - 20 = 80$).

The last step is to compute a best-fit LSQ line for the end- and mid-samples, and for the FDS, giving three pairs of slope and associated confidence intervals. The way the end- and mid-samples are defined in relation to the FDS (here for 103 points) is illustrated in Figure 5.3. Observe that the plots for the end-sample (20 points) and mid-sample (83 points) have been displaced downwards for clarity. The essential numerical features of Figure 5.3 are displayed in Table 5.4. Note that in Table 5.4, the value of pWR (3.3) for the mid-sample is only half of the value (6.6) for the end-sample and the FDS. One may find the outcomes of this exercise rather surprising (see Chapters 6, 21).

Intercepts calculated for a pre-assigned slope

When a slope (b) is near a *reference* slope (usually 0 or 1) one may want to estimate what the intercept would be if the slope were exactly 0 or 1. We calculate the revised intercept in such instances as follows. Denote the means of the logarithms of all the x- and all the y-values by $ML(x)$ and $ML(y)$, respectively. Then the logarithm of the intercept (denoted by $\log(a)$) is specified by:

$$\log(a) = ML(y) - b \cdot ML(x) \tag{5.8}$$

The value for the intercept in native coordinates is found by taking the antilog of $\log(a)$. If the assigned slope (b) is zero, then $\log(a) = ML(y)$. For example, assigning a slope of *zero* to the best-fit line for hemoglobin (Hb) concentration in the blood we find, after taking the antilog of $\log(a)$ that the mean adult Hb concentration for the specific sample is 15 g Hb per 100 cm^3 of blood (Chapter 9). Similarly, assigning a slope of 1 to the best-fit line for the fresh skeleton, we find (for the specific sample) that the skeleton comprises 9.5% of body mass, on average (Chapter 12).

Inherent limits on slopes greater than 1

It is shown in Chapters 3 and 24 that for slopes greater than one there is an inherent upper limit (i.e., b_{max}) to the value of the slope for any specific values of pWR and intercept (a). This limiting value is given by:

$$b_{\text{max}} = 1 - (1/\text{pWR}) \cdot \log(a) \quad 0 < a < 1 \tag{5.9}$$

An expression for a tentative minimum slope (for slopes less than 1), denoted by $b_{\text{reflected}}$, is conveyed by:

$$b_{\text{reflected}} = 2 - b_{\text{max}} \tag{5.10}$$

Invariance

Two forms of *invariance* were defined in Chapter 3: absolute and relative. Absolute invariance, which is common in mammalian scaling at the molecular and in some instances at the microscopic levels, means that the value of a y-variable is a constant, independent of body size. Relative invariance means that y-values scale in direct proportion to body size (x-values) on average.

Criteria for scaling at constant shape

If both total body length and total body surface area scale as the one-third and two-thirds power of body size, respectively, then we infer scaling at constant shape. If, however, an *organ* scales with body size raised to the power b (where $b < 1$ or $b > 1$), then the criterion for scaling at constant organ shape is that organ diameters or lengths scale with body size raised to the power $b/3$ (rather than 1/3, as is often assumed) (see Chapters 3, 24).

A repeating units model

If an organ, such as kidney or lung, consists of repeating units, then a stated model (see Chapters 3, 24) tells us that the volume and number of repeating units should scale as the 1/3 and 2/3 powers of *organ* volume, respectively.

Part II chapter organization

The chapters in Part II are organized for the most part systemically and alphabetically (circulatory to urinary system). Within chapters, the organization is based partly on structural size (smaller structures tend to be first) and partly on an alphabetical ordering. This flexible method of organization was chosen with a view to making it easier to locate topics of particular interest.

Standardized body weight table

For a discussion of the present *standardized body weight table* (SBT), see Chapter 4. Primarily the SBT was employed to screen body weights (or masses) prior to computing a best-fit LSQ line (see also Appendix B). However, there are a number of instances in Part II where an author supplied useful (and otherwise largely unavailable) values for a y-variable but no body size data. In those instances, I have (reluctantly) taken the needed body size data from the SBT, as available. Generally, one prefers to use sources where both y- and x-variables are specified in the same work.

Verbal classification of body size in adult mammals

It was suggested in Chapter 2 (Table 2.4) that adult mammals be classified into four verbal *size categories*: small (1–100 g), medium (10^2–10^4 g), large (10^4–10^6 g), and very large (10^6–10^8 g).

See also Chapter 7, Table 7.1. Note that each category spans a 100-fold range in body weight or mass.

Summary

Seven *evaluative* parameters are introduced. Four of these (N_{ord}, N_{sp}, pWR, and DI) are aimed at assessing how well any particular dataset represents mammals generally (Chapter 2). Another three parameters (MPD, 99% CL, and r^2) provide information as to how well a best-fit LSQ line describes the underlying data (Chapter 3). Taken together, these seven evaluative parameters provide a concise, consistent, and objective account of scaling datasets and their best-fit LSQ lines. In addition, a simple tactic based on the concept of end- and mid-samples (each a subset of a full dataset) is introduced for evaluating the sensitivity of slopes to sample size and body size range.

Table 5.2 provides suggested criteria for judging how confident we can be that a specific dataset is representative of mammals generally (columns a, b, d, e) and how well a best-fit LSQ line represents the underlying data (columns c, f, g). Other matters are reviewed, including how to compute the value of an intercept for a certain dataset when the value of the slope is pre-assigned. It should be kept in mind that the above inventory of parameters and methods of calculation would be largely pointless but for the fact that the analyses reviewed in Part II are based for the most part on *unearthing*, *screening*, and *re-analyzing* the original data as reported by many workers. This contrasts with the more usual approach of accepting the varied published data as given. A synopsis of symbols and abbreviations used throughout this study is provided below.

Abbreviations and symbols

a	Intercept of a best-fit line (usually the y-value when $x = 1$ for x, y in log–log coordinates)
AI	Absolute invariance
AR	Aspect ratio (e.g., ratio of length to diameter of a long bone)
b	Slope of a best-fit line or exponent in a power function

b_{max}	The maximum slope (for slopes > 1) expressed as a function of intercept and size range
BM	Body mass
BMR	Basal metabolic rate
BW	Body weight or mass
BW_{min}	Minimum body weight or mass in a set of body weights
BW_{max}	Maximum body weight or mass in a set of body weights
CI	Confidence interval(s)
CL	Confidence limit(s) on a slope or intercept
CO	Cardiac output
DI	Diversity index: indicates how representative a sample is of mammals generally
end-	End-sample; consists of the ten smallest and ten largest species in a dataset by body weight
FDS	Full dataset(s); contains all the records in a dataset; see end- and mid-
From Lit.	Data taken from the literature but explicit reference not given (owing to the large number of citations)
g, g	Grams or force of gravity
G.	Greek
GDP	Gross domestic product
Hb	Hemoglobin
L.	Latin
lin–lin	A plot that is linear in the x- and y-coordinates
log	Logarithm
log–log	A plot that is logarithmic in the x- and y-variables
LSQ	Least squares method for fitting a best-fit line to a dataset
mya	Million years ago
mid-	Mid-sample; consists of FDS less End-sample
MPD	Mean percent deviation(s)
N_d	Number of doublings
N_{ord}	Number of taxonomic orders represented in a given dataset
N_{pnts}	Number of points in a dataset
N_{sp}	Number of taxonomic species represented in a given dataset
PD	Percent deviation(s) computed in absolute form
ppm	Parts per million
pWR	Measure of weight range calculated as the logarithm of (BW_{max}/BW_{min})
r	Coefficient of correlation
r^2	Coefficient of determination
RI	Relative invariance
RO	Rank order
s_a	Standard error in intercept
s_b	Standard error in slope
s.e.	Standard error in intercept or slope
SBT	Standardized body weight or mass table

SMR	Standard metabolic rate
SQRT	Square root
STD	Standard deviation
STPD	Refers to a volume of gas at Standard Temperature Pressure Dry
ton	So-called "short ton" = 2,000 lb or 907.2 kg
tonne	1,000 kg
μm	Unit of length = 10^{-6} m
y'	Value of y-variable predicted by best-fit LSQ line for a given value of the x-variable
[ ]	Used to indicate my interpolations and/or clarifications in quotations and elsewhere
$\sum$	Summation sign

References

1. Corbet, G.B. and Hill, J.E. (1991). *A World List of Mammalian Species*, 3rd edn. Oxford: Oxford University Press.
2. Wilson, D.E. and Reeder, D.M. (eds.) (2005). *Mammal Species of the World: A Taxonomic and Geographic Reference*. Baltimore, MD: Johns Hopkins University Press.
3. Edwards, A.L. (1976). *An Introduction to Linear Regression and Correlation*. San Francisco, CA: W.H. Freeman.

Part II

Empirical analyses

6   Body length, girth, and surface area

Large and small animals are not similar geometrically but rather tend to be physiologically... similar.

Brody 1945

The assumption of geometric similarity [in adult mammals] *is not as unrealistic as one might expect.*

Schmidt-Nielsen 1984

It is often asserted that adult mammals of differing sizes are not geometrically similar (of constant shape) within or between species (see above quote by Brody). Commonly little or no empirical evidence is cited; perhaps the statement is assumed to be self-evident. In any case, the scaling of body length, girth, and surface area with body size stand to shed light on the scaling of external body shape in mammals. Most field guides to mammal identification give measurements of average body length (nose to tip of tail) and tail length. There are many fewer measurements available of body girth and surface area than of body length; still, these three measures afford the best evidence available on the scaling of external shape in mammals generally. Recall that an analysis of body shape based on only one parameter, such as body length or girth or body surface area, may give a misleading picture as to geometric similarity (Chapter 24). For example, the scaling of body length might be consistent with constant shape, but scaling of girth might not (see below). For a brief discussion of *elastic* similarity see Chapter 12.

Scaling of body length in mammals

The results of a LSQ analysis of body length versus body weight or mass in 1,266 species of adult mammals are summarized in Table 6.1, row nine and Figure 6.1 [1–13]. The body length data range from a shrew (*Suncus etruscus*) to a blue whale (*Balaenoptera musculus*). Note that the diversity index for these data is 76% (column e, row nine). The means of columns a to g are given in row ten.

The best-fit LSQ line for 1,266 species of adult mammals generally has a slope of 0.354 ± 0.001 (s.e.) (Table 6.1, row nine and Figure 6.1, panel a, dotted line). The slope of 0.354 ± 0.003 (99% CL) exceeds the *reference* slope of 1/3 (for geometrical

Table 6.1 Scaling of body length in mammals at large and in various subgroups

Column/Row	N_{ord} a	N_{sp} b	MPD c	pWR d	DI (%) e	r^2 f	Slope ± 99% CL g	Entity h
1	1	144	11	3	1.1	0.886	0.298 ± 0.022	Artiodactyla
2	1	180	10	3.6	1.3	0.923	0.307 ± 0.017	Carnivora
3	1	72	13	3.6	1.1	0.954	0.330 ± 0.023	Cetacea
4	1	161	8.5	2.7	1.0	0.917	0.362 ± 0.022	Chiroptera
5	15	1,000	11	6.4	47	0.979	0.335 ± 0.004	Land mammals
6	1	145	8	3.9	1.4	0.973	0.328 ± 0.0115	Marsupialia
7	1	108	9	3.4	1.1	0.925	0.285 ± 0.021	Primates
8	1	331	8	3.9	1.6	0.961	0.333 ± 0.010	Rodentia
9	19	1,266	13	7.9	76	0.982	0.354 ± 0.003	Mammals at large
10	4.6	379	10.2	4.3	14.6	0.944	0.326	Means

N_{ord} = number of orders; N_{sp} = number of species; MPD = mean percent deviation; pWR = logarithm of weight or mass ratio; DI = diversity index; r^2 = coefficient of determination.

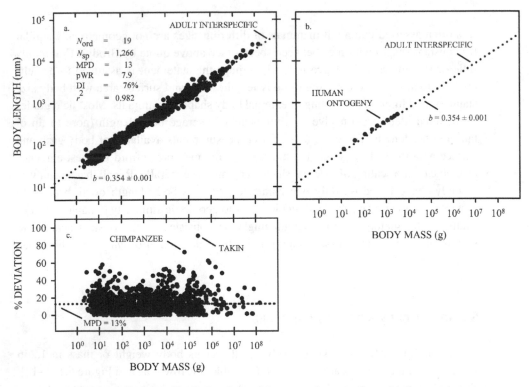

Figure 6.1 Scaling of body length in adult mammals generally and in human ontogeny.

similarity) by 0.021. This deviation from 1/3 implies that a large blue whale is 1.5 times longer than it would be if body length in adult mammals (for pWR = 8.0) scaled precisely in accord with geometric similarity (see Chapter 3, relation (3.5); $N_d = 26.6 \cdot \Delta b = 0.56; f = 2^{Nd} = 2^{0.56} = 1.5$). Observe that the numerical data in the upper left corner of

Figure 6.1a are the same as those in Table 6.1, row nine. (Note, though, that the confidence level (s.e.) on the slope shown in Figure 6.1 panel a is ±0.001, whereas Table 6.1, column g, row 9, gives the 99% CL, ±0.003.)

Jerison [14] reported a slope of 0.33 for body length on body weight in carnivores and ungulates. See also Prothero [15]. Humans [16] have a body length (height) 18% greater than predicted by the best-fit LSQ line for mammals at large and 41% greater than predicted for primates. (Incidentally, based on a small sample, I find from Greenewalt [17] that bat wing-length scales as the 0.304 ± 0.071 (99% CL) power of body weight (N_{sp} = 13, pWR = 2.6).) See also Norberg [8].

In Figure 6.1b, observe that body length during human ontogeny approaches the adult interspecific line from slightly above [18, 19]. In Figure 6.1c, the MPD between the data points and the best-fit LSQ line is 13%, with a standard deviation of 11%. (The species "takin" (*Budorcas taxicolor*), identified in panel c, refers to a goat-like member of artiodactyla found in China.)

Scaling of body length in subgroups of mammals

A detailed analysis of body length in various subgroups of adult mammals is given in Table 6.1, rows one to eight. The mean slope for nine cases (rows one to nine; artiodactyls to mammals at large) featured in Table 6.1, column g, bottom row, is 0.326 $\pm$ 0.024 (STD). The slopes range from 0.285 (primates) to 0.362 (bats). In an earlier paper I reported that body length scales as the 0.335 power of body size in 938 species of land mammals [15]. Compare this result with Table 6.1, column g, row five.

The slopes for body length versus body mass in cetaceans, land mammals, marsupials, and rodents are not statistically different from the *reference* slope of 1/3 (consistent with constant shape) (see Chapters 3, 24). The greatest deviation from the reference slope (Chapter 3, Table 3.5) of 1/3 is 0.048 (0.333 − 0.285 = 0.048) for primates.

The three major groups of mammals (bats, cetaceans, land mammals) with divergent adult body shapes have slopes (and 99% CL) for the scaling of body length as a function of adult body mass of 0.362 ± 0.022, 0.330 ± 0.023 and 0.335 ± 0.004 (Table 6.1, column g, rows four, three, and five, respectively). The slope of 0.362 ± 0.022 (99% CL) for bats (row four) differs from 1/3 by 0.03, an amount greater than the 99% CL of $\pm$ 0.022. Body length in bats appears not to scale strictly in agreement with constant shape.

Scaling of body girth

This analysis of body *girth* (measured at heart level) in land mammals is adapted from a prior study [15]. The girth data extend from the house mouse (*Mus musculus*) to the African elephant (*Loxodonta africanus*). For data on land mammals, see [20–24, 26–28], and for those on marine mammals, see [25, 29]. In this study (atypically) multiple points for some species (e.g., six points for the dog) were used in order to expand the dataset. The results of a LSQ analysis of girth in land mammals are summarized in

Table 6.2 Scaling of body length, girth, and surface area in adult land mammals

	N_{ord}	N_{pnts}	N_{sp}	MPD	pWR	DI (%)	r^2	Slope ± 99% CL	Entity
Column/Row	a	b	c	d	e	f	g	h	i
1	15	1,000	1,000	11	6.4	47	0.979	0.335 ± 0.004	Body length
2	6	43	26	6	5.3	7.4	0.993	0.365 ± 0.013	Body girth
3	7	30	16	14	5.5	7.6	0.994	0.670 ± 0.027	Body surface area

See Table 6.1. N_{pnts} = number of points in sample (column b); DI (column f) calculated from columns a, c, and e.

Table 6.2, row two. The slope of the best-fit line is 0.365 ± 0.013 (99% CL). This slope differs from a reference slope (Chapter 3, Table 3.5) of 1/3 by 0.032. Stahl and Gummerson [28] reported a slope of 0.37 for girth in five species of primates (pWR = 2.0); the 95% CL span the interval 0.35 to 0.38.

Scaling of body surface area

Data bearing on the scaling of external surface area as a function of body mass in 16 species of land mammals are summarized in the bottom row of Table 6.2 [15]. These data extend from the masked shrew (*Sorex cinereus*) to the black rhinoceros (*Diceros bicornis*). Multiple points were included for some species, such as the dog. The slope of the best-fit line (0.670 ± 0.027) (99% CL) is consistent with a reference slope of 2/3 called for by geometrical similarity. Keep in mind, however, that surface area is relatively insensitive to moderate changes in shape (Chapters 3, 24).

Scaling of end- and mid-samples compared with the FDS

In Table 6.1, each of nine groups (cetaceans, row three, excepted) is represented by more than 100 species. Hence eight body length datasets allow one to test how slopes and their CL vary as between end- and mid-samples and the full dataset (FDS) (see Chapter 5). The results of this analysis are shown in Figure 6.2.

In Figure 6.2, except for the cetaceans (entry e), each group is represented by a triplet of slopes, comprising the end-, mid-samples, and FDS, reading from left to right. The slopes are given together with their 99% CL. Each triplet is plotted in accord with rank order for the FDS. The main finding is that in each case except one (that for land mammals, item g) the 99% CL for the FDS include the slope determined for the end-sample (consisting of the ten smallest and ten largest species). If we were to use the slightly wider CL on the end-sample (rather than on the FDS), we would infer that none of the end-sample slopes differs significantly from the slope for the FDS.

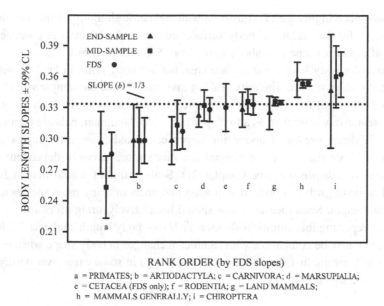

Figure 6.2 Slopes and 99% CL for adult body length as found in end- and mid-samples and FDS.

The instance of mammals generally (Table 6.1, row nine) is especially striking. The slope for the end-sample (20 species) is 0.357 ± 0.017 (99% CL) whereas the slope for the full dataset (1,266 species) is 0.354 ± 0.003 (99% CL). In this instance, there is no significant difference between a slope based on 20 species and one based on 1,266 species! The price paid for this economy is that the 99% CL on the end-sample slopes tend to be a little larger than those for the FDS. The mean absolute deviation between the 99% CL on the FDS and for the 99% CL on the end-samples is 0.012 ± 0.009 (STD).

In addition, I computed the absolute deviation ($|\Delta b|$) in slope between the end-sample and FDS for each of eight cases in Table 6.1 (cetaceans omitted; $N_{sp} < 100$). The mean of the eight absolute deviations is 0.0075 ± 0.0047 (STD). The findings of this study, except for bats (reported slope of 0.317), are broadly in agreement with those of Silva [30], whose analysis is based on a larger sample than used here. Unfortunately, this paper was not discovered until after the above analysis had been completed.

Summary

Scaling of body length was analyzed in 1,266 species of adult mammals drawn from 19 taxonomic orders (Table 6.1, row nine). The slope of the best-fit LSQ line is 0.354 ± 0.003 (99% CL). The slopes for body length in cetaceans ($b = 0.330$), land mammals (0.335), marsupials (0.328), and rodents (0.333) are consistent with scaling at constant body shape (Table 6.1). The range in slopes for body length in various subgroups of mammals extends from 0.285 in primates to 0.362 in bats. The slope (0.365) for girth in land mammals is higher than is consistent with geometric similarity. However, the slope ($b = 0.670$) for the scaling of body surface area in land mammals is consistent with scaling at constant shape (see above quote from Schmidt-Nielsen).

The data for body length and surface area, but not body girth, in land mammals are consistent with geometric similarity. Taking into account that the sample size ($N_{sp} = 26$, Table 6.2, row two) for body girth is quite restricted (relative to the body length data), we may tentatively infer that, as a first approximation, land mammals are geometrically similar. Evidence presented above for slopes calculated from end-samples and full datasets suggests that slopes in general are much more strongly dependent on *size range* than on *sample size* (see Chapter 21). Scaling studies would benefit from the determination of girth and body size in adult mammals in many more species and over wider size ranges. Such measurements should be relatively straightforward.

It bears repeating that empirical slopes near 1/3 for body length, and near 2/3 for body surface area, may be compatible with significant changes in body shape when evaluated over the full mammalian size range (pWR = 8.0) and in some cases over narrower size ranges (see Chapter 24).

References

1. Boitani, L. and Bartoli, S. (1982). *Simon & Schuster's Guide to Mammals*. New York: Simon and Schuster.
2. Dorst, J. and Dandelot, P. (1972). *A Field Guide to the Larger Mammals of Africa*, 2nd edn. London: Collins.
3. Eisenberg, J.F. (1981). *The Mammalian Radiations*. Chicago, IL: University of Chicago Press.
4. Flannery, T. (1995). *Mammals of New Guinea*. Ithaca, NY: Cornell University Press.
5. Garbutt, N. (1999). *Mammals of Madagascar*. New Haven, CT: Yale University Press.
6. Grzimek, H.C.B. (ed.) (1968). *Grzimek's Animal Life Encyclopedia: Mammals*. New York: Van Nostrand Reinhold, Vol. 10–13.
7. MacDonald, D. (1984). *The Encyclopedia of Mammals*. New York: Facts On File.
8. Norberg, U.M. (1981). Allometry of bat wings and legs and comparison with bird wings. *Philosophical Transactions of the Royal Society London*, **292**:359–398.
9. Nowak, R.M. and Paradiso, J.L. (1983). *Walker's Mammals of the World*, 4th edn. Baltimore, MD: Johns Hopkins University Press.
10. Strahan, R. (ed.) (1983). *The Australian Museum Complete Book of Australian Mammals*. London: Angus & Robertson Publishers.
11. van Zyll de Jong, C.G. (1983). *Handbook of Canadian Mammals: 1. Marsupials and Insectivores*. Ottawa: National Museums of Canada.
12. van Zyll de Jong, C.G. (1985). *Handbook of Canadian Mammals: 2. Bats*. Ottawa: National Museums of Canada.
13. Whitaker, J.O. Jr. (1980). *The Audubon Society Field Guide to North American Mammals*. New York: Alfred A Knopf.
14. Jerison, H.J. (1973). *Evolution of the Brain and Intelligence*. New York: Academic Press.
15. Prothero, J. (1992). Scaling of bodily proportions in adult terrestrial mammals. *American Journal of Physiology*, **262**:R492–R503.

16. Ogiu, N., Nakamura Y., Ijiri I., Hiraiwa K. and Ogiu T. (1997). A statistical analysis of the internal organ weights of normal Japanese people. *Health Physics*, **72**:368–383.

17. Greenewalt, C.H. (1962). Dimensional relationships for flying animals. *Smithsonian Miscellaneous Collections*, **144**:1–46.

18. Scammon, R.E. and Calkins, L.A. (1924). The relation between body-length and body-weight in the human embryo and fetus. *Proceedings of the Society for Experimental Biology and Medicine*, **21**:549–551.

19. Klein, A.D. and Scammon, R.E. (1929). Relations between surface area, weight and length of the human body in prenatal life. *Proceedings of the Society for Experimental Biology and Medicine*, **27**:456–461.

20. Benedict, F.G. (1936). *The Physiology of the Elephant*. Washington, DC: Carnegie Institution of Washington.

21. Brody, S. and Elting, E.C. (1926). Growth and development with special reference to domestic animals: II. A new method for measuring surface area and its utilization to determine the relation between growth in surface area and growth in weight and skeletal growth in cattle. *Missouri Agricultural Experimental Station, Research Bulletin*, **89**:4–18.

22. Cowgill, G.R. and Drabkin, D.L. (1927). Determination of a formula for the surface area of the dog together with a consideration of formulae available for other species. *American Journal of Physiology*, **81**:36–61.

23. Dempster, W.F. (1955). Space requirements of the seated operator: geometrical, kinematic, and mechanical aspects of the body with special reference to the limbs. *USAF Wright Air Development Center. Technical Report* 55–159.

24. Diack, S.L. (1930). The determination of the surface area of the white rat. *Journal of Nutrition*, **3**:289–296.

25. Kenyon, K.W. (1961). Cuvier beaked whales stranded in the Aleutian Islands. *Journal of Mammalogy*, **42**:71–76.

26. Ritzman, E.G and Colovos, N.F. (1930). Surface areas of sheep. *University of New Hampshire, Agricultural Experimental Station*, **32**:2–8.

27. Sachs, R. (1967). Liveweights and body measurements of Serengeti game animals. *East African Wildlife Journal*, **5**:24–36.

28. Stahl, W.R. and Gummerson, J.Y. (1967). Systematic allometry in five species of adult primates. *Growth*, **31**:21–34.

29. West, G.C., Burns, J.J. and Modafferi, M. (1979). Fatty acid composition of Pacific walrus skin and blubber fats. *Canadian Journal of Zoology*, **57**:1249–1255.

30. Silva, M. (1998). The scaling of mammal body length: elastic or geometrical similarity in mammalian design. *Journal of Mammalogy*, **79**:20–32.

7 Body size distribution in adult mammals

Animals come in different sizes, and the little ones are much more common than the big.

Colinvaux 1978

The standardized body weight table (SBT) (Chapter 4) is here used to examine three aspects of body size distribution in adult mammals. The first aspect concerns the distribution of body size over four size-classes (small, medium, large, and very large). See Chapter 2, Table 2.4.

Distribution of body sizes

Mean body sizes for 1,410 species were obtained either directly from the *standardized body weight table* (SBT) or by averaging minimum and maximum values in the SBT. See Chapter 4. The way that body sizes are distributed over the given four size-classes is summarized in Table 7.1. (Since pWR = 8.0 for mammals in general, dividing this into four size-classes each spanning a size range of pWR = 2 gives the most convenient division for scaling purposes.)

Note that of the 1,410 distinct species represented in Table 7.1, 1,080 (541 + 539 = 1,080) or some 76.6% (100·1,080/1,410 = 76.6) fall into the small and medium size-classes (column d, first two rows, brackets). Only 24% (21 + 3) of species are found in the large and very large size-classes. The diversity index (column f) ranges from 8.9 to 12.5% for the first three classes, but falls to 3.6% for the "very large" class [1]. Here we restrict our attention to data for the "general" class of mammals (i.e., gender not specified).

Ratio of maximum to minimum body weight by taxonomic order

Recall that in Chapter 2, Table 2.3 and in Figure 2.1, we identified – as nearly as permitted by the data on hand – the smallest and largest species for each taxonomic order. Here we look at another measure of the distribution of body sizes in adult mammals. For every species for which we have "general" (i.e., gender not specified)

Table 7.1 Number of species found in the "general" category of SBT in each of four size-classes

Size-class		N_{ord}	N_{sp} (%)	pWR	DI (%)
Verbal	Mass range (g)				
a	b	c	d	e	f
Small	10^0–10^2	10	541 (38)*	2	8.9
Medium	10^2–10^4	14	539 (38)	2	12.5
Large	10^4–10^6	12	292 (21)	2	9.7
Very large	10^6–10^8	7	38 (3)	2	3.6

N_{ord} = number of orders; N_{sp} = number of species; pWR = logarithm of weight range; DI = diversity index; * (100·541/1,410 = 38).

Table 7.2 Mean ratio of maximum to minimum body mass per taxonomic order

Order	Body mass ratio (max/min)	STD	N_{sp}	Order	Body mass ratio (max/min)	STD	N_{sp}
Artiodactyla	2.03	1.17	144	Lagomorpha	1.86	0.45	20
Carnivora	2.28	0.96	140	Marsupialia	2.71	2.42	113
Cetacea	1.71	0.74	32	Perissodactyla	1.48	0.31	7
Chiroptera	1.63	0.40	181	Pinnipedia	2.81	2.22	19
Edentata	1.62	0.36	13	Primates	1.88	0.66	60
Insectivora	2.06	0.79	64	Rodentia	2.00	0.92	318

See Table 7.1. STD = standard deviation.

minimal and maximal body size data in the SBT, there is a calculable ratio of maximal (max) to minimal (min) body size that is greater than one, if only slightly (see below).

An analysis of this body size *max/min* ratio for 12 different taxonomic orders is summarized in Table 7.2. One of the two smallest mammals in this dataset is a pygmy white-toothed shrew (*Suncus etruscus*) (1.5 g) (Chapter 2, Table 2.3); the largest is a black right whale (*Eubalaena glacialis*) (80×10^6 g). The *mean* value of this max/min ratio over 12 taxonomic orders and 1,111 species is 2.0, with a standard deviation of 0.40. The *maximum* value of this max/min ratio for any species in the sample is 18.9, for the eastern gray kangaroo (*Macropus giganteus*) (i.e., 66 kg/3.5 kg = 18.9); the *least* value of the max/min ratio for any species in the sample is 1.025, for the lesser flying fox (*Pteropus mahagamus*) (1,250 g/1,220 g = 1.025).

Among the 12 taxonomic orders represented in Table 7.2, Pinnipedia show the greatest max/min ratio (2.81 ± 2.22 (STD)) and Perissodactyla the smallest (1.48 ± 0.31 (STD)).

Sexual dimorphism

In most mammalian species for which data are available, adult males are larger than females, although in some species adult females are larger than males [2]. A study

Table 7.3 Dimorphic body size ratio by taxonomic order

Order	Dimorphic ratio (M/F)	STD	N_{sp}	Order	Dimorphic ratio (M/F)	STD	N_{sp}
Artiodactyla	1.58	0.44	47	Lagomorpha	0.91	0.09	10
Carnivora	1.43	0.33	41	Marsupialia	1.41	0.37	68
Cetacea	1.37	0.58	19	Pinnipedia	2.32	1.18	23
Chiroptera	0.99	0.27	32	Primates	1.48	0.30	45
Insectivora	1.08	0.18	15	Rodentia	1.07	0.21	72

See Table 7.1. STD = standard deviation.

(based on the SBT) of the *mean* dimorphic body size ratio (male/female) in each of ten taxonomic orders totalling 372 species is summarized in Table 7.3. The approach was to take mean body weights directly from the SBT, or to calculate mean body weights by averaging minimal and maximal body weights, for males and females separately. Then the ratio of mean body weight (or mass) for males relative to females was computed. The mean dimorphic ratios per order are reported.

The mean dimorphic ratio (M/F) for ten orders (N_{sp} = 372) is 1.37, with a STD of 0.54. On average, females are slightly larger than males in the orders Chiroptera (ratio = 0.99 $\pm$ 0.27 (STD)) and Lagomorpha (ratio = 0.91 $\pm$ 0.09 (STD)). Pinnipedia show both the highest ratio of maximal to minimal body weight (2.81) (Table 7.2) and the highest dimorphic ratio (2.32 $\pm$ 1.18 (STD)) (Table 7.3). Monogamous primate species exhibit little by way of sexual dimorphism [3].

Summary

The great majority (almost 80%) of adult mammals, as represented in the standardized body weight table (SBT), fall into the *small* and *medium* size-classes, ranging from 1 to 10,000 g. On average, for the given data, the maximal body size for any given adult species is *twice* the minimal size. Equivalently, on average, maximal size is four-thirds of the mean and minimal size is two-thirds of the mean body size. In ten taxonomic orders, the mean ratio of male to female body size was found to be 1.4 $\pm$ 0.39 (STD). Thus, sexual dimorphism may account for about 70% of the total variation seen in body size in the present sample ($100 \cdot (1.4)/2 = 70$). Despite the great overall range in body size amongst adult mammals, it is clear that the distributions in body mass within species and orders for a large sample are more regular than might have been expected.

References

1. Colinvaux, P. (1978). *Why Big Fierce Animals are Rare: An Ecologist's Perspective*. Princeton, NJ: Princeton University Press.
2. Ralls, K. (1976). Mammals in which females are larger than males. *Quarterly Review Biology*, **51**:245–276.
3. Leutenegger, W. (1978). Scaling of sexual dimorphism in body size and breeding systems in primates. *Nature*, **272**:610–611.

8 Body composition

The same minerals are present and they perform the same sorts of functions in the bodies of grasshoppers, mealworms, oysters, trout, duck, and foxes as they do in our own.

Widdowson 1968

There appears to be very little evidence... that the chemical composition of animals changes systematically with body size.

Kleiber 1961

Compositional studies pertain to the kind and quantity of components making up an animal or plant, without regard to how those components are organized in space and time. Composition may be regarded as the lowest level of analysis for scaling studies. Most studies of composition in mammals have focused on humans (but see above quote from Widdowson). Those interested in human body composition should consult the work by Heymsfield *et al.*, which provides an exhaustive account of current knowledge and understanding [1]. An analytical framework for compositional studies is formulated by Wang *et al.* [2]. They recognize five levels of analysis: atomic, molecular, cellular, tissue, and whole body. This framework is employed here, with minor modifications. In reading the following material it will be helpful to keep in mind the concept of *relative invariance* (Chapters 3, 5, 24). This term applies, in particular, when a *structural* slope is equal or nearly equal to 1, and the x- and y-variables have comparable dimensions (volume, weight or mass). It means that on average a y-variable is a constant fraction of the x-variable, in a given sample. In this chapter we analyze the LSQ results for seven y-variables, 11 instances, and 328 records, all in tabular form.

Origins of chemical composition studies

Before Robert Boyle (1627–1691) published "The Sceptical Chymist" in 1661, it was widely accepted that there exist only four "bodies" or elements (air, earth, fire, water); these candidate elements testify to the conceptual difficulty of distinguishing among physical states (gas, liquid, solid) or the process of oxidation on the one hand and chemical elements on the other. Boyle argued that an "element" should be "homogeneous" and not further "resolvable" into other "substances." By the time of Lavoisier

(1743–1794), 30 distinct elements were known, including calcium, carbon, chlorine, hydrogen, nitrogen, oxygen, phosphorus, and sulfur [3]. It was a short step then for Dalton (1766–1844) to propose that the elements combine in integral proportions to form chemical compounds. With some oversimplification, modern chemistry was born and never looked back.

A brief historical background to the study of body composition is provided by Heymsfield *et al.* [1], by Keys and Brozek [4], and by Siri [5]. The study of adult body composition in different mammalian species is mostly skirted in general works on scaling. The first analytical report bearing on the chemical composition of humans may be owing to the Dutch physiologist Jacob Moleschott (1822–1893), whose 1859 estimate of total body water (67.6%) is in rough agreement with current estimates of about 62.4% for adult males aged 25 to 46 (64.5% in adult mammals generally; see below). One of the first serious compilations of *elemental abundances* in humans is due to Hackh [6]. There followed a spate of analytical studies, by Mitchell *et al.* [7], Widdowson *et al.* [8], Widdowson and Dickerson [9], Forbes *et al.* [10, 11], and more recently Wang *et al.* [2] and Heymsfield *et al.* [1].

Elemental abundances

The chemical elements most widely distributed in terrestrial living creatures are the ones (apart from inert helium and neon) that are the commonest in the Universe – hydrogen, oxygen, carbon and nitrogen.

Trimble 1997

Thanks to Alpher, Bethe, and Gamow [12] and their "alphabetical" paper, we have a creation "myth" for our universe based on physics. In their ground-breaking paper, the authors argued that all the hydrogen and helium in our universe was created during the first three minutes of the Big Bang, an explosion of unimaginable intensity, thought to have occurred just before noon (EST), 13.8 billion years ago. Although these workers believed all the elements were created during the "nucleon" phase of the Big Bang, it is now known that the heavier elements (in particular those essential to life) were "cooked up" later in stars. The cosmic sequence of events is currently pictured as a Big Bang with the formation of hydrogen and helium nuclei, inflation, formation of black holes and then stars and galaxies, followed by the formation of the heavier elements, subsequently dispersed into the interstellar medium by massive stellar explosions. Our own star (sun) evidently formed in a region of the Milky Way where at least one supernova explosion occurred earlier. Barring a like explosion "nearby," life perhaps would not have evolved on earth.

Estimates of the elemental abundances in the universe at large, in the solar system, on the sun's surface, and in the adult human are summarized in Table 8.1.

The composition of the solar system and the sun's surface closely mirror the composition of the universe at large. These data show that while the essential elements in a mammal are among the more common elements in our universe (see above quote by Trimble), their detailed distribution is very different. Carbon is the "signature" element

Table 8.1 Selected elemental abundances by percent of total mass

Element	H	He	C	N	O	Fe	Source
Universe	73	25	0.4	0.1	0.9	0.1	[13]
Solar system	71	27	0.3	0.1	1.0	0.1	[14]
Sun's surface	73	25	0.3	0.1	0.8	0.3	[15]
Human being	10	-	23	2.6	61	0.005	[2, 6]

Table 8.2 Elemental abundances in the adult human as a percent of body mass

Element	Amount (%) (1919)	Amount (%) (1992)
Oxygen	62.43	61
Carbon	21.15	23
Hydrogen	9.86	10
Nitrogen	3.10	2.6
Calcium	1.90	1.4
Phosphorus	0.95	0.83
Sulfur	0.16	0.20
Potassium	0.23	0.20
Sodium	0.08	0.14
Chlorine	0.08	0.14
Magnesium	0.027	0.027
Total	100	99.54
Source	[6]	[2]

in humans and organisms generally (see below); it is about 60 (23/0.4 = 57.5) times more common in the adult human body than in the universe at large. Oxygen, chemically less adept than carbon, is even more common (Table 8.1) (61/0.9 = 68). Here we restrict our attention to baryonic (visible) matter, which constitutes only a fraction of the matter in our universe. Most of our universe is comprised of dark energy and dark matter, neither of which is yet understood.

A more detailed analysis of the *elemental* composition of adult humans is given in Table 8.2. The data are ranked in decreasing order by the entries in column three. The fact that two independent estimates for the more commonly occurring elements, published 73 years apart, are in reasonable agreement is striking. It is noteworthy that just five elements (oxygen, carbon, hydrogen, nitrogen, and calcium) rise above the 1% level. The first six elements in Table 8.2 account for 98 to 99% of human body weight or mass.

The outcomes of LSQ analyses of elemental levels (in g) for three elements in six species of adult mammals as a function of body mass are summarized in Table 8.3.

Taking into account the 99% CL, the slopes for these three elements lie between 0.885 and 1.11 (Table 8.3, column h). Recall that the total amount of any given element in the body varies with organ or tissue size, although some variation in elemental abundance (concentration) is seen from one organ or tissue to another. For example, in the adult human, we find that 60% of potassium is found in skeletal muscle, only 1% in

Table 8.3 LSQ analysis of calcium, potassium, and phosphorus levels in adult mammals

	Element	N_{ord}	N_{sp}	MPD	pWR	DI (%)	r^2	Slope ± 99% CL	Source
Column/ Row	a	b	c	d	e	f	g	h	i
1	Calcium	4	6	6.9	4.1	2.1	0.999	1.054 ± 0.055	[2, 16, 17, 18]
2	Potassium	4	6	5.7	3.4	1.7	0.999	0.944 ± 0.059	[2, 16, 17, 19]
3	Phosphorus	4	6	7.2	4.1	2.1	0.999	1.043 ± 0.061	[2, 16, 17, 18]

N_{ord} = number of orders; N_{sp} = number of species; MPD = mean percent deviation; pWR = logarithm of weight or mass ratio; DI = diversity index; r^2 = coefficient of determination; CL = confidence limit(s).

the lungs [20], reflecting in part differences in tissue and organ size (see Table 8.8). For a brief discussion of the functions of calcium, magnesium, potassium, sodium, and the hydrogen ion in the economy of animals see Lockwood [21].

Scaling of small molecules

Consider the reduction of an adult mammal to the level of small (relative to protein) molecules. These would be primarily amino acids, carbohydrates, lipids (mainly fatty acids, making up triacylglycerol), minerals (mostly hydroxyapatite), nucleic acids, and water. Most carbohydrate in the adult human is stored as glycogen, found primarily in liver and muscle (see below).

Free amino acids

The data of Munro [22] suggest that the *concentrations* (μmol/100 cm^3 plasma) of eight free amino acids (arginine, histidine, isoleucine, leucine, phenylalanine, threonine, tyrosine, and valine) found in blood plasma are independent of body size ($b \approx 0.0$) ($N_{ord} = 5$, $N_{sp} = 9$, pWR = 4.4, DI = 3.5%). Two other amino acids, lysine and possibly methionine, might have declining concentrations with increasing body size ($b = -0.13 \pm 0.15$ and -0.06 ± 0.22, respectively; 99% CL). In each of these ten instances the definitive slopes are quite uncertain because of the wide (>0.1) 99% confidence limits.

Ash

The oldest actual fossils that show direct evidence of biologically produced minerals occur in deposits formed 1.6 billion years ago.

Lowenstam and Weiner 1989

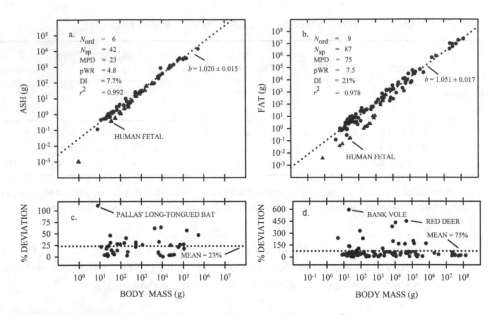

Figure 8.1 Scaling of ash and fat in adult mammals and the human fetus.

When a carcass is incinerated to constant weight in a muffle furnace at 600 °C or higher, the resultant ash consists mainly of the mineral components of bone, closely resembling hydroxyapatite, with a chemical formula of $Ca_5(PO_4)_3(OH)$ and a molecular weight of about 500 daltons. Hydroxyapatite occurs in bone mineral in the form of rod-like crystals about 50 Å in diameter and 200 Å long. The estimated surface area of the hydroxyapatite crystals is some 100 m^2 per gram. For the skeleton of an average adult male, this corresponds to a surface area equivalent to about 70 soccer fields. It is this enormous surface (relative to body surface) that makes possible the rapid exchange of minerals between bone and blood [23].

According to Blitz and Pellegrino [24], *calcium* and *phosphorus* comprise 38.3 ± 0.5 (STD) and 17.3 ± 0.4 (STD) percent of cortical bone ash (femur and tibia), respectively, in 11 species of mammals (including human) ranging in size from rat (*Rattus rattus*) to elephant (*Elephas maximus*) (body sizes taken from the standardized body weight table (Chapter 4)). The predicted amounts for these two elemental components of hydroxyapatite are 40 and 18%, respectively. However, the composition of bone mineral is rather more complex than this simple comparison with hydroxyapatite implies.

The most extensive study of ash content (in g) in the skeletons of various mammalian species is that reported by Pitts and Bullard [25]. From their data ($N_{sp} = 31$) a LSQ slope of 1.024 ± 0.052 (99% CL) is obtained for the line describing total ash content as a function of adult body weight. When ash data for another 11 species (drawn from the literature) are added to the data of Pitts and Bullard, the best-fit LSQ slope is 1.020 ± 0.040 (99% CL) (see Figure 8.1a and c). This slope is approximately consistent with *relative invariance* ($b = 1.0$). The percent deviations around the LSQ line are shown in Figure 8.1, panel c, on a linear scale. These data span the size range from bat

(*Glossophaga soricina*) to cattle (*Bos taurus*). The fetal human data (Figure 8.1a, triangles) are from Kelly [26].

The very large percent deviation (relative to other species) seen in Figure 8.1c, for Pallas' long-tongued bat (*Glossophaga soricina*) suggests that these data may be in error. Forcing a straight line (log–log) through the adult interspecific data with an *assigned* slope of 1 (rather than 1.02) (Chapter 5) gives an intercept of 0.033, implying that ash forms, on average, about 3.3% of adult body mass in the given sample. The findings for the scaling of ash are reasonably consistent with relative invariance (b = 1.0). See also slope (b = 1.019 $\pm$ 0.022 (99% CL)) for skeletal mass, Chapter 12, Table 12.5. We know from paleontology that minerals are of very great antiquity in animals (see above quote by Lowenstam and Weiner).

Carbohydrates

The primary carbohydrates in humans are glucose ($C_6H_{12}O_6$) and its polymer glycogen ($C_6H_{10}O_5)_n$. Glycogen stores in the adult human are about 1% of the fresh weight of skeletal muscle and 6 to 8% of liver weight. There is much more information available for blood sugar levels in mammals than for glycogen. Umminger [27] reported his findings from a LSQ analysis of data taken from the literature for blood sugar concentrations (mg sugar per 100 cm^3 of blood) in 73 species of mammals. He obtained a slope of –0.101 $\pm$ 0.029 (95% CL). Neither body size nor blood sugar data are given in tabular form. Carbohydrate reserves in the human and probably in adult mammals generally are much smaller than those for fat and protein (see Chapter 18, Table 18.2).

Fat

It is shown that all mammals studied have a similar pattern of distribution of larger and smaller adipose cells, regardless of the sex, age or fatness of the specimen.

Pond 1984

In the internal economy of higher vertebrates, the primary role of fat is as a medium of energy storage and energy transfer (e.g., lactation). In this regard fat is, per gram, more than *twice* as efficient as carbohydrate or protein. The actual ratios are about 2.25:1:1 for fat, protein, and carbohydrate, respectively. In addition, fat plays a mechanical role (as in the soles of the feet, or palms), in thermoregulation (e.g., in the neonates of some species), and in sexual selection (as in adult humans, where the fat distribution is "atypical") [28]. In marine mammals, fat plays a role in buoyancy.

Fat in most species of adult mammals consists of "white" fat; "brown" fat (so-named because of the presence of iron-containing mitochondria) occurs in some species, typically hibernators. Fat-containing cells aggregated into lobules are collectively referred to as *adipose* tissue, a type of connective tissue (see above quote from Pond). Adipose tissue also includes other connective tissue components, such as collagen (see below) and blood. Fat cells are unusual in that they are considered to be metabolically inert (most of the time), essentially spherical and exhibit a considerable size range (about 8,000-fold in volume) [29].

Fat is quantitatively the most variable of the major tissues in adult mammals [25]. Variations in fat content can occur within individuals of a species as a function of age, diet, exercise, gender, or season, as well as between species as a function of body size, habitat, and life style. For this reason it is often suggested that mammalian scaling studies would be better normalized to *fat-free body weight* than to body weight. This practice might reduce the scatter around LSQ lines, and also produce a modest increase in their slopes. Unfortunately, fat-free body weights have been determined in too few instances to make this a practical proposal for current scaling studies in mammals generally.

Most of the techniques for determination of fat content were first worked out for application to humans. Techniques that involve skin fold thickness or hydrostatic weighing have seldom been employed in other species. Two (destructive) techniques in common use involve weighing before and after ether extraction or manual dissection. Ether extraction is impractical on very large mammals. The largest example of fat determination by ether extraction used in this study is for a pig weighing about 150 kg. By contrast, manual dissection has been used over essentially the whole mammalian size range. Solvent extraction has the advantage that small dispersed fat samples will be included that might be missed in a manual dissection. Dissection actually focuses on adipose tissue, and is therefore anatomically more specific.

Pond and Mattacks [29] found that *collagen* in fat scales as 0.96 power of body weight in 46 species of land mammals, from guinea pig to horse (coefficient of determination $r^2 = 0.86$). Collagen makes up less than half a percent of adipose tissue by weight in land mammals [29]. Dissection and solvent extraction provide outcomes based, in small part, on qualitatively different materials.

Two datasets bearing on fat content (in g) were analyzed, one obtained by ether extraction, one by dissection. The analytical findings are shown in Table 8.4. When the two sets of data are analyzed separately, the slopes are 1.073 ± 0.081 and 1.123 ± 0.082, respectively (99% CL). These results were combined, as they agree within the 99% CL. No single species was common to both datasets. A LSQ line fitted to the combined dataset ($N_{sp} = 83$) is shown in Figure 8.1b. The best-fit LSQ line has a slope of 1.051 ± 0.045 (99% CL) (99% CI extends from 1.006 to 1.096). The data span shrew (*Sorex minutus*) to blue whale (*Balaenoptera musculus*). This slope implies that fat

Table 8.4 LSQ analysis of fat distribution in adult mammals as a function of body mass

N_{ord}	N_{sp}	MPD	pWR	DI (%)	r^2	Slope $\pm$ 99% CL	Method	Source
8	53	58	4.5	10	0.961	1.073 ± 0.081	Ether extraction	[7, 25, 30–42]
7	34	73	6.8	12	0.978	1.123 ± 0.082	Dissection	[43–61]
9	87	75	7.5	21	0.978	1.051 ± 0.045	Both	As above

See Table 8.3 for definition of terms.

content varies from 8.5% in the smallest mammal (1.5 g) to 22% in the largest mammal (150 tonnes). The percent deviation is shown in panel d. The two most deviant species in Figure 8.1d are the bank vole (*Clethrionomys glareolus*) and the red deer (*Cervus elephas*).

Notice in Figure 8.1b that the human fetal points (triangles) approach the adult interspecific line for fat from below [9]. The large MPDs (58 to 75%) seen in Table 8.4 and Figure 8.1d are consistent with the principle that fat is quantitatively the most variable of the major tissues in adult mammals [25].

Water

Water is one of the most important of all chemical substances... Its physical properties are strikingly different from those of other substances, in ways that determine the nature of the physical and biological world.

Pauling 1970

Water is the single most important small molecule making up life as we know it (see quote from Pauling). It is widely believed that life arose in the presence of water. In the absence of water we find an absence of life. That ice floats could account for the fact that the earth's oceans, through cycles of rising and falling global temperatures, have remained at least partly liquid since the earth's crust solidified about four billion years ago. This could have been a significant factor in the continued survival of metazoan life through successive Ice Ages.

Of course, no organism is solely water; most, however, are mainly water. According to Widdowson and Dickerson [9] a human fetus weighing less than 1 g is more than 90% water. Adult mammals are roughly two-thirds water (see below). As organisms mature, they dry out. This generalization seems to be true for a diverse sample of vertebrates and invertebrates, where dry weight increases faster than wet weight during development [62].

The simplest method of measuring *total body water* (TBW) is by *desiccation*. A carcass is placed in an oven and dried until a stable weight is achieved. Total body water is calculated as the difference between fresh wet weight and dry weight (or mass). Although desiccation tends to be regarded as the gold standard for estimating body water content, Kleiber [63] points out it is likely to remove other components as well as water. Desiccation seems not to have been applied to larger mammals. In the present sample (see below), TBW was determined by desiccation in 13 species; the largest specimen (a goat) weighed slightly less than 40 kg.

The more usual method is to inject a known quantity of isotope into an animal and then determine the radioactivity in a sample of body fluid drawn after a suitable time interval. Total body water is then calculated from the amount of isotope injected and the dilution factor. The isotopes most commonly used have been deuterium oxide (D_2O) and tritiated water (T_2O). Table 8.5 summarizes the findings of LSQ analyses of TBW (in g or cm^3) obtained by desiccation (row one) and by isotope dilution (row two) as a function of body mass in various species of adult mammals.

Table 8.5 LSQ analysis of TBW content as determined by desiccation and isotope dilution

N_{ord}	N_{sp}	MPD	pWR	DI (%)	r^2	Slope $\pm$ 99% CL	Method	Source
5	13	9	3.9	3.6	0.998	0.991 $\pm$ 0.039	Desiccation	[34, 36, 38, 64–69]
9	49	8	4.8	12	0.999	1.004 $\pm$ 0.0125	Isotope dilution	[35, 40, 41, 70–88]
11	62	8	5.1	16	0.999	1.002 $\pm$ 0.011	Combined	As above

See Table 8.3.

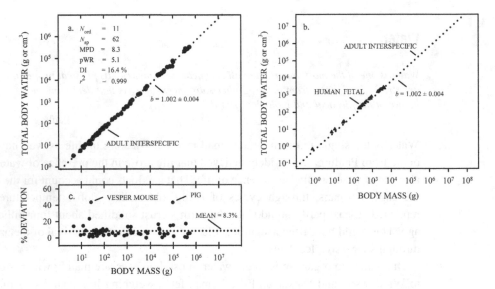

Figure 8.2 Total body water in adult mammals and in the human fetus.

In Table 8.5 (middle row), for 46 of 49 species the isotope used was tritiated water. Neither dataset (first and second rows) had any species in common. A reasonable inference from Table 8.5 is that the two methods give statistically indistinguishable outcomes. The results of combining both datasets are shown in the bottom row. The best-fit LSQ line computed for this combined dataset is shown in Figure 8.2a. The data include shrew (*Sorex minutus*) and zebu (*Bos indicus*). Cetaceans (and nine other taxonomic orders) are not represented in this dataset. After assigning a slope of 1 to the adult interspecific LSQ line, one finds that total body water comprises 64.5% of body weight or mass (in the given sample). Forbes [89] cites a figure of 62.4% (range from 56 to 68%) for the total water content derived from three male human cadavers (aged 25 to 46 years).

Data for the fetal human are shown in Figure 8.2b (triangles), relative to the adult interspecific line (dotted) [9]. During ontogeny, total body water in the human approaches the adult interspecific line from slightly above. In panel c, two deviant species are the vesper mouse (*Calomys lepidus*) and the pig (*Sus scrofa*). Combining the above figures for total body water (62.4%) and for carbon (23%) (Table 8.1) in the human we see that these two substances account for about 85% of adult human body mass. Forbes *et al.* [89] found that in the adult human water content in brain, cartilage, kidney, lungs, muscle (cardiac, skeletal, smooth), pancreas, and spleen varied between 76 and 83%, but averaged only 14% in bone. See also [90]. Since water molecules are all of the same size and shape, and since total body water scales as the first power of body weight (or mass), it follows that the *number* of water molecules in an adult mammal must, on average, also scale as the first power of body weight ($b = 1.0$).

MACROMOLECULES

Bulk proteins

We start from the premise that proteins are at the heart of all living processes, uniquely versatile in their capability to do whatever is needed... there is no limit to the adaptability of proteins.

Tanford and Reynolds 2001

...we have had access for many years to data that show proteins can't be the whole story.

Carey 2012

The marked complexity of living organisms resides chiefly in the diverse properties of their constituent proteins. Think of chess: there are two players; the game is played in a 2D space of 64 squares (8 × 8); there are two classes of pieces, black and white; each class consists of 16 pieces, of six different types: eight pawns, one king, one queen and two bishops, two knights, and two rooks. Each piece (bishop, king, knight, pawn, queen, rook) has its own unique pattern of legal moves. A game thus simply defined, teachable to a young child, nevertheless turns out to be one of mind-bending complexity.

By contrast, a mammal consists of perhaps 10,000 different proteins, these possessing, in the aggregate, a large number ($n > 20$) of distinctive properties (e.g., adhesivity, allosterism, antigenicity, binding, buffering, catalysis, charge transfer, conformational change, elasticity, folding, force generation, hormonal activity, insulation, light absorption, mechanical support, mutability, pigmentation, polymerization, pumping, reception, regulation, renaturation, repression, transcription, signal transduction, translation, transport). These varied properties "play" out in 3-D and co-vary with the microenvironment and time. It is difficult to think of a function of organisms that does not involve proteins. No other class of macromolecules is so central to the diverse functions of living organisms (see above quotes from Tanford and Reynolds and from Carey).

Proteins are polymers of amino acids; they consist chiefly of carbon, hydrogen, oxygen, nitrogen, and in some cases sulfur. The amount of protein in an organism is usually determined indirectly. The approximate elemental composition of proteins is given in Table 8.6 [63]. Carbon accounts for more than half of protein by mass.

The first four elements in Table 8.6 are among the most common in the universe (Table 8.1). Recall, however, that the amount of protein found in different tissues is highly variable. Skeletal muscle contains 47% of the adult human body's total protein, skeleton 18%, and skin 11%. These differences reflect, in major part, variations in organ and tissue size (see Table 8.8). Liver, lungs, nerve tissue, and gastrointestinal tract are

Table 8.6 Approximate elemental composition of proteins

Element	Atomic weight	% of total protein mass	Element	Atomic weight	% of total protein mass
Carbon	12	53	Hydrogen	1	7
Oxygen	16	23	Sulfur	32	1
Nitrogen	14	16	-	-	-

See Table 8.3.

each less than 3% [1, 2, 11]. Almost two-thirds (64.7%) of total body protein in the adult human male is located in skeletal muscle and bone [11].

A LSQ analysis was carried out of whole body protein as a function of body mass in 11 species drawn from five taxonomic orders [16, 34, 39, 40, 64, 91–95]. In this small sample MPD = 11, pWR = 4.6, DI = 3.9%, and r^2 = 0.998. The slope, possibly consistent with relative invariance, is 0.985 $\pm$ 0.046 (99% CL) (see Figure 8.3, item d). The fetal human data (not shown) approach the adult interspecific line from below [9]. Assigning a slope of one to the adult interspecific best-fit line, one finds an intercept of 0.165, implying that bulk protein constitutes some 16.5% of body mass in the given sample (see Chapter 5).

Analyzing a rather similar database (N_{ord} = 4, N_{sp} =10, MPD = 10.5, pWR = 4.5, DI = 2.9%, r^2 = 0.999) taken from the literature, it was found that bulk protein scales as the 0.997 $\pm$ 0.042 (99% CL) power of *fat-free* body mass, possibly consistent with relative invariance (b = 1). For an analysis of the scaling of hemoglobin concentration in adult mammals see Chapter 9. Note that bulk protein (16.5%) and water (64.5%) in mammals generally together account for about 81% of adult body size.

RNA and DNA

Munro and Gray [96] provide data on RNA and DNA concentration (mg/100 g fresh tissue) in the posterior thigh muscles of six different species (mouse to horse). Inferences from LSQ analyses of their data are summarized in Table 8.7. It might be anticipated (since muscle scales directly with body weight – see Chapter 12) that the slopes for RNA and DNA would be close to 1. Needed are larger datasets with greater diversity indices and narrower CL.

Extracellular fluid compartment

The ECF is more dilute than present-day sea-water, but its composition closely resembles that of the primordial oceans in which, presumably, all life originated.
                                                                    Ganong 1965

In the course of biological evolution, multicellular organisms may have "sought" to preserve the environment provided by the early oceans (see above quote from Ganong).

Table 8.7 Concentration of RNA and DNA in thigh muscles of six adult mammalian species

N_{ord}	N_{sp}	MPD	pWR	DI (%)	r^2	Slope ± 99% CL	Macromolecule
5	6	6	4.4	2.8	1.000	0.909 ± 0.043	Muscle RNA
5	6	6	4.4	2.8	0.999	0.923 ± 0.0565	Muscle DNA

See Table 8.3.

Rubey [97] and Holland [98] provide useful discussions of the evolution of the oceans. From the perspective of cells, the extracellular compartment is the environment. Because of its large size, the extracellular space is of considerable importance from a scaling standpoint; it is, however, very heterogeneous. The primary components of the extracellular space are: blood plasma, the non-cellular and non-solid portions of connective tissue (especially of bone and cartilage), glandular secretions like thyroidal colloid, interstitial fluid, lymph, and various liquid or semi-liquid compartments (cerebrospinal fluid, gut water, urine, vitreous humor). What is usually measured is not the extracellular "space," but rather the size of the *extracellular water* (ECW) compartment. Thiocyanate has often been employed (using the dilution principle) to estimate its size; this has been largely supplanted by bromide solutions and more recently by bio-impedance measurements [1, 99].

Of the estimates of ECW (in g or cm^3) in eight species analyzed here, five were obtained using thiocyanate [100–106]. For these data, $N_{ord} = 6$, $N_{sp} = 8$, MPD = 9, pWR = 2.6, and DI = 2.3%. The slope of the best-fit LSQ line is 0.996 ± 0.074 (99% CL), possibly consistent with relative invariance ($b = 1.0$). Assigning a slope of 1 to this best-fit line gives an intercept of 0.225, implying that the extracellular water compartment of adult mammals (in this small sample) constitutes 22.5% of body mass on average.

Intracellular water (ICW) compartment

It was found above that *total body water* in 62 species of adult mammals is 64.5% of body mass. As just discussed, the extracellular water compartment constitutes 22.5% of body mass in eight species. This suggests that intracellular water (ICW) accounts for 42% of body mass (64.5 – 22.5 = 42). Wang *et al.* [2] set the ICW compartment in adult humans at 34% of body mass. Further data are needed to better establish this aspect of body composition in adult mammals generally.

Size of organs and tissues in an adult human

Organs, organ systems, and tissues represent, in the main, a natural hierarchical endpoint for scaling studies in adult mammals. Beyond them we have the whole body. A summary of organ and tissue masses in an adult male human is given in Table 8.8 [11]. The fourth and last columns of Table 8.8 give cumulative sums.

Table 8.8 Organs and tissues in an adult male human (of 60 years) weighing 73.5 kg

Organ/tissue	Mass (g)	% body mass	Sum (%) [a]	Organ/tissue	Mass (g)	% body mass	Sum (%) [a]
Striated muscle	29,600	40.2	40	Nerve tissue	1,570	2	90
Adipose tissue	16,000	21.7	62	GI tract [b]	1,110	1.5	92
Skeleton	11,000	15.0	77	Heart	440	0.6	92
Skin	4,800	6.6	83	Kidneys	316	0.4	93
Liver	1,750	2.4	86	Spleen	74	0.1	93
Lungs	1,620	2.2	88	Other	5,220	7.1	100

[a] Cumulative sum; [b] GI = gastrointestinal.

Table 8.9 Component best-fit slopes with 99% CL ordered approximately by component size

Component	Example	Slope ± 99% CL
Elements	Phosphorus	1.043 ± 0.061
	Potassium	0.944 ± 0.059
	Calcium	1.054 ± 0.055
Small molecules	Water	1.002 ± 0.011
	Fatty acids	1.051 ± 0.045
	Ash	1.020 ± 0.040
Macromolecules	RNA	0.909 ± 0.043
	DNA	0.923 ± 0.074
	Bulk protein	0.985 ± 0.046
Compartment	ECF	0.996 ± 0.074

Summary

Glancing back to Table 8.1, we see that at the elemental level the universe is dominated by hydrogen and helium, whereas mammals (e.g., humans) are dominated by oxygen and carbon. In the following, body components are expressed as a percent of body mass. For small molecules, adult mammals are composed primarily of minerals (3.3%), fat (8.5 to 22%), and water (64.5%). Macromolecules are chiefly proteins (16.5% of body mass). At the tissue and organ level the mammalian body is, on average, mostly skeletal muscle (42.6%) and skeleton (9.5%) (42.6 + 9.5 = 52.1) (see Chapter 12).

In Table 8.9, the various component slopes are ordered roughly by component size. With the possible exceptions of RNA and DNA, there does not seem to be any significant correlation between component size and slope.

A simple overall picture emerges when we plot the slopes discussed above against their rank order (see Figure 8.3).

The slopes for ten y-variables relating to body composition tend to cluster roughly around one (relative invariance). The mean of these ten slopes encountered in this

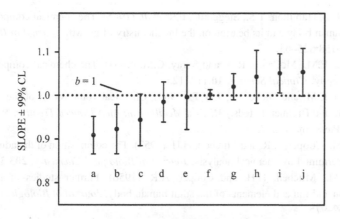

RANK ORDER

a = muscle RNA concentration; b = muscle DNA concentration;
c = potassium content; d = total body protein; e = ECW; f = TBW;
g = ash; h = phosphorus content; i = fat content; j = calcium content

ECW = extracellular water; TBW = total body water

Figure 8.3 Summary of slopes for body composition by rank order.

chapter is 0.993, with a standard deviation of 0.050. Note that seven of the slopes are given in tabular form, and three slopes (see Figure 8.3, items d, e, g) are embedded in the text. The slopes for ash (1.020 ± 0.040), extracellular water (0.996 ± 0.074), total body protein (0.997 ± 0.040), and total body water (1.002 ± 0.011) (all 99% CL) are suggestive of *relative invariance* (but for a discussion of "special pleading" see Chapter 23).

Collectively, the above findings derived from the scaling of body composition testify to the potential importance of relative invariance as a major aspect of scaling in adult mammals (see also Tables 8.3, 8.4, and 8.5, and above quote by Kleiber).

References

1. Heymsfield, S.B., Lohman, T.G., Wang, Z. *et al.* (2005). *Human Body Composition*, 2nd edn. Champaign: Human Kinetics.
2. Wang, Z.-M., Pierson, R.N. Jr., Heymsfield, S.B. *et al.* (1992). The five-level model: a new approach to organizing body-composition research. *American Journal of Clinical Nutrition*, **56**:19–28.
3. Mason, B. (1966). *Principles of Geochemistry*. New York: John Wiley & Sons.
4. Keys, A. and Brozek, J. (1953). Body fat in adult man. *Physiological Reviews*, **33**:245–325.
5. Siri, W.E. (1961). Body composition from fluid spaces and density: an analysis of methods. In: Brozek, J. and Henschel, A. (eds.) *Techniques for Measuring Body Composition*. Washington, DC: National Academy of Sciences, pp. 223–244.
6. Hackh, I.W.D. (1919). Bioelements: the chemical elements of living matter. *Journal of General Physiology*, **1**:429–433.

7. Mitchell, H.H., Hamilton, T.S., Steggerda, F.R. *et al.* (1945). The chemical composition of the adult human body and its bearing on the biochemistry of growth. *Journal of Biological Chemistry*, **158**:625–637.

8. Widdowson, E.M., McCance, R.A. and Spray, C.M. (1951). The chemical composition of the human body. *Clinical Science*, **10**:113–125.

9. Widdowson, E.M. and Dickerson, J.W.T. (1964). Chemical composition of the body. In: Comar, C.L. and Bronner, F. (eds.) *Mineral Metabolism: An Advanced Treatise*. New York: Academic Press, pp. 1–217.

10. Forbes, R.M., Cooper, A.R. and Mitchell, H.H. (1953). The composition of the adult human body as determined by chemical analysis. *Journal of Biological Chemistry*, **203**:359–366.

11. Forbes, R.M., Mitchell, H.H. and Cooper, A.R. (1956). Further studies of the gross composition and mineral elements of the adult human body. *Journal of Biological Chemistry*, **223**:969–975.

12. Gamow, G. (1965). *The Creation of the Universe*. New York: Bantam Books.

13. Reeves, H. (1984). *Atoms of Silence: An Exploration of Cosmic Evolution*. Cambridge, MA: MIT Press.

14. Trimble, V. (1997). Origin of the biologically important elements. *Origins of Life and Evolution in the Biosphere*, **27**:3–21.

15. Kitchin, C.R. (1987). *Stars, Nebulae and the Interstellar Medium. Observational Physics and Astrophysics*. Bristol: Adam Hilger.

16. Spray, C.M. (1950). A study of some aspects of reproduction by means of chemical analysis. *British Journal of Nutrition*, **4**:354–360.

17. Spray, C.M. and Widdowson, E.M. (1950). The effect of growth and development on the composition of mammals. *British Journal of Nutrition*, **4**:332–353.

18. Meigs, E.B. (1935). The effects, on calcium and phosphorus metabolism in dairy cows, of feeding low-calcium rations for long periods. *Journal of Agricultural Research*, **51**:1–26.

19. Williams, R.E., Hughes, D., Lee, P. *et al.* (1964). The measurement of whole-body potassium by gamma-ray spectrometry and its relation to total body water in dogs. *Clinical Science*, **27**:305–312.

20. Lloyd, R.D. and Mays, C.W. (1987). A model for human body composition by total body counting. *Human Biology*, **59**:7–30.

21. Lockwood, A.P.M. (1963). *Animal Body Fluids and their Regulation*. Cambridge, MA: Harvard University Press.

22. Munro, H.N. (1969). Evolution of protein metabolism in mammals. In: Munro, H.N. (ed.) *Mammalian Protein Metabolism*. New York: Academic Press, pp. 133–182.

23. McLean, F.C. and Urist, M.R. (1968). *Bone: Fundamentals of the Physiology of Skeletal Tissue*. Chicago, IL: University of Chicago Press.

24. Blitz, R.M. and Pellegrino, E.D. (1969). The chemical anatomy of bone: I. A comparative study of bone composition in sixteen vertebrates. *Journal of Bone Joint Surgery,* **51-A**:456–466.

25. Pitts, G.C. and Bullard, T.R. (1968). Some interspecific aspects of body composition in mammals. In: *Body Composition in Animals and Man*. Washington, DC: National Academy of Sciences, pp. 45–70. (Body weights by personal communication from the late W.A. Calder.)

26. Kelly, H.J., Sloan, R.E., Hoffman, W. and Saunders, C. (1951). Accumulation of nitrogen and six minerals in the human fetus during gestation. *Human Biology*, **23**:61–74.

27. Umminger, B.L. (1975). Body size and whole blood sugar level concentrations in mammals. *Comparative Biochemistry and Physiology*, **52A**:455–458.

28. Pond, C.M. (1978). Morphological aspects and the ecological consequences of fat deposition in wild vertebrates. *Annual Review Ecology Systematics*, **9**:519–570.

29. Pond, C.M. and Mattacks, C.A. (1989). Biochemical correlates of the structural allometry and site-specific properties of mammalian adipose tissue. *Comparative Biochemistry and Physiology*, **92A**:455–463.

30. Churchfield, S. (1981). Water and fat contents of British shrews and their role in the seasonal changes in body weight. *Journal of Zoology*, **194**:165–173.

31. Dawson, N.J. (1970). Body composition of inbred mice (*Mus musculus*). *Comparative Biochemistry and Physiology*, **37**:589–593.

32. Doornenbal, H., Asdell, S.A. and Comar, C.L. (1962). Relationship between Cr^{51}-determined total red cell volume and lean body mass in rats. *Journal of Applied Physiology*, **17**:737–740.

33. Goyal, S.P., Ghosh, P.K. and Prakash, I. (1981). Significance of body fat in relation to basal metabolic rate in some Indian desert rodents. *Journal of Arid Environments*, **4**:59–62.

34. Hayward, J.S. (1965). The gross body composition of six geographic races of Peromyscus. *Canadian Journal of Zoology*, **43**:297–308.

35. Kodama, A.M. (1971). In vivo and in vitro determination of body fat and body water in the hamster. *Journal of Applied Physiology*, **31**:218–222.

36. McNab, B.K. (1968). The influence of fat deposits on the basal rate of metabolism in desert homoiotherms. *Comparative Biochemistry and Physiology*, **26**:337–343.

37. Millar, J.S. (1975). Tactics of energy partitioning in breeding Peromyscus. *Canadian Journal of Zoology*, **53**:967–976.

38. Myrcha, A. (1969). Seasonal changes in caloric value, body water and fat in some shrews. *Acta Theriologica*, **14**:211–227.

39. Panaretto, B.A. and Till, A.R. (1963). Body composition in vivo: II. The composition of mature goats and its relationship to the antipyrine, tritiated water, and N-acetyl-4-aminoantipyrine spaces. *Australian Journal of Agricultural Research*, **14**:926–943.

40. Reilly, J.J. and Fedak, M.A. (1990). Measurement of the body composition of living gray seals by hydrogen isotope dilution. *Journal of Applied Physiology*, **69**:885–891.

41. Rumpler, W.V., Allen, M.E., Ullrey, D.E. *et al.* (1987). Body composition of white-tailed deer estimated by deuterium oxide dilution. *Canadian Journal of Zoology*, **65**:204–208.

42. Shields, R.G., Mahan, D.C. and Graham, P.L. *et al.* (1983). Changes in swine body composition from birth to 145 kg. *Journal of Animal Science*, **57**:43–54.

43. Bjarnason, I. and Lingaas, P. (1954). Some weight measurements of whales. *Norsk Hvalfangstted*, **43**:8–11.

44. Bryden, M.M. and Erickson, A.W. (1976). Body size and composition of Crabeater seals (*Lobodon carcinophagus*), with observations on tissue and organ size in Ross seals (*Ommatophoca rossi*). *Journal of Zoology*, **179**:235–247.

45. Bryden, M.M. (1972). Body size and composition of elephant seals (*Mirounga leonina*): absolute measurements and estimates from bone dimensions. *Journal of Zoology*, **167**:265–276.

46. Davis, D.D. (1962). Allometric relationships in lions vs. domestic cats. *Evolution*, **16**:505–514.

47. Hamilton, J.E. (1949). Weight, etc. of Elephant seal. *Nature*, **163**:536.

48. Hock, R.J. (1960). Seasonal variations in physiologic functions of arctic ground squirrels and black bears. *Bulletin of the Museum of Comparative Zoology*, **124**:155–169.

49. Kamiya, T. and Yamasaki, F. (1974). Organ weights of *Pontoporia blainvillei* and *Platanista gangetica* (Platanistidae). *Scientific Reports Whales Research Institute*, **26**:265–270.

50. Kasuya, T. (1972). Some information on the growth of the Ganges dolphin with a comment on the Indus dolphin. *Scientific Reports Whales Research Institute*, **24**:87–108.

51. Latimer, H.B. and Sawin, P.B. (1955). Morphogenetic studies of the rabbit XII. Organ size in relation to body weights in adults of small sized race X. *Anatomical Record*, **123**:81–102.

52. Nishiwaki, M. (1950). On the body weight of whales. *Scientific Reports Whales Research Institute*, **4**:184–209.

53. Ohno, M. and Fujino, K. (1952). Biological investigation on the whales caught by the Japanese Antarctic whaling fleets, season 1950/51. *Scientific Reports Whales Research Institute*, **7**:125–188.

54. Omura, H. (1957). Report on two right whales. *Norsk Hvalfangst-Tidende*, **46**:374–378.

55. Pond, C.M. and Mattacks, C.A. (1986). Allometry of the cellular structure of intra-orbital adipose tissue in eutherian mammals. *Journal of Zoology*, **209**:35–42.

56. Pond, C.M. and Mattacks, C.A. (1985). Body mass and natural diet as determinants of the number and volume of adipocytes in eutherian mammals. *Journal of Morphology*, **185**:183–193.

57. Pond, C.M. and Mattacks, C.A. (1994). The anatomy and chemical composition of adipose tissue in wild wolverines (*Gulo gulo*) in northern Canada. *Journal of Zoology*, **232**:603–616.

58. Pond, C.M., Mattacks, C.A. and Colby, R.H. (1992). The anatomy, chemical composition, and metabolism of adipose tissue in wild polar bears (*Ursus maritimus*). *Canadian Journal of Zoology*, **70**:326–341.

59. Simpson, A.M., Webster, A.J.F., Smith, J.S. and Simpson, C.A. (1978). Energy and nitrogen metabolism of Red deer (*Cervus elaphus*) in cold environments; a comparison with cattle and sheep. *Comparative Biochemistry and Physiology*, **60**:251–256.

60. Winston, W.C. (1950). The largest whale ever weighed. *Natural History*, **59**:392–398.

61. Zenkovic, B.A. (1937). Weighing of whales. *Comptes Rendus Académie Sciences USSR*, **16**:177–182.

62. Needham, J. (1934). Chemical heterogony and the ground-plan of animal growth. *Biological Review*, **9**:79–109.

63. Kleiber, M. (1961). *The Fire of Life: An Introduction to Animal Energetics*. New York: John Wiley & Sons.

64. Bailey, C.B., Kitts, W.D. and Wood, A.J. (1960). Changes in the gross chemical composition of the mouse during growth in relation to the assessment of physiological age. *Canadian Journal of Animal Science*, **40**:143–155.

65. Cizek, L.J. (1954). Total water content of laboratory animals with special reference to volume of fluid within the lumen of the gastrointestinal tract. *American Journal of Physiology*, **179**:104–110.

66. Holleman, D.F. and Dietrich, R.A. (1975). An evaluation of the tritiated water method for estimating body water in small rodents. *Canadian Journal of Zoology*, **53**:1376–1378.

67. Pace, N. and Rathbun, E.N. (1945). Studies on body composition. III. The body water and chemically combined nitrogen content in relation to fat content. *Journal of Biological Chemistry*, **158**: 685–691.

68. Panaretto, B.A. (1963). Body composition in vivo. III. *Australian Journal of Agricultural Research*, **14**:944–952.

69. Panaretto, B.A. (1963). Body composition in vivo. I. The estimation of total body water with antipyrene and the relation of total body water to total body fat in rabbits. *Australian Journal of Agricultural Research*, **14**:594–601.

70. Bell, G.P., Bartholomew, G.A. and Nagy, K.A. (1986). The roles of energetics, water economy, foraging behavior, and geothermal refugia in the distribution of the bat, *Macrotus californicus*. *Journal of Comparative Physiology*, **B156**:441–450.

71. Bradshaw, S.D., Morris, K.D., Dickman, C.R., Withers, P.C. and Murphy, D. (1994). Field metabolism and turnover in the Golden Bandicoot (*Isoodon auratus*) and other small mammals from Barrow Island, Western Australia. *Australian Journal of Zoology*, **42**:29–41.

72. Costa, D.P. (1982). Energy, nitrogen, and electrolyte flux and sea water drinking in the sea otter *Enhydra lutris*. *Physiological Zoology*, **55**:35–44.

73. Dawson, T.J., Denny, M.J.S., Russell, E.M. and Ellis, B. (1975). Water usage and diet preferences of free ranging kangaroos, sheep and feral goats in the Australian arid zone during summer. *Journal of Zoology*, **177**:1–23.

74. Denny, M.J.S. and Dawson, T.J. (1975). Comparative metabolism of tritiated water by macropodid marsupials. *American Journal of Physiology*, **228**:1794–1799.

75. Hansard, S.L. (1963). Radiochemical procedures for estimating body composition in animals. *Annals of the New York Academy of Science*, **110**:229–245.

76. Holleman, D.F. and Dietrich, R.A. (1973). Body water content and turnover in several species of rodents as evaluated by the tritiated water method. *Journal of Mammalogy*, **54**:456–465.

77. Hulbert, A.J. and Dawson, T.J. (1974). Water metabolism in perameloid marsupials from different environments. *Comparative Biochemistry and Physiology*, **47A**:617–633.

78. Kamis, A.B. and Latif, N.B.T. (1981). Turnover and total body water in Macaque (*Macaca fascicularis*) and Gibbon (*Hylobates lar*). *Comparative Biochemistry and Physiology*, **70A**:45–46.

79. Kennedy, P.M. and Heinsohn, G.E. (1974). Water metabolism of two marsupials – the brush-tailed possum, *Trichosurus vulpecula* and the rock-wallaby, *Petrogale inornata* in the wild. *Comparative Biochemistry and Physiology*, **47A**:829–834.

80. Knox, K.L., Nagy, J.G. and Brown, R.D. (1969). Water turnover in mule deer. *Journal of Wildlife Management*, **33**:389–393.

81. Kodama, A.M. (1970). Total body water of the pig-tailed monkey, *Macaca nemestrina*. *Journal of Applied Physiology*, **29**:260–262.

82. Leon, B., Shkolnik, A. and Shkolnik, T. (1983). Temperature regulation and water metabolism in the elephant shrew *Elephantulus edwardii*. *Comparative Biochemistry and Physiology*, **74A**:399–407.

83. Luft, U.C., Cardus, D., Lim, T.P.K., Anderson, E.C. and Howarth, J.L. (1963). Physical performance in relation to body size and composition. *Annals of the New York Academy of Science*, **110**:795–808.

84. MacFarlane, W.V. and Howard, B. (1972). Comparative water and energy economy of wild and domestic mammals. *Symposium of the Zoological Society London*, **31**:261–296.

85. McLean, J.A. and Speakman, J.R. (1999). Energy budgets of lactating and non-reproductive Brown long-eared bats (*Plecotus auritus*) suggest females use compensation in lactation. *Functional Ecology*, **13**:360–372.

86. Richmond, C.R., Langham, W.H. and Trujillo, T.T. (1962). Comparative metabolism of tritiated water by mammals. *Journal of Cellular Comparative Physiology*, **59**:45–52.

87. Wade, L. and Sasser, L.B. (1970). Body water, plasma volume, and erythrocyte volume in sheep. *American Journal of Veterinary Research*, **31**:1375–1378.

88. Yousef, M.K., Johnson, H.D., Bradley, W.G. and Seif, S.M. (1974). Tritiated water-turnover rate in rodents: desert and mountain. *Physiological Zoology*, **47**:153–161.

89. Forbes, G.B. (1987). *Human Body Composition: Growth, Aging, Nutrition and Activity*. New York: Springer-Verlag.

90. Davies, M. (1961). On body size and tissue respiration. *Journal of Cellular Comparative Physiology*, **57**:135–147.

91. Chinn, K.S.K. (1967). Prediction of muscle and remaining tissue protein in man. *Journal of Applied Physiology*, **23**:713–715.

92. Doornenbal, H. and Martin, A.H. (1965). The evaluation of blood volume and total red cell mass as predictors of gross body composition in the pig. *Canadian Journal of Animal Science*, **45**:203–210.

93. Fedyk, A. (1977). Seasonal changes in the water content and level in the bank vole against the background of other gross body components. *Acta Theriologica*, **22**:355–363.

94. Haecker, T.L. (1920). Investigations in beef production. *Minnesota Agricultural Experimental Station Bulletin*, **193**:1–111.

95. Hatai, S. (1917). Changes in the composition of the entire body of the albino rat during the life span. *American Journal of Anatomy*, **21**:23–37.

96. Munro, H.N. and Gray, J.A.M. (1969). The nucleic acid content of skeletal muscle and liver in mammals of different body size. *Comparative Biochemistry and Physiology*, **28**:897–905.

97. Rubey, W.W. (1951). Geologic history of sea water: an attempt to state the problem. *Bulletin of the Geological Society America*, **62**:1111–1148.

98. Holland, H.D. (1984). *The Chemical Evolution of the Atmosphere and Oceans*. Princeton, NJ: Princeton University Press.

99. Frankenfield, D.C., Cooney, R.N., Smith, J.S. and Rowe, W.A. (2007). Bioelectrical impedance plethysmographic analysis of body composition in critically injured and healthy subjects. *American Journal of Clinical Nutrition*, **69**:426–431.

100. Cameron, R.D. and Luick, J.R. (1972). Seasonal changes in total body water, extracellular fluid, and blood volume in grazing reindeer. *Canadian Journal of Zoology*, **50**:107–116.

101. Denny, M.J.S. and Dawson, T.J. (1975). Effects of dehydration on body-water distribution in desert kangaroos. *American Journal of Physiology*, **229**:251–254.

102. Fernandez, L.A., Rettori, O. and Mejía, R. (1966). Correlation between body fluid volumes and body weight in the rat. *American Journal of Physiology*, **210**:877–879.

103. Harrison, H.E., Darrow, D.C. and Yannet, H. (1936). The total electrolyte content of animals and its probable relation to the distribution of body water. *Journal of Biological Chemistry*, **113**:515–529.

104. MacFarlane, W.V., Morris, R.J.H., Howard, B. and Budtz-Olsen, O.E. (1959). Extracellular fluid distribution in tropical Merino sheep. *Australian Journal of Agricultural Research*, **10**:269–286.

105. McCance, R.A. and Widdowson, E.M. (1951). A method of breaking down the body weights of living persons into terms of extracellular fluid, cell mass and fat, and some applications of it to physiology and medicine. *Proceedings of the Royal Society of London*, **138**:115–130.

106. Sheng, H-P. and Huggins, R.A. (1973). Body cell mass and lean body mass in the growing beagle. *Proceedings of the Society for Experimental Biology and Medicine*, **142**:175–180.

9   Circulatory system

The greatest physiological advance of the seventeenth century, and perhaps of all times, was the discovery of the circulation of the blood.

Ackerknecht 1968

The discovery of the blood circulation is due to William Harvey (1578–1657), whose work was published in 1628. His "Anatomical Treatise on the Movement of the Heart and Blood in Relation to Animals" was a major stimulus to physiology and to a deterministic conception of biology (see above quote from Ackerknecht). The closed circulatory system, consisting of four pumps in series linked to virtually all parts of the body by branching roughly cylindrical tubes of greatly varying diameter, is among the most obviously physical of the body's various systems [1]. The general *macroscopic* configuration of the circulatory system, as inferred mostly from dog and man, is illustrated in Figure 9.1.

The primary function of the circulatory system in mammals is to transport heat and chemical substances, chiefly oxygen, carbon dioxide, and glucose. This system features "long-haul" arterial and venous subsystems, operating over distances of centimeters in small mammals to tens of meters in very large mammals, mediated by convection, driven by the heart. In addition, we have local "drop-off and pick-up" capillary subsystems, operating over microns to millimeters, mediated by diffusion, and driven by concentration gradients between the blood and surrounding tissues. No doubt transport costs for blood increase as some function of distance from the heart. Metabolically expensive organs, such as brain, kidneys, and liver, tend to be located nearer to the heart than less expensive organs and tissues such as bone and fat.

The number and relative locations of named large blood vessels (e.g., descending aorta) typically do not vary with body size or species, whereas the total number of unnamed small vessels increases progressively with body size (Figure 9.1).

The mammalian heart consists of two pairs of pumps. The right heart is a low-pressure system: the right atrium pumps venous blood drawn from the systemic circulation into the right ventricle, which in turn drives it through the lungs and into the left atrium. The left atrium forces blood into the left ventricle, which pumps it at a relatively high pressure throughout the body. The transport function of the vascular system as a whole can be understood in part through an analysis of its geometry and physical properties. Some important aspects of the *microscopic* and *macroscopic* geometry of the vascular system in mammals are summarized in Figure 9.2.

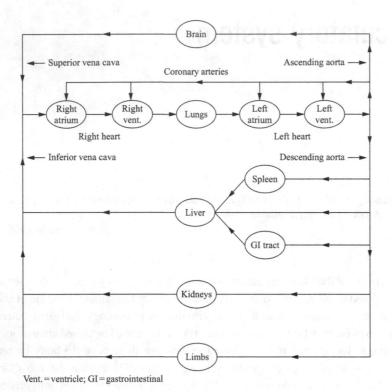

Vent. = ventricle; GI = gastrointestinal

Figure 9.1 Mammalian circulatory system.

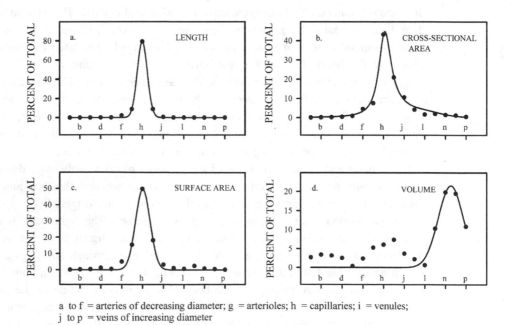

a to f = arteries of decreasing diameter; g = arterioles; h = capillaries; i = venules;
j to p = veins of increasing diameter

Figure 9.2 Geometry of the mammalian circulatory system.

The data on which Figure 9.2 is based (see solid circles) were gathered by Schleier [2] from the dog mesentery; see also Landis and Hortenstine [3] and Wiedman [4]. The smooth curves shown in Figure 9.2 are arbitrary mathematical functions selected to give a reasonable fit to the empirical data (less so for panel d). In each panel of Figure 9.2, a y-variable is expressed as a percent of the total. It is evident from Figure 9.2 that nature has "contrived" to aggregate most of total blood vessel length (panel a), cross-sectional area (panel b), and surface area (panel c) in vessels about the size of capillaries. Blood volume (panel d) is more unevenly distributed, with the largest portion being in the veins. The large cross-sectional area (Figure 9.2, panel b) represented by the capillaries collectively implies that the impedance (resistance to flow) is relatively low in this portion of the circulatory system. Furthermore, the large surface area (Figure 9.2, panel c) of the combined capillaries is ideal for diffusive transport (e.g., of oxygen) from the blood into the tissues.

The relatively smaller total volume of the capillaries (see item h in Figure 9.2d) implies that the advantages of the capillary circulation (large cross-sectional area and large surface area) are achieved at low cost in terms of tissue volume (e.g., in the human cerebral cortex, the capillaries occupy 2.3% of tissue volume) [5]. Total capillary blood volume in the adult human may account for as little as 0.5% of body volume [6].

It may be objected, with some justification, that the above findings, based on one vascular bed (mesentery) in one species (dog), should not be extrapolated to other vascular beds or other species. However, it is known from the data of Mall [7] that a broadly similar picture holds for the dog small intestine, stomach, adrenal gland, spleen, lung, and liver. In Figure 9.3 it is shown how blood pressure and blood velocity (panel a) vary with vessel diameter (panel b). Vessel diameter is approximately symmetric on either side of the capillaries; it is true, nonetheless, that vessel diameter decreases more slowly on the arterial side and increases more rapidly on the venous side.

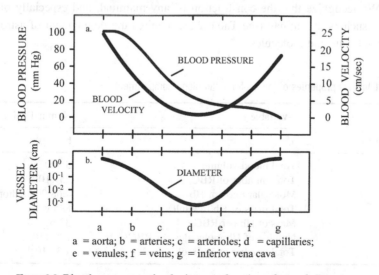

a = aorta; b = arteries; c = arterioles; d = capillaries;
e = venules; f = veins; g = inferior vena cava

Figure 9.3 Blood pressure and velocity as a function of vessel diameter.

The blood velocity data are taken from Burton [8], and the remaining data from the literature. For the most part these data are from humans and dogs. Blood velocity, as expected, is lowest in the capillaries [9] with roughly symmetrical values in larger arteries and veins, being a little lower in the vena cava than in the aorta (Figure 9.3, panel a). The relatively low blood velocity in the capillaries implies a relatively longer transit-time, favoring diffusive transport. Observe that blood pressure falls continuously throughout the vascular system, the drop being steepest in the small arteries and arterioles. The blood pressure, on average, in a young adult human, is about 100 mm Hg in the aorta and 5 mm Hg in the vena cava.

These data may not be representative of diving mammals, where a pronounced bradycardia (slowing of heart rate) and other differences are observed during activity [10]. The following discussion considers 20 scaling y-variables pertaining to the circulatory system, of which six are functional, the remainder structural. For these 20 y-variables there are 39 instances. The total number of records analyzed in this chapter is 1,776. These 20 y-variables are, for the most part, treated here in isolation from one another. A snapshot of how these 20 respective slopes are collectively distributed is provided in the summary (see Figure 9.11).

Blood

Some general properties of the blood compartment in the adult human as derived from the literature are displayed in Table 9.1 [11, 12]. Total blood volume (row 1) in the adult human is less than 8% of body mass ($100 \cdot 5{,}500/70{,}000 = 7.9$). The total number of red blood cells (row two) is about 10^{13}, whereas the total number of hemoglobin molecules (row 7) is nearly 10^{22} (about the same as the estimated number of stars in the observable universe). Various properties of hemoglobin are summarized in Table 9.1, rows three to seven. We recognize that the construction of any mammal, and especially of a large mammal such as the human (see Table 2.5), requires the deployment of astronomical numbers of cells and molecules.

Table 9.1 Various properties of the adult human blood compartment

Column/Row	y-variable a	Nominal values in adult b
1	Total blood volume	5,500 cm^3
2	Total number of RBC	25×10^{12}
3	Molecular mass of Hb	$\approx 68{,}000$ daltons
4	Total mass of Hb in adult	825 g
5	Mass of Hb per RBC	33 pg
6	Number of Hb molecules per RBC	300×10^6
7	Total number of Hb molecules	7.3×10^{21}

RBC = red blood cell(s); Hb = hemoglobin; pg = picogram (10^{-12}g).

Hemoglobin

Hemoglobin thus increases the O_2 capacity of the blood by 100 times, and as a result the O_2 content of the blood is the same as that of air.

                                                              Weibel 1984

The function of hemoglobin is to transport oxygen efficiently from the lungs to the tissues. Hemoglobin is a protein, consisting of two α- and two β-subunits, each subunit having a molecular weight of about 17,000 daltons. The hemoglobin molecule is roughly spherical, with a mean diameter of 55 Å. The subunits are organized in a tetrahedral array [13]. The α- and β-subunits closely resemble myoglobin, probably the ancestral molecule from which hemoglobin is descended [14]. Hemoglobin is widely distributed across vertebrates, invertebrates, plants, fungi, and bacteria.

Myoglobin and each subunit of hemoglobin contain a heme prosthetic group (i.e., a non-protein portion), in the form of a ring structure, with a molecular weight of about 600 daltons, at the center of which is an iron atom in the ferrous state. An oxygen molecule binds weakly, and reversibly, to the iron atom. Hemoglobin also facilitates the transport of carbon dioxide (mainly in the form of bicarbonate ion, HCO_3-, a waste product), from the tissues to the lungs. Myoglobin is located intracellularly, chiefly in skeletal muscle, where it stores oxygen; it is most abundant in diving mammals.

Myoglobin and hemoglobin were the first proteins whose atomic structures were determined (by John Kendrew and Max Perutz and their colleagues, at the Medical Research Council Laboratory in Cambridge, England) thereby throwing open the doors to the age of molecular biology. X-ray diffraction analysis showed that both myoglobin and the subunits of hemoglobin each have about 70% alpha helical structure. This degree of alpha helicity [15] is now known to be higher than that in proteins generally. It is curious that the very first analysis of protein structure at the atomic level pointed to an evolutionary relationship between two proteins – myoglobin and hemoglobin. That is, the application of physical methods to a biological problem may often raise questions which physics alone cannot answer (see Box 9.1).

Box 9.1 Reductionism

The ultimate aim of the modern movement in biology is in fact to explain all biology in terms of physics and chemistry... Thus eventually one may hope to have the whole of biology "explained" in terms of the level below it, and so on right down to the atomic level.

                                                              Crick 1966

...DNA can be regarded as a set of instructions for how to make a body...

                                                              Dawkins 1978

...to try to derive the behaviour of plants and people from first principles would be folly.

                                                              Cox and Forshaw 2011

Box 9.1 (*cont.*)

The ability to reduce everything to simple fundamental laws does not imply the ability to start from those laws and reconstruct the universe... the constructionist hypothesis breaks down when confronted with the twin difficulties of scale and complexity... at each level of complexity entirely new properties appear ... in this case we can see how the whole becomes not only more but very different from the sum of its parts.

                                                                    Anderson 1972

Prior to the first half of the twentieth century, physics was strongly reductionist in character. Reductionism expresses the conviction that progress in science is best made by breaking systems of interest into parts, and then studying the properties and interactions of these parts. Much of the triumph of classical physics is attributable to reductionism. But this success was purchased at a high price – that of restricting attention to those systems that tend to exhibit simple geometry, homogeneous composition, and linearity. In such systems the whole is no more than the sum of its parts.

Given its success in physics, it is not surprising that reductionism affected other branches of science, including biology. It was strong amongst molecular biologists, particularly in the period 1950–80 (see first two quotes above from Crick and Dawkins). But in the second half of the twentieth century, physics underwent a sea change. It began with an accidental discovery in the early 1960s, by Edward Lorenz. Using a computer to model weather, he found that the relevant non-linear physical equations were astonishingly sensitive to initial conditions. A change in the third decimal place in a parameter value could lead to significant changes in weather predictions. This is often referred to as the Butterfly Effect, the idea being that a butterfly flapping its wings may affect the weather at later times and remote distances. We now know that such "chaotic" non-linear systems are more common in the real world than are linear systems. Even a simple dripping faucet may exhibit chaotic behavior.

A key question is the degree to which physics can explain and predict biological phenomena. To a limited extent this is surely possible. But much of biology involves evolution. Few would maintain that physical arguments can predict evolutionary outcomes. Nature has many ways of responding to physical constraints (see Box 3.1). Different forms of life find different solutions. Biological systems are complex, heterogeneous, hierarchical, interactive, and non-linear. As Nobel Prize-winning physicist Paul Anderson argued, one cannot assume that higher-level properties in such systems are necessarily predictable from lower levels. Complexity, non-linearity, and size may be of crucial importance in understanding scaling and many other phenomena seen across multiple disciplines. In this long-term enterprise, a strictly reductive approach may prove to be of limited usefulness (see above quote from Cox and Forshaw).

Table 9.2 Scaling of hemoglobin (g per 100 cm^3 blood) as a function of body mass

N_{ord}	N_{sp}	MPD	pWR	DI (%)	r^2	Slope $\pm$ 99% CL	Source
12	113	12	4.7	19	0	0.012 ± 0.0165	[16]
-	171	-	-	-	0	0.003	[17]
11	86	14	6	21	0	-0.010 ± 0.015	From Lit.

N_{ord} = number of orders; N_{sp} = number of species; MPD = mean percent deviation; pWR = log (BW_{max}/BW_{min}); DI = diversity index; r^2 = coefficient of determination; CL = confidence limit(s); Lit. = literature.

Because of the presence of hemoglobin, blood is able to carry (at a maximum) some 70 times more oxygen than would an equivalent volume of water at the same temperature and partial pressure of oxygen (as noted by Weibel in above quote). In effect, blood is able to carry oxygen up to the same concentration as in the atmosphere at sea-level. The results of three LSQ analyses of the scaling of hemoglobin in adult mammals are brought together in Table 9.2. The analysis in row one draws on Hawkey [16], who does not provide body sizes (here taken from the standardized body weight table; see Chapter 4). The data in Table 9.2, row two, are from Promislow [17], who also drew on Hawkey [16].

The account presented in Table 9.2 (bottom row) is based on sources other than Hawkey [16]. One infers from Table 9.2 that hemoglobin concentration in the blood of adult mammals is probably independent of body size in the given samples (i.e., shows absolute invariance, $b = 0.0$). *Ontogenetic* data for the dog (not shown) are provided by Deavers *et al.* [18]. The data employed to construct the bottom row of Table 9.2 are shown in Figure 9.4 (panels a and b). Observe that the data for the armadillo (*Dasypus novemcinctus*), cat (*Felis silvestris*), and sheep (*Ovis aries*) deviate substantially from the best-fit LSQ line (Figure 9.4b).

Assigning a slope of 0 (rather than -0.010) (Chapter 5) to the line fitted to the data of Table 9.2 (bottom row), one obtains an intercept of 16 g Hb/100 cm^3 blood (Chapters 5 and 24, and Figure 9.4a). The normal values for adult women and men are 14 g and 16 g per 100 cm^3 blood, respectively. A further discussion of the properties of hemoglobin is presented below.

Red blood cell diameter

Among the vertebrate classes, mammals have the "smallest and most numerous" red blood cells [19]. The fruits of a LSQ analysis of red blood cell diameter for three datasets are presented in Table 9.3. The slopes for these three instances are possibly consistent with absolute invariance ($b = 0.0$) of red blood cell diameter expressed as a function of body size. Calder [22] reported that red blood cell diameter scales as the -0.02 power of body size. Assigning a slope of 0 to the best-fit line for each dataset

Table 9.3 Scaling of red blood cell diameter (μm) in adult mammals

N_{ord}	N_{sp}	MPD	pWR	DI (%)	r^2	Slope ± 99% CL	Source
6	21	11	3.3	4.3	0.008	0.006 ± 0.041	[20] [a]
3	17	2	1.4	0.9	0.509	0.034 ± 0.026	[21]
11	78	17	4.7	16	0.002	−0.004 ± 0.030	[16]

See Table 9.2. [a] Body sizes from standardized body weight table (Chapter 4).

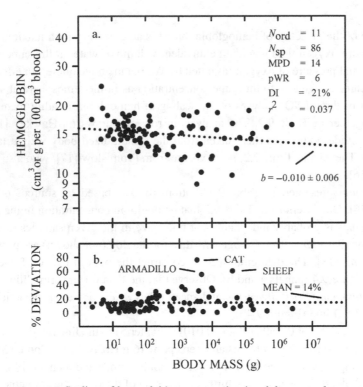

Figure 9.4 Scaling of hemoglobin concentration in adult mammals.

(Table 9.3, rows one to three), the mean diameters for red blood cells are found to be 7.4, 6.4, and 6.0 μm, respectively. The data from which the bottom row of Table 9.3 is computed are shown in Figure 9.5, panels a and c.

While diameter appears on average to be independent of body size (i.e., exhibits absolute invariance), there is nevertheless considerable variation among species. According to Slijper [23], the largest known mammalian red blood cells (9.5 μm in diameter) are found in the sperm whale (*Physeter catodon*). Red blood cells with diameters as small as 1.5–2.1 μm have been reported in the lesser mouse-deer mouse (*Tragulus javanicus*) [24].

Table 9.4 LSQ analysis of mean red blood cell volume in adult mammals

N_{ord}	N_{sp}	MPD	pWR	DI (%)	r^2	Slope $\pm$ 99% CL	Source
3	27	12	2.9	2	0.703	0.139 ± 0.050	[25]
12	109	32	4.7	19	0.0007	-0.005 ± 0.044	[16] [a]

See Table 9.2. [a] Body size data from standardized body weight table (Chapter 4).

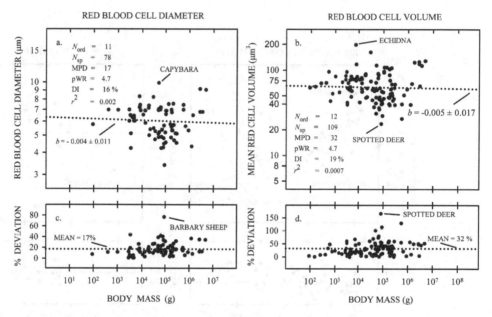

Figure 9.5 Scaling of red blood cell diameter and volume in adult mammals.

Mean volume of red blood cells

The phrase "red cell volume" is open to misinterpretation; it may refer to the volume of all red blood cells, usually expressed as a percent of total blood volume (i.e., hematocrit or packed cell volume (PCV)) (see below) or to the mean volume of individual red cells or corpuscles.

Here the concern is with corpuscular volume. The outcomes of LSQ analyses of mean red blood cell volume data from two different laboratories are reported in Table 9.4 (rows 1, 2). See also Sealander [21]. The data for row two of Table 9.4 are displayed in Figure 9.5, panels b and d. Promislow [17] found a slope of -0.009 for mean red cell volume versus body mass. These estimates of mean red cell volume (Table 9.4, rows one, two) were not measured directly, but rather were computed from measurements of hematocrit and red cell density (number/mm^3).

A plausible inference to draw from row two of Table 9.4 and from Figure 9.5, panel b, is that mean red cell volume is independent of body size ($b = 0.0$) in this sample [16].

Table 9.5 LSQ analysis of the red blood cell concentration in adult mammals

N_{ord}	N_{sp}	MPD	pWR	DI (%)	r^2	Slope ± 99% CL	Source
12	113	32	4.7	19	0.000	−0.001 ± 0.042	[16] [a]
-	173	-	-	-	-	0.001	[17] [b]
10	62	22	5.7	17	0.453	−0.080 ± 0.030	From Lit.

See Table 9.2. [a] Body size data from SBT (Chapter 4); [b] red blood cell concentration data taken from Hawkey [16].

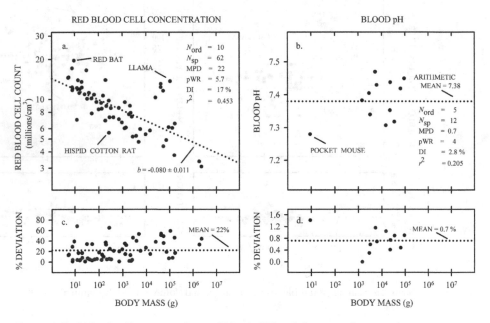

Figure 9.6 Red blood cell concentration and blood pH in adult mammals.

Assigning a slope of 0 (for absolute invariance) to the best-fit line yields an intercept of 63 μm^3. The arithmetic mean of the red cell volume data is 68 μm^3.

Red blood cell concentration

The outcomes of LSQ analyses of red blood cell concentration (millions of cells per mm^3) in three different datasets are shown in Table 9.5. The data used to compute the bottom row of Table 9.5 are illustrated in Figure 9.6, panels a and c.

The data presented in Table 9.5 (first two rows) are consistent with absolute invariance of red blood cell concentration. But the data presented in Table 9.5 (bottom row) and Figure 9.6a suggest there may be, in this sample, a tendency (slope = −0.08 ± 0.03) (99% CL) for the number of red blood cells per unit volume to decline with increasing

Table 9.6 LSQ analysis of the scaling of red blood cell lifespan in adult mammals

N_{ord}	N_{sp}	MPD	pWR	DI (%)	r^2	Slope $\pm$ 99% CL	Source
6	15	17	4.3	5	0.818	0.161 $\pm$ 0.064	From Lit.

See Table 9.2.

body size. The common name in Figure 9.6a for llama refers to *Llama glama*, for red bat to *Lasiurus borealis*, and for hispid cotton rat to *Sigmodon hispidus*.

Red blood cell lifespan

In the adult human, a red blood cell spends about 110–120 days in the circulation before being destroyed by phagocytes in the bone marrow, liver, and spleen. The results of a LSQ analysis of a small dataset bearing on red blood cell lifespan in adult species of mammals as a function of body mass are itemized in Table 9.6.

The range in lifespan for red blood cells in the present dataset extends from 25 days in the mouse (*Mus musculus*) to 113 in the human and to 147 in the Barbary sheep (*Ammotragus lervia*). Red blood cell lifespan in adult mammals scales as the 0.161 $\pm$ 0.064 (99% CL) power of body size (see Table 9.6) in the given sample.

A slope of 0.161, if applicable over the whole mammalian size range, implies that red blood cells in the blue whale (*Balaenoptera musculus*; body mass = 10^8 g) would survive for 408 days (assuming an intercept of 21 days, then $21 \cdot 10^{(8 \cdot 0.161)} = 408$). In Chapter 17 we find that maximal lifespan for mammals scales as the 0.187 $\pm$ 0.023 (99% CL) power of body mass.

Blood pH

The normal pH of adult human blood plasma is 7.4 (range of 7.35 to 7.45), being slightly alkaline. A plasma pH of 7.0 or less (and 8.0 or greater) may be lethal [26] (but see also Osnes and Hermansen [27] and Chapter 18). A summary of the findings obtained here (data from the literature) is shown in Figure 9.6, panels b and d. The *arithmetic* mean pH is 7.38. It is seen from Figure 9.6d that the mean deviation (from 7.38) is only 0.7%. Note, however, that the pH data (panel b) are badly distributed, with the pocket mouse (*Perognathus longimembris*) being an outlier. If one omits the data for the pocket mouse, the arithmetic mean is 7.395. When, for consistency, one does a LSQ analysis of log(pH) (rather than pH) as a function of log(BW), a slope 0.002 $\pm$ 0.003 (99% CL) and a coefficient of determination (r^2) of 0.205 are obtained. Whichever inference one prefers, normal plasma hydrogen ion concentration varies little from one species to another or with body size in this small sample, consistent with absolute invariance ($b = 0.0$).

Blood P_{50}

At sea-level the atmosphere exerts a mean pressure of 760 mm Hg. Oxygen constitutes 20.9% (by volume) of the (dry) atmosphere and exerts a partial pressure $pO_2 = 159$ mm Hg ($0.209 \cdot 760 = 159$) at the earth's surface. However, this is not the partial pressure of oxygen in the lungs, owing to the presence of water vapor and carbon dioxide. Water vapor at 37 °C exerts a partial pressure of 47 mm Hg. Correcting for water vapor implies a partial pressure of oxygen of 149 mm Hg (($760 - 47) \cdot 0.209 = 149$). Then correcting for pCO_2 (40 mm Hg) in the alveoli gives a partial pressure of oxygen in the alveoli at 37 °C of 109 mm Hg($149 - 40 = 109$). The partial pressure of oxygen in the arteries is commonly taken as 100 mm Hg. In the tissues it is found to be 30 mm Hg or less, so that the net driving partial pressure for oxygen between arterial blood and the tissues is about 60–70 mm Hg, or at most 100 mm Hg in very active tissues where the local cell pO_2 may approach zero.

It is important that the hemoglobin dissociation curve (degree of oxygen saturation vs pO_2) is sigmoid, so that oxygen is unloaded more rapidly with decreasing pO_2 than would be the case if the dissociation curve were linear. Moreover, the dissociation curve as a whole is shifted left or right in different species of mammals. This shift, measured by determining the pO_2 at which hemoglobin is 50% saturated, is called P_{50} or T_{50}, or the half-saturation pressure. P_{50} is usually measured at normal body temperature and blood pH. P_{50} provides a measure of hemoglobin's affinity for oxygen; smaller values of P_{50} imply a higher affinity for oxygen and vice versa. The results of the present LSQ analysis are exhibited in Figure 9.7, panels a and c (data from the literature).

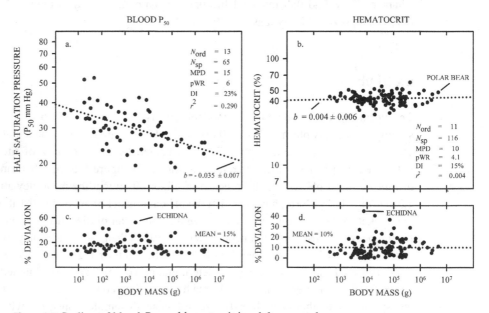

Figure 9.7 Scaling of blood P_{50} and hematocrit in adult mammals.

The slope of the best-fit LSQ line in Figure 9.7a, for P_{50} is -0.035 ± 0.018 (99% CL). (The intercept for this line is 39.2 mm Hg.) The point for the echidna (*Tachyglossus setosus*) deviates noticeably from the best-fit line (panel c). Assigning a slope of zero to the best-fit line furnishes an intercept of 30 mm Hg (Chapter 5). The arithmetic mean of the values of P_{50} shown in Figure 9.7a is 31 mm Hg. In a prior study a slope of -0.03 ± 0.007 (s.e.) for P_{50} as a function of adult body size was reported [28]. These findings are reasonably consistent with the hypothesis of absolute invariance ($b = 0.0$) for the scaling of P_{50} as a function of adult body size in the given sample.

Bohr effect and oxygen capacity

Carbon dioxide is probably the most important physiological modifier of tissue oxygen delivery.

Adamson and Finch 1975

One thinks of the hemoglobin molecule as evolved primarily to take up oxygen in the lungs. But hemoglobin is only effective if it also unloads oxygen readily in the tissues. What is wanted is a molecule with high affinity for oxygen in the lungs and low affinity in the tissues. At the same time, blood needs to have a high affinity for carbon dioxide in the tissues and a lower affinity in the lungs. To a surprising extent, nature has been able to achieve these seemingly conflicting goals.

The mechanism of unloading oxygen in the tissues takes advantage of the fact that energy metabolism produces carbon dioxide. This ambient carbon dioxide diffuses into the red blood cells and is rapidly converted by intracellular carbonic anhydrase into hydrogen and bicarbonate ions. The effect of hydrogen ions binding to the hemoglobin molecule, thereby lowering the effective pH, is to shift the hemoglobin dissociation curve to the right, thus lowering the affinity for oxygen, and so facilitating the unloading of oxygen to the tissues. Free carbon dioxide molecules also bind directly to hemoglobin, again lowering the affinity for oxygen. These combined effects of carbon dioxide (see above quote from Adamson and Finch) and pH on oxygen affinity are termed the *Bohr effect*. In highly active tissues, such as contracting muscle, the greater consumption of oxygen tends to increase the gradient between the blood and the tissues and also to raise tissue and blood temperature. These several factors facilitate the unloading of oxygen.

The Bohr effect is measured as a change in the logarithm of the oxygen pressure (denoted by "$\Delta \log (P_{50})$") divided by the associated change in pH (denoted by "ΔpH"), measured at constant temperature (e.g., body temperature), at a specified hemoglobin oxygen saturation and at varying CO_2 tensions. That is, the Bohr effect measures the magnitude of the change in the oxygen affinity of hemoglobin as a function of alteration in pH or the partial pressure of carbon dioxide.

Riggs [29] determined the magnitude of the Bohr effect in hemoglobin solutions derived from blood samples taken from mammals of varying body size (mouse to elephant). He found that hemoglobin taken from small mammals, with higher relative

Table 9.7 Scaling of Bohr effect and oxygen capacity in adult mammals

N_{ord}	N_{sp}	MPD	pWR	DI (%)	r^2	Slope $\pm$ 99% CL	Entity	Source
7	9	6.7	5	5.5	0.859	-0.095 ± 0.051	Bohr effect	[29]
9	23	8.6	6	12	0.496	0.032 ± 0.020	Bohr effect	From Lit.
11	30	15	5.6	15	0.010	-0.006 ± 0.031	O_2 capacity	From Lit.

See Table 9.2.

rates of energy metabolism, had larger Bohr effects (shifts). From his data, one obtains a LSQ slope of -0.0955 ± 0.051 (99% CL) for the Bohr effect as a function of adult body mass (see Table 9.7, row one). For pWR = 5.0, one also finds from his empirical data that the Bohr effect is 2.5 times as large in a mouse as it is in an elephant.

In a separate analysis of the Bohr effect one obtains a LSQ slope of 0.032 ± 0.020 (99% CL) (Table 9.7, row two). From these data, one finds that the Bohr effect is 1.9 times as large in a shrew as it is in a hippopotamus, for pWR = 5.4. These outcomes are roughly consistent with the findings of Riggs [29]; see also Pietschmann *et al.* [30]. It is likely that the Bohr effect shows at most a weak association with increasing adult body size, or scales in harmony with absolute invariance ($b = 0.0$).

Consider next the blood *oxygen capacity*. Each of the four subunits of Hb can combine, at most, with one molecule of oxygen. Taking the molecular weight of a hemoglobin tetramer to be about 68,000 daltons, then the mean weight of one subunit is 68,000/4 or 17,000 daltons. At STPD (denoting a volume of dry gas at zero degrees centigrade and a pressure of 760 mm Hg) this amount of Hb can combine with 1 gram mole or 22.4 liters of oxygen. It follows that 1 g of Hb can combine with 22,400/17,000 or $1.32 \, cm^3 \, O_2$ [31]. If one takes 16 g Hb/100 cm^3 of blood as a mean figure for the concentration of hemoglobin (see above) in the blood of adult mammals, then the maximum oxygen capacity would be $1.32 \cdot 16 = 21 \, cm^3 \, O_2$ per 100 cm^3 of blood, corresponding to the atmospheric concentration of oxygen (21% by volume).

The end results of a LSQ analysis of the oxygen capacity (expressed in *volume percent*) of blood from 30 different species and 11 different taxonomic orders are shown in Table 9.7, bottom row. The LSQ slope (-0.006) is not significantly different from zero (absolute invariance). Assigning a slope of zero to the best-fit line, one obtains an intercept of 19.7 volume percent. At the same time, recall that seals, no doubt as an adaptation to diving, have unusually high oxygen capacities [32].

It is commonly, and in part correctly, accepted that blood oxygen in mammals is mostly found in the arterial blood. The ratio of the oxygen saturation of arterial blood to venous blood in an adult human at rest is about 1.3 [11]. Hence, there is roughly one-quarter to one-third more oxygen per cm^3 of arterial blood than per cm^3 of mixed venous blood. However, the ratio of systemic venous blood volume to the volume of the aorta and systemic arterial blood volume in the adult human is 3.7 [6]. It follows that in the adult human there is much more oxygen in the veins than in the arteries (by a factor of about 2.8 (3.7/1.3 = 2.8)). This is a counter-intuitive (rather than functional) aspect of the way the circulatory system has evolved.

Table 9.8 Summary of LSQ analyses of blood (BV) and plasma volume (PV) in adult mammals

N_{ord}	N_{sp}	MPD	pWR	DI (%)	r^2	Slope ± 99% CL	CL	Entity	Source
3	3	-	4 (est.)	0.9	0.994	0.987 ± 0.017	s.e.	BV	[35]
11	35	20	4.9	13.5	0.992	0.996 ± 0.043	99%	BV	From Lit.
9	26	19	4.9	10.2	0.992	0.978 ± 0.049	99%	PV	From Lit.

See Table 9.2. est. = estimated; BV = total blood volume; PV = plasma volume; both in cm^3; s.e. = standard error.

Blood and plasma volumes

Blood volume is usually determined as the sum of plasma volume (PV) and red cell volume. The use of double labels, one (e.g., the isotopes ^{51}Cr or ^{32}P) for red cells and another (e.g., Evan's blue (T-1824) or ^{131}I) for plasma, is capable of giving reproducible accuracies of 5% in the human [33]. The use of a single label, perhaps together with hematocrit measurements, gives a lower accuracy of about 10% [34]. In this analysis of total blood volume, about 30% of the studies involved T-1824, and another 34% used hematocrit. The yields of a LSQ analysis of blood volume in 35 species representing 11 orders are shown in Table 9.8 (row two). The LSQ slope ($b = 0.996 \pm 0.043$ (99% CL)) for blood volume on adult body size is close to that reported earlier by Brody [35] (see Table 9.8, row one) and Stahl [36]. Both slopes are reasonably consistent with relative invariance ($b = 1.0$).

Assigning a slope of 1 (rather than 0.996) to the best-fit line for the scaling of total blood volume (Table 9.8, row two) on adult body mass gives an intercept of 0.077, implying that on average blood volume is 7.7% of body volume (mass). In a prior study [28] an empirical slope of 1.00 and an intercept of 0.076 were reported. Because blood volume scales nearly as the first power of body mass (Table 9.8, rows one and two), and hemoglobin concentration in the blood scales nearly as the zero power of body mass (Table 9.2), it follows that the *number* of hemoglobin molecules must scale as, or nearly as, the first power of body volume or mass (relative invariance).

Here one finds that plasma volume (PV) in the adult mammal scales as the 0.978 ± 0.049 (99% CL) power of adult body mass (see Table 9.8, bottom row). In 62% of the cases Evan's blue was used as the marker; in another 23% either ^{125}I or ^{131}I was employed. After assigning a slope of 1 (rather than 0.978) to the best-fit LSQ line, an intercept of 0.046 is obtained, implying that on average PV is 4.6% of adult body mass. In a previous study [28] a LSQ slope of 1.00 and an intercept of 0.044 were found for PV in adult mammals. Blood volume and possibly plasma volume apparently scale in accord with relative invariance ($b \approx 1.0$).

Hematocrit

The hematocrit, or packed cell volume (PCV), is the fraction of blood volume comprised of red blood cells. In the past, hematocrit was determined by filling a calibrated

Table 9.9 Scaling of hematocrit (or PCV) in adult mammals as a function of body mass

N_{ord}	N_{sp}	MPD	pWR	DI (%)	r^2	Slope ± 99% CL
11	116	10	4.1	15	0.004	0.004 ± 0.016

See Table 9.2.

glass tube up to a specified level with a fresh sample of blood and centrifuging it. The red blood cells sink to the bottom, leaving the blood plasma on top. A thin buffy coat on top of the red blood cells comprises the white cells. The hematocrit is simply read off the scale as a percent of the length of the centrifuge tube. The mean values of hematocrit in normal adult men and women are usually stated as 45% and 42%, respectively. Elevated hematocrit may by indicative of polycythemia (increased proportion of red blood cells), or it may be lowered by say internal hemorrhage. In exact work, measurements of hematocrit, as determined by centrifugation, are corrected for trapped plasma between the red blood cells.

In current practice, hematocrit may be inferred from independent measurements of the red cell count (number of cells per unit volume of blood) multiplied by the directly measured mean cell volume (see above). The packed cell volume data of Hawkey [16] together with body masses extracted from the SBT (see Chapter 4) were employed to compute the results seen in Table 9.9; see also Figure 9.7, panels b and d. The largest species represented in panel b is the polar bear (*Ursus arctos*).

If we assign a slope of 0 (rather than 0.004) to the best-fit LSQ line (Table 9.9), we obtain an intercept of 42.2%. On the other hand, the *arithmetic* mean of the hematocrits in this sample of 116 species of mammals is 42.6%. The ontogenetic hematocrit in the dog [37] (not shown) follows a similar pattern to that seen for hemoglobin in the dog [18]. Promislow [17] reported a slope of –0.009 for the scaling of hematocrit on adult body mass. These several findings are reasonably consistent with absolute invariance ($b = 0.0$). By coincidence, in Table 9.9, r^2 and the slope are numerically the same (0.004).

Organ blood volume

It is known that while a few components (e.g., calcium, fat) of the body may scale faster than adult body size ($b > 1.0$) (see Chapter 8), numerous other components (e.g., brain, kidney, and liver) scale more slowly than body size ($b < 1.0$), whereas blood volume, for example, scales almost as the first power of body mass ($b = 0.996 \pm 0.043$) (99% CL) (Table 9.8). Thus, it is of interest to know how blood volume per organ scales with body size. Few studies were found where organ blood volume was determined in multiple organs and in several different species. The main results, expressed as percent of total blood volume per organ, are depicted in Table 9.10. It should be noted that the data provided by Hansard [38] are only approximate, owing to variations in the figures offered for body mass. Otherwise, these few findings suggest that organ blood

Table 9.10 Organ blood volume as a percent of total blood volume at rest in adult mammals

Species/Organ	Mouse	Rat	Pig	Sheep	Cow	Horse
Heart	-	-	0.2	0.3	0.3	0.36
Kidney	4.97	2.61	0.5	0.2	0.3	0.84
Liver	-	8.1	2.5	1.8	3.0	4.80
Lungs	2.98	-	3.5	4.5	3.1	4.76
Spleen	0.56	1.84	0.5	1.5	1.6	-
TBV (cm^3)	2.57	26.1	2,910	3,160	23,400	-
Body mass (g)	21.2	320	46,300	53,500	41,000	-
						-
Method	PV, HCT	PV, HCT	^{32}P labeled RBC, TBV [a]			[b]
Source	[39]	[40]	[38]	[38]	[38]	[41]

RBC = red blood cell; PV = plasma volume; HCT = hematocrit; TBV = total blood volume. [a] Applies to pig, sheep, and cow. [b] Tissue blood content calculated by comparing the activity of ^{51}Cr-labeled RBC with the activity of a labeled reference sample characteristic of the total blood volume.

volume as a percent of total blood volume is broadly similar across species for heart, lungs, and spleen. Observe that the percent blood volumes for mouse and rat kidney and rat liver are substantially higher than for larger species. If blood volume were evenly distributed across all organs and tissues, one would expect to find each being 7.7% blood (see above). More data obtained with a consistent and precise methodology are needed.

BLOOD VESSELS

Capillary diameter

Considering that capillaries are the primary site of exchange between the blood and the tissues, it is surprising how little quantitative work has been carried out on capillary dimensions in mammals differing substantially in body size. This may reflect in part the morphological complexity of the capillary circulation. Much of the reported work has been carried out in the myocardium. Kassab and Fung [42] cited data from the literature (12 papers from 11 laboratories for four species – dog (four points), human (one), pig (one), rat (11) – giving a total of 17 data points, of which 15 are taken from the left ventricle, and two from the right). The mean capillary diameter for the 17 cases was 5.5 µm, with a STD of ±0.84. No data for body size were cited. Kayar *et al.* [43] reported estimates of capillary diameter based on their own measurements as well as some taken from the literature. For dog (*Canis familiaris*), goat (*Capra aegagrus*), steer (*Bos indicus*), and horse (*Equus ferus*) skeletal muscle, the capillary diameters varied from 4.1 to 4.6 µm, with a mean of 4.5 ± 0.2 µm (STD), and gave no evident correlation with body size or energy metabolism. As matters stand, the consensus holds that there is little or no variation in capillary diameter in either skeletal or cardiac muscle with increasing body size (i.e., $b \approx 0.0$).

Capillary length

Mean capillary length is problematic in many tissues in part because capillaries need not be straight. Various estimates of capillary length in different species and tissues are summarized in Table 9.11.

The estimates of mean capillary length seen in Table 9.11 lie between 0.5 and 1 mm, with a mean of 0.75 ± 0.22 mm (STD). There is no evidence of significant variations in capillary length with body size or energy metabolism in this small sample.

Total pulmonary capillary surface area and volume

The most extensive estimates of *capillary surface area* and *volume* in a variety of mammalian species I found are those of Gehr *et al*. [48] made by stereological methods in the lung. From a *subset* of their data, one finds that pulmonary capillary surface area scales as the 0.888 ± 0.056 (99% CL) power and capillary volume as the 0.970 ± 0.050 (99% CL) power of *lung volume* (see Chapter 3, "A repeating units model"). The LSQ analytical inferences are displayed in Table 9.12. Weibel [49] gives values of 126 m^2 and 213 cm^3 for the capillary surface area and volume, respectively, of the adult human lungs.

Capillary number

Mall [7] offers for the total number of capillaries in the crypts and villi of the dog's small intestine a figure of 51 million, and for the dog's stomach 23 million. Weibel and

Table 9.11 Nominal values of mean capillary length in a few species of adult mammals

Species	Organ/Tissue	Mean capillary length (mm)	Source
Cat	Tenuissimus muscle	1.0	[44]
Cow, dog, goat, horse, rabbit, rat	Renal glomeruli	0.7	[45]
Rat	Gracilis muscle	1.0	[46]
Cat, dog, pig, rabbit, rat	Left and right ventricles	0.5	[42]
Human	Renal glomeruli	0.5	[47]

Table 9.12 Total pulmonary capillary surface area and volume as a function of lung volume

N_{ord}	N_{sp}	MPD	pWR	DI (%)	r^2	Slope $\pm$ 99% CL	Entity
6	20	3.2	4.4	5.6	0.991	0.888 ± 0.056	Capillary surface area
6	20	22	4.4	5.6	0.994	0.970 ± 0.050	Capillary volume

See Table 9.2.

Gomez [50] state that on average the adult human lung has 277 billion capillary "segments," each segment joined to other essentially identical segments forming an array of hexagonal meshes.

Capillary density

...there is little if any variation in minimum capillary size and capillary density among mammals of a wide range of sizes.

Promislow 1991

Blood flow through capillaries (in the frog) was first described by Malpighi (1628–1694) in 1661, following Harvey's discovery of the circulation in 1628 (see above). August Krogh (1874–1949) appears to have been the first to make a quantitative study of *capillary density* in mammals of different sizes (dog (*Canis familiaris*), guinea pig (*Cavia porcellus*), horse (*Equus ferus*)). His studies were motivated by a desire to better understand how the blood supply to the tissues meets the demand or "call" for oxygen. He found that in relaxed muscle few capillaries are open, but that as activity increases, the number of open capillaries also increases [51]. He further showed that the increased number of open capillaries was not due to increased blood pressure. For his pioneering work, Krogh was awarded the Nobel Prize for Medicine in 1920.

Plyley and Groom [52] in their survey of the literature observed that capillary density measurements made before 1950 tended to be three times as high as those reported afterwards. It is also known that capillary density may vary with the type of skeletal muscle cell. Type I muscle fibers are oxidative, contract relatively slowly, tire gradually, and have substantial quantities of myoglobin, making them red in color. Type IIb muscle cells are glycolytic, contract and fatigue quickly, have little myoglobin and are therefore white in color. Type IIa fibers are intermediate between I and IIb. Table 9.13 reports a LSQ calculation made from the white fiber capillary density data of Schmidt-Nielsen and Pennycuik [53] in the gastrocnemius muscle (a locomotor muscle located on the posterior aspect of the lower or hind limb, with both red and white fibers). These data may (or given the wide CL may not) suggest that capillary density scales only weakly or possibly not at all with body size ($b = 0.0$). See above quote from Promislow.

Table 9.13 Scaling of capillary densities (number of capillaries per unit cross-sectional area) in the gastrocnemius muscle

N_{ord}	N_{sp}	MPD	pWR	DI (%)	r^2	Slope $\pm$ 99% CL	Entity	Source
4	8	15	4.2	2.5	0.537	-0.063 ± 0.088	Gastrocnemius [a]	[53]

See Table 9.2. [a] A point for a bat species appears to be a substantial outlier and was omitted from the calculations.

Additional data on capillary density in the vicugna (*Vicugna vicugna*) left ventricle are provided by Jürgens *et al.* [54]. From the above and other data one finds that the mean capillary density in the hearts of various adult species is about 3,000 capillaries per mm^2, implying a mean separation between capillaries of about 20 μm (square root of $(10^6/3,000) = 18.3$).

Aortic diameter

Since body area scales as the 2/3 power of the body mass, any characteristic internal area of the organism should scale likewise. Such is the case, for instance, with the cross-sectional area of the aorta.

Schepartz 1980

Apart from its intrinsic interest, one wants to know how aortic diameter scales mainly because cardiac output (CO) (see below) may be defined by the relation:

$$CO = \text{aortic cross-sectional area} \times \text{mean aortic blood velocity} \qquad (9.1)$$

If it were the case that aortic cross-sectional area scales as does cardiac output, one would infer from relation (9.1) that mean aortic blood velocity is independent of body size. The present findings (see Table 9.14) suggest that aortic cross-sectional area scales as the 0.84 ± 0.14 (99% CL) power of body mass (as computed from the square of aortic diameter on body mass). Resting cardiac output (in small samples) scales as the 0.769 ± 0.070 (99% CL) power of body mass (see Table 9.16). Because of the *wide* 99% confidence *interval* (0.699 to 0.839) one cannot infer that aortic blood velocity is invariant at a resting heart rate. If one chooses to ignore the confidence interval, one might infer that blood velocity in the aorta at rest scales as the -0.07 power of body mass ($0.77 - 0.84 = -0.07$). (The diameter of the aorta of a blue whale weighing around 100 tons has been estimated at 23 cm; the aortic diameter predicted from Table 9.14, bottom row (see also Table 19.2) is about twice as large.)

Rat *ontogenetic* data on aortic diameter are available from Dreyer *et al.* [57]. It has, on occasion, been asserted (e.g., see above quote from Schepartz) that aortic cross-sectional area should scale as the two-thirds power of body size; the data of Table 9.14 (see larger sample sizes in top and bottom rows) cast doubt on this assumption. These slopes (0.41 and 0.418) imply that aortic cross-sectional area scales as the 0.82 to 0.84 power of body weight or mass.

Table 9.14 Aortic diameter as a function of adult body mass

N_{ord}	N_{sp}	MPD	pWR	DI (%)	r^2	Slope $\pm$ CL	CL	Source
9	21	17	6.75	13	0.984	0.41 ± 0.02	99%	[55] [a]
5	7	-	4.6	3.2	0.98	0.3 ± 0.02	95%	[56]
8	13	24	5.9	8.6	0.968	0.418 ± 0.071	99%	From Lit.

See Table 9.2. [a] There are 42 points and 21 species represented in this sample.

HEART

Sarcomere length

Furthermore the Z to Z distance [sarcomere length] *proved to be the same in a wide variety of animals.*

Linzbach 1960

The mammalian heart is composed chiefly of contractile cells termed myofibers. These myofibers may be branched, forming an anastomosing 3-D network. Each myofiber is in turn composed of longitudinally oriented myofibrils, organized into sarcomeres, these forming the contractile unit of the cell. A sarcomere is an example of an intracellular repeating structural unit (see Chapter 3) found in striated (cardiac and skeletal) muscle. A sarcomere consists of a highly ordered longitudinal array of protein filaments, chiefly actin and myosin. In the sliding filament model of muscle contraction, shortening of a sarcomere occurs when myosin and actin fibers slide past each other owing to make-and-break rowing-like contacts between the two classes of filaments. The length of cardiac sarcomeres varies with the state of contraction and also with location in the heart. Data for mean sarcomere lengths, determined partly under different conditions in rat, cat, dog, and human, are 2, 2.2, 2.1, and 2.0 µm, respectively [58, 59, 60, 61]. These few results are consistent with the view that sarcomere length is an absolute invariant ($b = 0.0$) (see above quote from Linzbach).

Cardiac myofiber diameter and length

They [cardiac muscle cells] *are usually 50–100 µm in length and do not often exceed 20 µm in breadth.*

Garven 1965

It was found (data from the literature) that cardiac myofiber diameter scales as the 0.015 ± 0.077 (99% CL) power of *heart mass* ($N_{ord} = 5$, $N_{sp} = 5$, MPD = 8, pWR = 5.3, DI = 3%, $r^2 = 0.30$). Mean cardiac myofiber *length* in cells isolated from the rat heart is given as 94 µm [62]. A morphometric study in the dog heart yielded a mean myofiber length of 71 µm [63]. Myofiber diameter may scale consistent with absolute invariance, but the CL are too wide to be certain. See above quote from Garven.

Heart size

Interest in how heart size varies with body size goes back at least a century. A likely motivation for much of this work comes from the fact that energy metabolism is a function of cardiac output, which in turn is a function of heart size. A portion of this earlier work is summarized in Table 9.15 (rows one to three). See also Prothero [64]. The conclusions as presented in Table 9.15, rows one to three may be viewed as

Table 9.15 Scaling of heart mass in adult mammals as a function of body weight or mass

N_{ord}	N_{sp}	MPD	pWR	DI (%)	r^2	Slope ± CL	CL	Source
10	29	31	6.4	15	0.989	1.010 ± 0.040 [a]	99%	[55]
7	11	27	5.7	7	0.949	1.014 ± 0.096	99%	[35]
6	9	30	3	3	0.975	1.080 ± 0.065	s.e.	[66]
14	126	33	7.7	37	0.991	0.953 ± 0.022	99%	From Lit.

See Table 9.2. [a] There are 54 points and 29 species represented in this dataset.

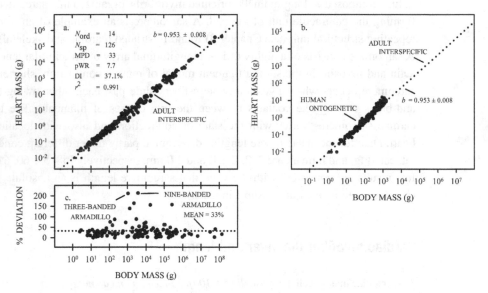

Figure 9.8 Scaling of heart mass in mammals as a function of body mass.

consistent with heart mass scaling directly with body mass. However, the 99% CL limits for these three rows make this doubtful (see "Special pleading," Chapter 23).

The outcomes of a study of heart mass in 126 species of adult mammals are set forth in Table 9.15 (bottom row). The range in heart size in this sample is from 0.0292 g in the pygmy white-toothed shrew (*Suncus etruscus*) to 430 kg in the blue whale (*Balaenoptera musculus*), for a ratio of 15 million. (The largest reported heart mass appears to be 698.5 kg from a female blue whale.) The end results of the LSQ analysis in Table 9.15 (bottom row) are illustrated in Figure 9.8a. The slope of 0.953 ± 0.022 (99% CL) implies that the largest mammals have hearts smaller by a factor of about 2.4 than would be the case if the heart scaled uniformly with body size (($0.047 \cdot 26.6 = 1.25$) doublings or a factor (*f*) of 2.4; $f = 2^{1.25} = 2.4$). This finding suggests that heart weight or mass scales with an exponent less than accords with relative invariance ($b \approx 1.0$).

Five species in Table 9.15 (bottom row) and Figure 9.8, panel c, exhibit heart masses that deviate from the LSQ line by more than 100%, all on the low side (recall that what is shown is absolute percent deviation: see Chapter 5). The most deviant points are for two species of armadillo (*Tolypeutes matacus, Dasypus novemcinctus*). The hearts in eight species of marsupials are larger by a factor of 1.20 than in placental mammals (as calculated from the ratio of the respective intercepts). Dawson [65] reported a factor of 1.33. For cetacean hearts relative to those of non-cetacean mammals, one finds a ratio of 1.45:1. However, this calculation involves a substantial implicit extrapolation, since the largest non-cetacean mammals are much smaller than the largest whales. If one compares cetacean hearts with those of mammals generally (including cetaceans), one obtains a ratio of 1.41:1.

The human *ontogenetic* data shown in Figure 9.8b are from Shepard [67]. These ontogenetic data appear to be nearly collinear with the adult interspecific line (see Chapter 3).

Cardiac output

Cardiac output in mammals is a physiological y-variable of major importance. Both resting and maximal energy metabolism are functions of cardiac output. A summary of three LSQ analyses of resting cardiac output is presented in Table 9.16. The inferences in Table 9.16, first row, have been quoted by some workers. Nevertheless, a slope of near 1 for resting cardiac output over nearly two orders of magnitude in body size is inherently implausible. Cardiac output is the product of stroke volume and heart rate. Stroke volume is likely to be proportional to heart size (e.g., Tenney [69]), which in turn scales as the 0.953 ± 0.022 (99% CL) power of body size (see Table 9.15, bottom row). Hence, it is plausible to assume that stroke volume scales roughly in proportion to body size. Few will think it likely that a pump weighing about 100 times more than another and constructed to the same plan and of the same materials will beat sustainably at the same relatively high rate (e.g., Hill [70]). Thus a slope of 1 for resting steady-state cardiac output as a function of body mass is improbable over any substantial size range. In this regard, the slopes reported in Table 9.16, rows two and three, are much more in line with what is physiologically reasonable.

The present LSQ analysis of cardiac output is shown in Table 9.16, bottom row. The exponent of 0.769 ± 0.070 (99% CL) for resting cardiac is compatible with the earlier findings ($b = 0.81$) reported in row two of Table 9.16. Dawson [65] found exponents for resting cardiac output of 0.87 and 0.81 for marsupials and placental

Table 9.16 Resting cardiac output as a function of adult body mass in prior and present studies

N_{ord}	N_{sp}	MPD	pWR	DI (%)	r^2	Slope ± CL	CL	Source	Year
4	11	-	1.8	1.2	-	0.999	-	[68]	1965
-	-	-	-	-	0.96	0.81 ± 0.01	s.e.	[36]	1967
8	21	23	4.4	7.5	0.981	0.769 ± 0.070	99%	From Lit.	-

See Table 9.2. s.e. = standard error.

Table 9.17 Scaling of resting and maximal heart rate in adult mammals as a function of body mass

ROW	N_{ord}	N_{sp}	MPD	pWR	DI (%)	r^2	Slope ± 99% CL	Heart rate (beats/min)	Source
1	9	19	27	5.8	11	0.894	−0.240 ± 0.058	Resting	[55]
2	7	13	24	5.2	6.6	0.909	−0.248 ± 0.073	Resting	[35]
3	14	65	31	6.1	26	0.830	−0.220 ± 0.033	Resting	From Lit.
4	8	29	13	5.3	10	0.91	−0.146 ± 0.025	Maximal	From Lit.

See Table 9.2.

mammals, respectively. Bartels [71] gave an exponent of 0.80 in 15 species of placental mammals for pWR = 6.8. Data for more diverse species taken over greater size ranges are needed.

Heart rate

The value of 1511 [beats] *per min recorded in* Suncus etruscus *is the highest heart rate ever reported for a mammal.*

Jürgens et al. 1996

Heart rate (HR) is of particular interest from the standpoint of whole-body energy metabolism, since cardiac output (see above) is the product of stroke volume and heart rate. The results of LSQ analyses of HR using data from Clark [65] published in 1927 and from Brody [35] in 1945 are shown in Table 9.17, rows one and two, respectively. The two slopes (−0.240 and −0.248) derived from the data of these authors are not significantly different. In addition, the best-fit LSQ lines for resting and maximal HR were computed using data taken from the literature. The fruits of this work are seen in Table 9.17 (rows three, four) and in Figure 9.9. The resting HR for echidna (*Tachyglossus aculeatus*), mole rat (*Spalax ehrenbergi;* not labeled), pygmy mouse (*Baiomys taylori;* not labeled), and southern coati (*Nasua nasua*) deviate from the best-fit LSQ line by 103% to 123%.

Data for the humpback whale (*Megaptera novaeangliae*) from Meijler et al. [72] were not included in any of the LSQ calculations (Table 9.17, Figure 9.9b) as these data represent a substantial outlier. (The conditions under which HR was measured in the humpback whale may not correspond to either resting or maximal heart rate.) The data on which rows three and four of Table 9.17 are based are shown in Figure 9.9 (panels a, b). Note that the slope (−0.146 ± 0.025, 99% CL) for maximal HR is significantly less steep (closer to zero) than the slope (−0.220 ± 0.033, 99% CL) for resting HR. However, maximal heart rates are only sustainable for short periods of time relative to those for, say, resting metabolic rate. The most deviant point in Figure 9.9d represents the wombat (*Lasiorhinus latifrons*). The highest heart rate (1,511 beats per minute) so far reported is that of a shrew (see above quote from Jürgens et al.).

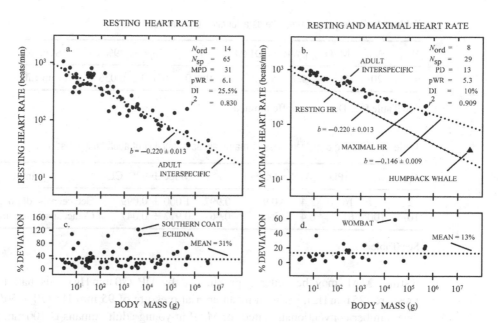

Figure 9.9 Resting and maximal heart rate (HR) in adult mammals.

Blood pressure

An understanding of blood flow in adult mammals is chiefly concerned with pressure differences rather than absolute pressure. In terrestrial (non-diving) mammals, the reference pressure is atmospheric. The arterial driving pressure is equal to the difference between the pressure at the entrance to the aorta and that at the end of the vena cava. This pressure difference is numerically equal to the transmural pressure difference across the aortic wall, where the pressure in the tissues is taken to be zero (i.e., atmospheric). Normally, in land mammals, this tissue pressure will be a few mm Hg above zero. (It may turn out that prolonged exposure to microgravity, as in space travel, will produce deleterious effects due to altered tissue pressures.) The concept of transmural pressure is important, because the overall pattern of blood flow in the vascular system is strongly dependent on wall compliance (ease of displacement), which is low in the arteries and high in the veins.

Because of the rhythmic nature of cardiac contraction, there is a peak pressure at maximal contraction, called systolic (SP), and a minimal pressure (during relaxation) called diastolic (DP). The approximate mean arterial pressure (MAP) in peripheral arteries may be calculated as follows:

$$MAP = DP + (1/3) \cdot PP \tag{9.2}$$

where PP is the pulse pressure, equal to SP − DP. The reason for the 1/3 factor is that systole only lasts for about one-third of the resting cardiac cycle. In a healthy young adult human, the systolic/diastolic ratio is usually stated as 120/70. According to relation (9.2), MAP should be 87 mm Hg (70 + 50/3 = 87). At higher heart rates than

Table 9.18 Resting blood pressure as a function of adult body mass

N_{ord}	N_{sp}	MPD	pWR	DI (%)	r^2	Slope ± 99% CL	Source
9	20	11	4.3	8.2	0.040	0.010 ± 0.036	From Lit.

See Table 9.2. Data for giraffe omitted.

Table 9.19 Scaling of spleen mass in mammals as a function of adult body mass

N_{ord}	N_{sp}	MPD	pWR	DI (%)	r^2	Slope ± 99% CL	Sample
2	7	16	3.4	0.9	0.997	1.000 ± 0.099	Cetaceans + dugong only
10	100	61	6.4	21	0.967	0.972 ± 0.047	Cetaceans + dugong omitted

See Table 9.2.

resting, MAP may be better approximated by DP (70 mm Hg) plus half the pulse pressure (25 mm Hg), giving a mean arterial pressure of 95 mm Hg (70 + 50/2 = 95). The number conventionally cited for MAP in young adult humans is 100 mm Hg.

The outcome of the present analysis of resting blood pressure is depicted in Table 9.18. Data are taken from the literature. In 12 of the 20 cases considered, a Statham pressure gauge, capable of measuring both pulse and mean arterial blood pressure, was employed. The arithmetic mean blood pressure (omitting data for the giraffe) is 115 mm Hg. (The resting blood pressure for the giraffe is 237.5 mm Hg [73].) Again we see general consistency with absolute invariance ($b = 0.0$).

Human ontogenetic blood pressure data are provided by Krishna *et al.* [74].

Spleen

The spleen is the largest mass of lymphoid tissue in the body and the only large lymphoid structure in the vascular circuit, where it is ideally placed to filter blood. It plays a number of different roles in the economy of mammals. In some mammals (such as the dog), the spleen is an important reservoir of replacement blood in the event of hemorrhage. The spleen eliminates ageing red blood cells (see above). Iron derived from the breakdown of hemoglobin is stored as ferritin in the liver and elsewhere. Other breakdown products contribute to the formation of bile in the liver. In the fetal human the spleen produces a proportion of the red blood cells.

The spleen weighs about 170 g in the young adult human and decreases to about 120 g in the elderly. Splenectomy in human adults is not life-threatening. The outcomes of a LSQ analysis of the scaling of spleen size in various species of mammals are summarized in Table 9.19 and Figure 9.10. Data are taken from the literature.

It may be seen in Table 9.19, row one, that the slope for the LSQ best-fit line for spleen size in adult cetaceans and dugong is 1.000 ± 0.099 (99% CL). Figure 9.10, panel a, suggests that the best-fit line for these species lies well below that for non-cetaceans

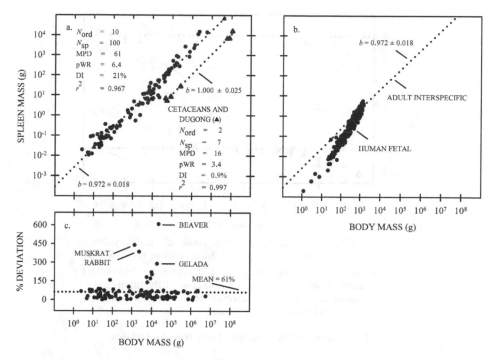

Figure 9.10 Scaling of spleen size as a function of adult body size.

(including Pinnipedia). The log (intercept) and 99% CL for the adult interspecific line are −2.563 ± 0.174 (see Chapter 19, Table 19.4), while those for the cetacean–dugong line are −3.942 ± 0.157 (99% CL), suggesting that the two intercepts may be statistically different.

Assigning a slope of 1 (rather than 0.972) to the LSQ best-fit line for the data of the bottom row of Table 9.19 implies that the spleen comprises about 0.2% of adult body mass, on average. Observe that the human *fetal* spleen data (Figure 9.10b), approach the adult interspecific line from below [67]. Take note in Figure 9.10c that spleen sizes for two species, beaver (*Castor canadensis*) and muskrat (*Ondatra zibethicus*), deviate from the best-fit LSQ line by more than 600 and 400%, respectively. Data for the rabbit (*Oryctolagus cuniculus*) and gelada (*Theropithecus gelada*) are also highly deviant. Note that Figure 9.10c does not include data for cetaceans and dugong.

Summary

What is more puzzling... is why red cell size and capillary diameter... are body-size independent.

Schmidt-Nielsen 1984

By plotting the various slopes derived above (by LSQ analysis in terms of body size) in rank order, one obtains a new perspective on the way the circulatory system scales with

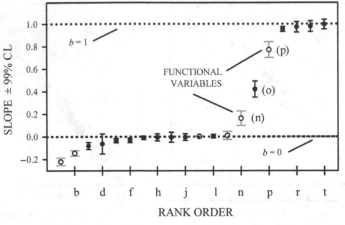

a = HRrest; b = HRmax; c = RBC count; d = capillary density;
e = P_{50}; f = Bohr effect; g = Hb concentration; h = blood O_2
capacity; i = RBC volume; j = RBC diameter; k = blood pH; l =
hematocrit; m = resting blood pressure; n = RBC lifespan; o =
aortic diameter; p = resting cardiac output; q = heart weight;
r = spleen mass; s = plasma volume; t = blood volume

HR = heart rate; RBC = red blood cell; Hb = hemoglobin

Figure 9.11 Summary of LSQ slopes for 20 circulatory y-variables.

increasing body size. See Figure 9.11. Note that the solid circles stand for 14 structural
y-variables and the open circles represent six functional y-variables (a, b, k, m, n, p).

A striking feature of Figure 9.11 is that eight of the 20 slopes lie within one
99% confidence interval of the reference slope of 0 (absolute invariance) (see d, g, h,
i, j, k, l, m). Another three slopes lie within one 99% confidence interval of 1 (relative
invariance) (see r, s, t). Two other slopes (c and q) are close to absolute or relative
invariance, respectively. The key point is that 55% to 65% of the slopes in this sample
are close to either absolute (see above quote from Schmidt-Nielsen) or relative invariance (see Chapter 3, Table 3.5; and Chapters 19, 20).

Observe that the six or seven slopes scaling in (or nearly in) accord with absolute
invariance (see items d to j) refer to *microscopic* and *macromolecular* structures whereas
the three structures (items r, s, t) scaling in accord with relative invariance are *macroscopic* (see Chapter 3). Six slopes in Figure 9.11 (a, b, c, n, o, p) are inconsistent with
either absolute or relative invariance. Compare Figure 9.11 with Figure 8.4, where
ten slopes lie between 0.909 and 1.054. Taken together, these findings are consistent
with the inference that absolute and relative invariance are of considerable importance in
arriving at an understanding of structural and functional scaling in adult mammals.

One might argue that Figure 9.11 compares apples and oranges, in the sense that
different slopes may be associated with different dimensions. That is true, but the slopes
(or exponents) per se are non-dimensional. Moreover, a power function can always be
cast in non-dimensional form, without altering the numerical value of the exponent
(Chapter 24). Recall that in the everyday commerce of life we have little difficulty in

comparing a host of *qualitatively* different objects in terms of varying prices. Thus the objection based on qualitative differences has only limited force.

We will see in coming chapters that the patterns of slopes appearing in the chapter summaries vary significantly as we move from the circulatory system to the digestive, musculoskeletal, and respiratory systems (see Chapter 20, Figure 20.2). Not all of the systems reviewed in Part II differ markedly with respect to these system-wide arrays of slopes, but some do. The circulatory system has long been a focus of scaling studies, partly because of its numerous physical aspects. The results shown in Figure 9.11 may imply that this will continue to be the case well into the future.

References

1. Stacy, R.W., Williams, D.T., Worden, R.E. and McMorris, R.O. (1955). *Essentials of Biological and Medical Physics*, New York: McGraw-Hill.
2. Schleier, J. (1918). Der energieverbrauch in der blutbahn. *Pflügers Archiv*, **173**:172–204.
3. Landis, E.M. and Hortenstine, J.C. (1950). Functional significance of venous blood pressure. *Physiological Reviews*, **30**:1–32.
4. Wiedeman, M.P. (1963). Dimensions of blood vessels from distributing artery to collecting vein. *Circulation Research*, **12**:375–378.
5. Meier-Ruge, W., Hunziker, O., Schulz, U., Tobler, H.-J. and Schweizer, A. (1980). Stereological changes in the capillary network and nerve cells of the aging human brain. *Mechanisms of Ageing and Development*, **14**:233–243.
6. Bazett, H.C. (1949). A consideration of the venous circulation. In: Zweifach, B.W. and Shorr, E. (eds.) *Factors Regulating Blood Pressure*, New York: Josiah Macy Jr. Foundation, pp. 53–81.
7. Mall, F.P. (1906). A study of the structural unit of the liver. *American Journal of Anatomy*, **5**:227–228. (No data provided for vessel lengths.)
8. Burton, A.C. (1966). *Physiology and Biophysics of the Circulation: An Introductory Text*, 2nd edn. Chicago, IL: Year Book Medical Publishers.
9. Prothero, J.W. and Burton, A.C. (1962). The physics of blood flow in capillaries: II. The capillary resistance to flow. *Biophysical Journal*, **2**:199–212.
10. Elsner, R. and Gooden, B. (1983). *Diving and Asphyxia: A Comparative Study of Animals and Man*. Cambridge: Cambridge University Press.
11. Ganong, W.F. (1965). *Review of Medical Physiology*. Los Altos, CA: Lange Medical Publications.
12. Dickerson, R.E. and Geis, I. (1969). *The Structure and Action of Proteins*. New York: Harper & Row.
13. Prothero, J.W. and Rossmann, M.G. (1964). The relative orientation of molecules of crystallized human and horse oxyhaemoglobin. *Acta Crystallographica*, **17**:768–769.
14. Scouloudi, H. and Prothero, J.W. (1965). The nature and configuration of the mercuri-iodide ion in the seal myoglobin derivative. *Journal of Molecular Biology*, **12**:17–26.
15. Prothero, J.W. (1968). A model of alpha-helical distribution in proteins. *Biophysical Journal*, **8**:1236–1255.
16. Hawkey, C.M. (1975). *Comparative Mammalian Haematology*. London: William Heinemann Medical Books.

17. Promislow, D.E.L. (1991). The evolution of mammalian blood parameters: patterns and their interpretation. *Physiological Zoology*, **64**:393–431.
18. Deavers, S., Smith, E.L. and Huggins, R.A. (1971). Changes in red cell volume, venous hematocrit, and hemoglobin concentration in growing beagles. *Proceedings of the Society for Experimental Biology and Medicine*, **137**:299–303.
19. Andrew, W. and Hickman, C.P. (1974). *Histology of the Vertebrates: A Comparative Text*. Saint Louis, MO: C.V. Mosby.
20. Ponder, E. (1948). *Hemolysis and Related Phenomena*. New York: Grune & Stratton.
21. Sealander, J.A. (1964). The influence of body size, season, sex, age and other factors upon some blood parameters in small mammals. *Journal of Mammalogy*, **45**:598–616.
22. Calder, W.A. (1984). *Size, Function, and Life History*. Cambridge, MA: Harvard University Press.
23. Slijper, E.J. (1962). *Whalers*. New York: Basic Books.
24. Weathers, W.W. and Snyder, G.K. (1977). Hemodynamics of the lesser mouse deer, *Tragulus javanicus*. *Journal of Applied Physiology*, **42**(5):679–681.
25. Dunaway, P.B. and Lewis, L.L. (1965). Taxonomic relation of erythrocyte count, mean corpuscular volume, and body-weight in mammals. *Nature*, **205**:481–484.
26. Lockwood, A.P.M. (1963). *Animal Body Fluids and their Regulation*. Cambridge, MA: Harvard University Press.
27. Osnes, J-B. and Hermansen, L. (1972). Acid–base balance after maximal exercise of short duration. *Journal of Applied Physiology*, **32**:59–63.
28. Prothero, J. (1980). Scaling of blood parameters in mammals. *Comparative Biochemistry and Physiology*, **67A**:649–657.
29. Riggs, A. (1960). The nature and significance of the Bohr effect in mammalian hemoglobins. *Journal of General Physiology*, **43**:737–752.
30. Pietschmann, M. Bartels, H. and Fons, R. (1982). Capillary supply of heart and skeletal muscle of small bats and non-flying mammals. *Respiratory Physiology*, **50**:267–282.
31. Dhindsa, D.S., Metcalfe, J., Hoversland, A.S. and Hartman, R.A. (1974). Comparative studies of the respiratory functions of mammalian blood. X. Killer Whale (*Orcinus orca Linnaeus*) and Beluga Whale (*Delphinapterus leucas*). *Respiratory Physiology*, **20**:93–103.
32. Scholander, P.F. (1940). Experimental investigations on the respiratory function in diving mammals and birds. *Hvalrådets Skrifter*, **22**:1–131.
33. Dagher, F.J., Lyons, J.H., and Finlayson, D.C. *et al.* (1965). Blood volume measurement: a critical study. *Advances Surgery*, **1**:69–109.
34. Gillett, D.J. and Halmagy, F.J. (1970). Accuracy of single-label blood volume measurement before and after corrected blood loss in sheep and dogs. *Journal of Applied Physiology*, **28**:213–215.
35. Brody, S. (1945). *Bioenergetics and Growth: With Special Reference to the Efficiency Complex in Domestic Animals*. New York: Reinhold. (The given value of heart weight of 2,200 g for the elephant on p. 642 is mistaken. The correct value is 26,100 g.)
36. Stahl, W.R. (1967). Scaling of respiratory variables in mammals. *Journal of Applied Physiology*, **22**:453–460.
37. Smith, E.L. (1972). Absolute and relative residual organ blood volumes and organ hematocrits in growing beagles. *Proceedings of the Society for Experimental Biology and Medicine*, **140**:285–290.
38. Hansard, S.L. (1956). Residual organ blood volume of cattle, sheep, and swine. *Proceedings of the Society for Experimental Biology and Medicine*, **91**:31–34.

39. Kaliss, N. and Pressman, D. (1950). Plasma and blood volumes of mouse organs, as determined with radioactive iodoproteins. *Proceedings of the Society for Experimental Biology and Medicine*, **75**:16–20.

40. Lewis, A.E., Goodman, R.D. and Schuck, E.A. (1952). Organ blood volume measurements in normal rats. *Journal of Laboratory Clinical Medicine*, **39**:704–710.

41. Weaver, B.M.Q., Staddon, G.E. and Pearson, M.R.B. (1989). Tissue blood content in anaesthetised sheep and horses. *Comparative Biochemistry and Physiology*, **94A**:401–404.

42. Kassab, G.S. and Fung, y-C.B. (1994). Topology and dimensions of pig coronary capillary network. *American Journal of Physiology*, **267**:H319–H325.

43. Kayar, S.R., Hoppeler, H., Armstrong, R.B. *et al.* (1992). Estimating transit time for capillary blood in selected muscles of exercising animals. *Pflügers Archiv*, **421**:578–584.

44. Eriksson, E. and Myrhage, R. (1972). Microvascular dimensions and blood flow in skeletal muscle. *Acta Physiologica Scandinavica*, **86**:211–222.

45. Holt, J.P. and Rhode. E.A. (1976). Similarity of renal glomerular hemodynamics in mammals. *American Heart Journal*, **92**:465–472.

46. Honig, C.R., (1977). Capillary lengths, anastomoses, and estimated capillary transit times in skeletal muscle. *American Journal of Physiology*, **233**:H122–H129.

47. Vimtrup, B.J. (1928). On the number, shape, structure, and surface area of the glomeruli in the kidneys of man and mammals. *American Journal of Anatomy*, **41**:123–151.

48. Gehr, P., Mwangi, DK, Ammann, A. *et al.* (1981). Design of the mammalian respiratory system: V. Scaling morphometric pulmonary diffusing capacity to body mass: wild and domestic mammals. *Respiratory Physiology*, **44**:61–86.

49. Weibel, E.R. (1984). *The Pathway for Oxygen*. Cambridge, MA: Harvard University Press.

50. Weibel, E.R. and Gomez, D.M. (1962). Architecture of the human lung. *Science*, **137**:577–585.

51. Krogh, A. (1929). *The Anatomy and Physiology of Capillaries*. New Haven, CT: Yale University Press.

52. Plyley, M.J. and Groom, A.C. (1975). Geometrical distribution of capillaries in mammalian striated muscle. *American Journal of Physiology*, **228**:1376–1383.

53. Schmidt-Nielsen, K. and Pennycuik, P. (1961). Capillary density in mammals in relation to body size and oxygen consumption. *American Journal of Physiology*, **200**:746–750.

54. Jürgens, K.D., Pietschmann, M., Yamaguchi, K. and Kleinschmidt, T. (1988). Oxygen binding properties, capillary densities and heart weights in high altitude camelids. *Comparative Physiology*, **B158**:469–477.

55. Clark, A.J. (1927). *Comparative Physiology of the Heart*. New York: MacMillan. (Apparently the units given for aortic cross-sectional area are square inches rather than square centimeters, as stated.)

56. Holt, J.P. and Rhode, E.A. (1981). Geometric similarity of aorta, venae cavae, and certain of their branches in mammals. *American Journal of Physiology*, **241**:R100–R104.

57. Dreyer, G., Ray, W. and Ainley Walker, E.A. (1912). Size of the aorta in warm-blooded animals and its relationship to the body weight and to the surface area expressed in a formula. *Proceedings of the Royal Society of London*, **86**:39–65.

58. Grimm, A.F., Katele, K.V., Klein, S.A. and Lin, H-L. (1973). Growth of the rat heart. Left ventricular morphology and sarcomere lengths. *Growth*, **37**:189–208.

59. Spiro, D. and Sonnenblick, E.H. (1965). The structural basis of the contractile process in heart muscle under physiological and pathological conditions. *Progress in Cardiovascular Diseases*, **7**:295–335.

60. Poole, D.C. and Mathieu-Costello, O. (1990). Analysis of capillary geometry in rat subepicardium and subendocardium. *American Journal of Physiology*, **259**:H204–H210.

61. James, T.N. and Sherf, L. (1978). Ultrastructure of the myocardium. In: Hurst, J.W., Logue, R.B., Schlant, R.C. and Wenger, N.K. (eds.) *The Heart. Arteries and Veins*. New York: McGraw-Hill, pp. 57–70.

62. Bishop, S.P. and Drummond, J.L. (1979). Surface morphology and cell size measurement of isolated rat cardiac myocytes. *Journal of Molecular Cellular Cardiology*, **11**:423–433.

63. Laks, M.M. (1967). Myocardial cell and sarcomere lengths in the normal dog heart. *Circulation Research*, **21**:671–678.

64. Prothero, J. (1979). Heart weight as a function of body weight in mammals. *Growth*, **43**:139–150.

65. Dawson, T.J. (1989). Responses to cold of monotremes and marsupials. *Advances in Comparative Environmental Physiology*, **4**:255–288.

66. Holt, J.P., Rhode,, E.A. and Kines, H. (1968). Ventricular volumes and body weight in mammals. *American Journal of Physiology*, **215**:704–715.

67. Professor T.H. Shepard. Personal communication.

68. Patterson, J.L., Goetz, R.H., Doyle, J.T. *et al.* (1965). Cardiorespiratory dynamics in the ox and giraffe, with comparative observations on man and other mammals. *Annals of the New York Academy of Science*, **127**:393–413. (The scatter in the data these authors provide is very considerable. No CL on slope are given.)

69. Tenney, S.M. (1967). Some aspects of the comparative physiology of muscular exercise in mammals. *Circulation Research*, **20**:I7–I14.

70. Hill, A.V. (1950). The dimensions of animals and their muscular dynamics. *Science Progress*, **38**:209–230.

71. Bartels, H. (1980). Aspects of respiratory gas transport in mammals with high weight specific metabolic rates. *Verhandlungen Deutschen Zoologischen Gesellschaft*, 188–201.

72. Meijler, F.L., Wittkampf, F.H.M., Brennen, K.R. *et al.* (1992). Electrocardiogram of the humpback whale (*Megaptera novaeangliae*), with special reference to atrioventricular transmission and ventricular excitation. *Journal of the American College of Cardiology*, **20**:475–479.

73. Goetz, R.H., Warren, J.V., Gauer, O.H. *et al.* (1960). Circulation of the giraffe. *Circulation Research*, **8**:1049–1058.

74. Krishna, P., PrasannaKumar, K.M., Desai, N. and Thennarasu, K. (2006). Blood pressure reference tables for children and adolescents of Karnataka. *Indian Pediatrics*, **43**:491–501.

10 Digestive system

The basic pattern of the digestive system is essentially the same in all vertebrates, with some special modifications imposed by varied diets.

Andrew and Hickman 1974

All free-living organisms do work, whether chemical, electrical, or mechanical, day-shift or night-shift. This work requires an ongoing flux of energy. Plants obtain energy directly from sunlight by photosynthesis; animals by consuming animal or plant matter. Animals have evolved several semi-independent subsystems for obtaining needed substances and excreting waste products. These subsystems may be defined by the states of matter. In mammals the respiratory system deals with gases, chiefly oxygen, carbon dioxide, and water vapor, whereas the digestive system treats liquids and solids. Liquids (mainly water and nitrogenous breakdown products) are excreted via the urinary system as well (see below). The overall organization of the digestive system is much the same in all vertebrates (see above quote from Andrew and Hickman).

The process of digestion converts large, complex (and possibly insoluble) materials such as proteins, fats, and carbohydrates into small molecules, for example amino acids, fatty acids, and glucose, which are readily absorbed across the gut wall. The stomach, not considered essential to life, temporarily stores semi-digested food (chyme) before transferring it to the small intestine. Most of the process of digestion and absorption of nutrients occurs in the small intestine. Digestion is associated with the release of large quantities of water into the lumen of the gastrointestinal tract, water which is reabsorbed, mostly in the small intestine, but also in the large intestine. Excreted material consists of semi-solids formed by cells eroded from the gut lining, undigested food and bacteria. Liquid is normally excreted mainly by the urinary system (Chapter 16) and also by the integument (Chapter 11) and the respiratory system (Chapter 15) (see below). The gut lining is exposed to wide variations in pH, to proteolytic enzymes, and to mechanical abrasion. It must be shielded from these hazards and repaired as needed.

In the following account, *structural* y-variables with the dimensions of length (six instances), area (three), and mass (12) are introduced. In addition *functional* data are presented for scaling of daily water intake (one instance). The slopes for 14 of these 22 instances are displayed in rank order in Figure 10.5. This chapter is based on the LSQ analysis of 970 records in total. The above figures refer to data expressed in tabular form.

Teeth

Logarithmic transformation of tooth crown area and body weight yields a linear model of slope 0.67 as an isometric [geometric] baseline for study of dental allometry.

Gingerich *et al.* 1982

Digestion begins in the mouth, partly owing to the secretion of digestive enzymes by the salivary glands, and partly owing to the cutting and crushing action of the teeth, actuated by the masticatory muscles. Most mammalian species (unlike reptiles), have teeth (incisors, canines, molars) of varying shapes (heterodont) and only two successive dentitions (diphyodont) [1]. Given the importance of chewing in mammalian physiology and the crucial role that teeth play in paleontology, there is a scarcity of quantitative information on tooth size and shape as a function of body size in different mammals [2].

Pilbeam and Gould [3] reported the results of a LSQ analysis of tooth *crown* area as a function of body weight (they derived a sum of areas for the first premolar through third molar for each specimen). These workers obtained a LSQ slope of 0.70 for 14 species of rodents, and of 0.78 in 10 species of male primates. No confidence limits were given.

A more extensive analysis of tooth data abstracted from the literature was reported by Fortelius [4]. He employed several statistical methods, including ordinary LSQ analysis. He concluded that the scaling of *tooth area* (as for lower and upper molars) scales consistent with geometric similarity (i.e., slopes near 2/3). From the LSQ portion of his analysis in male and female primates and bovids (e.g., cattle, sheep, goats) one finds that tooth area scales as a function of body size in these groups with a mean slope of 0.62 (range = 0.52–0.70) and a standard deviation (STD) of ±0.055.

Gingerich and Smith [5] took tooth area data (lower and upper first molars) in primates from Swindler [6] and their own prior work together with body weight data from the literature. They reported a slope of 0.67 for tooth crown area (calculated from width×length) in 43 species of male and female primates as a function of body size. They interpreted this result as supporting tooth scaling at constant shape (geometrical similarity) (see above quote from Gingerich *et al.*). Using a *subset* (e.g., males only) of the data provided by Gingerich and Smith [5] for primates one obtains the LSQ inferences for tooth length, width, and area shown in Table 10.1 and Figure 10.1 (panels a and c). Note the small value of pWR (2.7 vs 8.0 for mammals generally).

Table 10.1 Scaling of tooth crown parameters in adult male primates as a function of body mass [5]

Tooth variable	N_{ord}	N_{sp}	MPD	pWR	DI (%)	r^2	Slope ± 99% CL
Length	1	39	9	2.7	0.7	0.908	0.298 ± 0.042
Width	1	39	11	2.7	0.7	0.839	0.248 ± 0.048
Area	1	39	18.3	2.7	0.7	0.897	0.546 ± 0.083

N_{ord} = number of orders; N_{sp} = number of species; MPD = mean percent deviation; pWR = log (BW_{max}/BW_{min}); DI = diversity index; r^2 = coefficient of determination; CL = confidence limit(s).

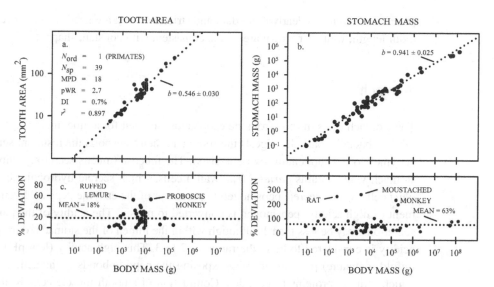

Figure 10.1 Scaling of tooth area in adult male primates and stomach mass in adult mammals.

From Table 10.1, rows one and two, one finds that tooth length ($b = 0.298 \pm 0.048$) (99% CL) might be construed (see Chapter 23, "Special pleading") as just compatible with constant shape (assumed reference slope of 1/3); however, the width data ($b = 0.248 \pm 0.048$) (99% CL) are *incompatible* with constant shape. Observe that for this sample the best-fit LSQ slope for *tooth area* (bottom row), taking into account the 99% CL, lies between 0.463 and 0.629. This range of slopes provides little support for arguments based on geometrical similarity with a presumptive reference slope of 2/3 for areas (Chapters 3, 24). As expected, the best-fit LSQ slope for tooth area is exactly the sum of the slopes for length and width ($0.298 + 0.248 = 0.546$) (Table 10.1, row three). The MPD for the scaling of tooth area (row three) is roughly the sum of the MPD for rows one and two; likewise the CLs (± 0.083) for tooth area (row three) are roughly the sum of the CLs seen in rows one and two ($0.042 + 0.048 = 0.090$). Keep in mind (Chapters 3, 24) that the way tooth crown area scales with body size is not in any case, by itself, an adequate criterion for appraising scaling at constant crown shape. A simpler (but not infallible) criterion for geometric similarity of crown area is that the *arithmetic* ratio of crown tooth length to width be constant over some range of tooth sizes, within the given CL.

This criterion bypasses the issue of scaling of tooth dimensions with body size. If, however, one prefers to express tooth width (with exponent b_1) and length (with exponent b_2) as power functions of body size, then the criterion for constant shape is $b_1 = b_2$. This criterion does not require that b_1 or b_2 have a specific value, such as 1/3, only that the two slopes be equal (Chapters 3, 24). Taking into account the 99% CL, the slopes for tooth length lie between 0.256 and 0.340, and those for tooth width between 0.200 and 0.296. The mean of both slopes (i.e., 0.248 and 0.298) is 0.273. The two deviant species in Figure 10.1c are the ruffed lemur (*Varecia variegata*) and the proboscis monkey

(*Narsalis larvatus*). Clearly more data and stricter criteria are needed to resolve whether tooth length, width, or crown area scale consistent with constant shape.

Stomach

The stomach takes in food from the esophagus, stores it for a time, mixes it, destroys or inhibits bacteria, partially digests the food, and then transports the resultant semi-liquid "chyme" to the duodenum (by peristalsis). The mucosa (mucus-secreting lining) of the stomach wall contains *parietal* (L. "wall") cells, which secrete hydrochloric acid, and *chief* cells, which secrete pepsinogen (a precursor of the enzyme pepsin). In an environment of high acidity, pepsinogen is converted to pepsin. The acidity of the parietal cell secretions may be less than 1, although within the lumen of the stomach, during a meal, acidity is considered to be in the range of 2 to 3. This high acidity (low pH) serves to unfold (denature) proteins, thereby exposing the peptide bonds to the action of pepsin which reduces proteins to peptides. Contraction of smooth muscle cells in the gastric mucosal wall serves to churn the gastric contents. Small molecules (e.g., alcohol and water) are absorbed directly across the stomach wall.

Ruminants (mammals that regurgitate, re-chew, and re-swallow their food) and cetaceans (excepting beaked whales) exhibit multi-chambered stomachs. In adult humans the stomach lining turns over about every three days. This rapid shedding of cells (relative to, say, epidermis) may be a partial explanation of the ability of the stomach wall to persist intact in a highly acidic environment. In addition, the alkaline mucus secreted by the stomach wall is acid-resistant, and also acts as a lubricant (thereby minimizing mechanical abrasion). *Stomach size* as a function of adult body size was analyzed using data from the literature (of these 58 records, 30 are from Chivers and Hladik [7]). The findings of this LSQ analysis are presented in Table 10.2 and Figure 10.1, panels b and d.

Stomach mass scales as the 0.941 ± 0.066 (99% CL) power of body mass. Taking into account the 99% CL, the best-fit slopes lie between 0.875 and 1.007, possibly (but not convincingly) consistent with relative invariance ($b = 1.0$). For eight different species, drawn from six orders (Carnivora, Insectivora, Perissodactyla, Pinnipedia, Primates, Rodentia) the MPD exceeds 100%. If these eight records are omitted, the MPD drops from 63 to 50% and the slope drops to 0.931 ± 0.058 (99% CL). Disconcertingly, another five records then appear with percent deviations greater than 100%. The two very deviant species noted in Figure 10.1d are the rat (*Rattus norvegicus*) and the moustached monkey (*Cercopithecus cephus*).

Table 10.2 Scaling of stomach mass as a function of body mass in adult mammals

N_{ord}	N_{sp}	MPD	pWR	DI (%)	r^2	Slope $\pm$ 99% CL
13	58	63	7.2	27	0.963	0.941 ± 0.066

See Table 10.1.

Small intestine

The small intestine is the primary site of digestion and absorption of nutrients and water. It plays an essential role in both water and acid–base balance. Considered lengthwise, it has three major consecutive parts: duodenum, jejunum, and ileum. The most distinctive structural feature of the small intestine is that of villi, made up mainly of absorptive cells, which project into the lumen and together with microvilli and macroscopic folds greatly increase the absorptive surface area. The mucus-secreting Brunner's glands of the duodenum, situated in the submucosa, between and at the base of the villi, are considered to be peculiar to mammals (but are not included in Table 2.1). Chyme, transferred from the stomach to the small intestine, is neutralized by the alkaline mucus present in the small intestine. Enzymes, derived mainly from the pancreas (see below), promote digestion. Absorption of the products of digestion (e.g., amino and fatty acids, electrolytes, sugars, water) begins in the duodenum.

Consider first scaling of intestinal *length* in the small intestine. In the adult human the post-mortem small intestine is about 5.5 m in length. The outcomes of a LSQ analysis of small intestinal length as a function of adult body size are brought together in Table 10.3.

Note that in the bottom row of Table 10.3, the data for the Indian elephant (*Elephas maximus*) fall 215% below the best-fit LSQ line and those for the black right whale (*Eubalaena glacialis*) 135%. The inclusion of the elephant and whale data (bottom row) in the calculations almost doubles the diversity index ($17/8.9 \approx 2$), mainly the result of increasing pWR and the N_{ord}. The findings reported in rows two and three of Table 10.3 are *inconsistent* with scaling of small intestinal length to the one-third power of body mass (see also Calder [9]). Table 10.4 brings together the inferences from a LSQ

Table 10.3 Scaling of small intestinal length as a function of adult body mass

N_{ord}	N_{sp}	MPD	pWR	DI (%)	r^2	Slope ± 99% CL	Comment	Source
4	29	23	2.7	2.6	0.883	0.399 ± 0.077	-	[8]
6	60	23	5.1	8.9	0.973	0.439 ± 0.025	Elephant and whale omitted	From Lit.
8	62	29	7.3	17	0.962	0.418 ± 0.028	Elephant and whale included	From Lit.

See Table 10.1. Lit. = literature.

Table 10.4 Scaling of small intestine volume or mass as a function of adult body mass

N_{ord}	N_{sp}	MPD	pWR	DI (%)	r^2	Slope ± 99% CL	Source
12	72	63	7	26	0.965	0.878 ± 0.053	From Lit.

See Tables 10.1, 10.3.

analysis of the scaling of small intestine mass (or volume) as a function of adult body mass in 72 species of mammals.

Some of the small intestinal masses used in the calculations for Table 10.4 may include ingesta. If one restricts the calculation to apparently empty small intestines, one obtains a slope of 0.946 ± 0.071 (99% CL), not significantly different from a slope of 0.878 (Table 10.4).

Large intestine

The outcomes of LSQ analyses of *large intestinal length* and *mass* as a function of adult body mass appear in Table 10.5. The first row indicates that large intestinal *length*, in the given sample, scales as the 0.439 ± 0.047 (99% CL) power of adult body mass. This is not significantly different from the slope of 0.418 ± 0.028 (99% CL) for small intestine length (Table 10.3, bottom row). Row two shows that large intestinal *mass* scales as the 0.873 ± 0.056 (99% CL) power of body mass, close to the slope of 0.878 ± 0.053 for the small intestine (see Table 10.4). Some of the intestinal masses used to compute the bottom row of Table 10.5 may include ingesta. If one confines the LSQ analysis to cases where, as far as one can judge, the large intestine is "empty," one finds that $b = 0.925 \pm 0.080$ (99% CL), not significantly different from a slope of 0.873 (Table 10.5, bottom row).

Goetz and Budtz-Olsen [10] state that the intestines of a giraffe (body size not given) measured 85 m in length; also, those of a black right whale weighing 65 tonnes measured 99 m [11].

Gut mass

Brody [12] reported that the weight of the stomach plus intestines scales as the 0.941 power of body weight. Calder [9] gives exponents ranging from 0.89 to 1.16 for various subgroups of mammals (carnivora, eutheria, rodents, ruminants). From the data of Crile and Quiring [13] one finds a slope of 1.047 ± 0.092 (99% CL) (Table 10.6, row one). The way in which gut mass scales with body mass and with *fat-free body weight* (FFBW) (or mass), based on the data of Pitts and Bullard [14], is set forth in Table 10.6, rows two and three.

Table 10.5 Scaling of large intestinal length and mass as a function of adult body mass

N_{ord}	N_{sp}	MPD	pWR	DI (%)	r^2	Slope $\pm$ 99% CL	Entity	Source
4	49	38	5.3	5.9	0.931	0.439 ± 0.047	Length	From Lit.
13	70	63	7.0	27.5	0.961	0.873 ± 0.056	Mass	From Lit.

See Tables 10.1, 10.3.

Table 10.6 Scaling of gut mass as a function of fat-free body weight or mass or adult body mass

N_{ord}	N_{sp}	MPD	pWR	DI (%)	r^2	Slope $\pm$ 99% CL	x-variable	Source
7	30	48	5.5	9.3	0.972	1.047 ± 0.092 [a]	Body mass	[13]
4	35	40	2.8	2.8	0.965	0.977 ± 0.088	FFBW[b]	[14]
4	35	34	2.8	2.8	0.974	0.962 ± 0.075	Body mass	[14]
13	48	33	6.8	24	0.987	0.932 ± 0.042 [a]	Body mass	From Lit.

See Tables 10.1, 10.3. [a] Stomach + intestines only; [b] FFBW = fat-free body weight.

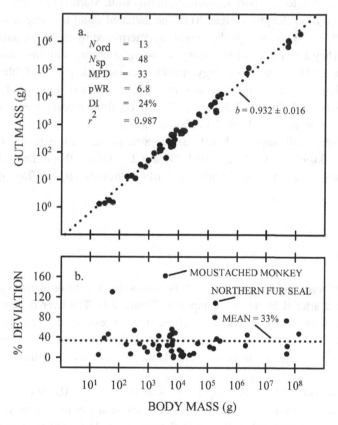

Figure 10.2 Scaling of gut mass as a function of adult body mass.

The present LSQ analysis of the scaling of gut mass as a function of adult body mass, based on data from the literature, is displayed in Table 10.6 (bottom row) and Figure 10.2. Note that 46% of the records making up this latter dataset are from Chivers and Hladik [7]. From these data, one finds that gut mass (i.e., stomach plus intestines) scales as the 0.932 ± 0.042 (99% CL) power of adult body mass (Table 10.6, bottom row) in 48 species. This exponent is inconsistent with scaling in accord with relative invariance ($b = 1.0$).

The first and third most deviant points in Figure 10.2 (panel b) are for the moustached monkey (*Cercopithecus cephus*) and northern fur seal (*Callorhinus ursinus*). The second most deviant point (not labeled) at 130% in Figure 10.2b, with a body mass of 180 g, is Demidoff's galago (*Galagoides demidoff*).

Small and large intestinal surface areas

The *surface area* of the gastrointestinal tract, and especially of the small intestine, is of importance in relation to transport across the mucosal wall. Martin *et al.* [15] found that the *basal surface area* (length × breadth) of the flattened small intestine scales as the 0.75 ± 0.05 (95% CL) power of body weight, as determined by the major axis method (Chapter 23). They interpreted their findings as being in accord with the conventional reference slope (3/4) for resting energy metabolism (see Chapter 3, Table 3.5 and Chapter 17). Here we recount the outcomes of a LSQ analysis of a *subset* of the basal surface area data of Snipes and Kriete [16] derived from measurements made on flattened tissue segments (see Table 10.7, bottom row).

In these rather small samples, basal surface area scales with exponents ranging from 0.749 ± 0.099 to 0.782 ± 0.083 (99% CL). Given the rather wide 99% CL, these results are consistent with the earlier findings (b = 0.75) of Martin *et al.* [15].

Liver weight

The liver is the largest compact organ in the human body. It is the only gland to receive both venous and arterial blood (see Chapter 9, Figure 9.1). The liver is involved in a complex array of biochemical functions. These include regulation of amino acid and glucose levels, bile production, clearing bilirubin, detoxification of abnormal substances, glycogen storage, inactivation of circulating hormones, and synthesis of plasma proteins and urea.

Brody [12] reported that liver weight scales as the 0.867 ± 0.009 (s.e.) power of body weight (see Appendix C). The findings of a LSQ analysis of data from Crile and Quiring [13] are summarized in Table 10.8 (row one). The LSQ analyses of liver mass as a function of FFBW and as a function of body mass, based on the data of

Table 10.7 Scaling of basal surface area in the small and large intestine as a function of adult body mass

N_{ord}	N_{sp}	MPD	pWR	DI (%)	r^2	Slope ± 99% CL	Organ [a]	Source
4	12	24	4.2	3	0.989	0.782 ± 0.083	S.I.	[16]
4	12	28	4.2	3	0.983	0.749 ± 0.099	S.I. + L.I.	[16]

See Table 10.1. [a] S.I. = small intestine; L.I. = large intestine.

Table 10.8 Scaling of liver mass as a function of adult body mass and fat-free body weight

Row	N_{ord}	N_{sp}	MPD	pWR	DI (%)	r^2	Slope ± 99% CL	x-variable	Source
1	9	36	28	4.9	11.2	0.982	0.891 ± 0.057	BM	[13]
2	4	35	28	2.8	3.4	0.980	0.965 ± 0.065	FFBW	[14]
3	4	35	25	2.8	3.4	0.985	0.949 ± 0.055	BM	[14]
4	13	134	27	7.7	35	0.992	0.895 ± 0.018	BM	From Lit.

See Tables 10.1, 10.3. BM = body mass; FFBW = fat-free body weight.

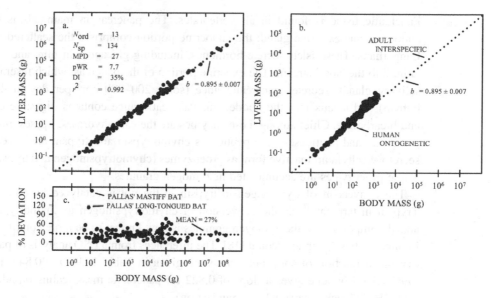

Figure 10.3 Scaling of liver mass in mammals.

Pitts and Bullard [14], are given in Table 10.8, rows two and three, respectively. The data (drawn from the literature) from which the bottom row of Table 10.8 is constructed are shown in Figure 10.3. Comparing the entries from rows one, three, and four, one finds that the best-fit LSQ slopes vary between 0.891 ± 0.0.057 and 0.949 ± 0.055 (both 99% CL). The slope based on the largest sample (row four) is 0.895 ± 0.018 (99% CL).

This latter result ($b = 0.895$) implies that the liver in the largest mammal (a blue whale) is smaller by 2.8 doublings (or a factor $f = 7$) than it would be if liver mass scaled as the first power of body mass ($N_d = 26.6 \cdot 0.105 = 2.8$; $f = 2^{2.8} = 7$) (Chapter 3). In a prior study a best-fit LSQ slope for liver mass on body mass of 0.886 ± 0.01 (s.e.) was reported [17]. One sees in Figure 10.3b that the human ontogenetic points (data of T. Shepard, personal communication) fall mostly above the adult interspecific line (i.e., 94% of the fetal points lie above the adult interspecific line). The two most deviant species in panel c are the Pallas' mastiff bat (*Molossus molossus*) and the Pallas' long-tongued bat (*Glossophaga soricina*).

Table 10.9 Scaling of pancreatic mass as a function of adult body mass

N_{ord}	N_{sp}	MPD	pWR	DI (%)	r^2	Slope $\pm$ 99% CL	Source
12	41	47	7.3	23	0.977	0.847 $\pm$ 0.057	From Lit.

See Table 10.1. Lit. = literature.

Pancreatic size

Pancreatic tissue is found in all vertebrates. The pancreas in mammals is both an endocrine and exocrine gland: the endocrine portion comprises the scattered Islets of Langerhans. These islets secrete hormones, including glucagon, insulin, and somatostatin, into the blood stream. The exocrine portion of the pancreas, which resembles the salivary glands, secretes pancreatic juice (some 200–800 cm^3 per day in the adult human) and mucus into the duodenum. Pancreatic juice contains digestive enzymes and bicarbonate. Chief among these enzymes are the carbohydrases, lipases, nucleases, proteases, and peptidases. The peptidases chymotrypsin and trypsin are stored in the secretory cells in an inactive form as proenzymes (chymotrypsinogen and trypsinogen). These proenzymes are accumulated in zymogen granules.

The conversion of trypsinogen to trypsin in the gut is catalyzed by enterokinase. Trypsin in turn catalyzes the conversion of chymotrypsinogen to chymotrypsin. The actual composition of the zymogen granules is now known to be more complex than is implied by this synoptic account [18]. The inferences from a LSQ analysis of pancreatic mass as a function of adult body mass are given in Table 10.9 ($b = 0.847 \pm 0.057$) (99% CL). Hörnicke gives a slope of 0.822 for pancreatic mass scaling on adult body mass (H. Hörnicke, personal communication).

Observe that liver scales with a slope of 0.895 $\pm$ 0.018 (99% CL), and pancreas with a slope of 0.847 $\pm$ 0.057 (99% CL). These slopes are not significantly different at the 99% CL.

Water intake

A man can survive weeks without food, but only a few days without water.

Ladell 1965

All adult mammals studied thus far are about two-thirds water (see Chapter 8). Daily water loss (of about 2.5 liters) in the adult human is normally through the kidneys (60%), skin (24%), lung (12%), and rectum (4%). This water loss is, on average, made up in the adult human through drinking (40%), food intake (48%), and metabolic oxidation (12%). Water consumption in many species increases substantially when ambient temperature rises (see Chapter 18). It is often said that three days without water in the desert is fatal for most adult humans (see above quote from Ladell; see also

Table 10.10 Scaling of water intake (cm³/day) as a function of adult body mass

N_{ord}	N_{sp}	MPD	pWR	DI (%)	r^2	Slope ± 99% CL
13	106	49	4.6	20	0.957	0.862 ± 0.047

See Table 10.1.

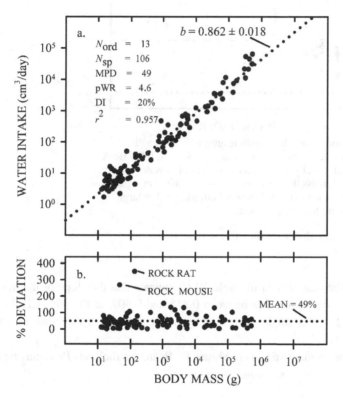

Figure 10.4 Scaling of water intake in adult mammals.

Chapter 18). However, some species, such as the desert kangaroo-rat, can meet all of their water needs through food consumption and oxidation alone. These species tend to be nocturnal, thereby avoiding the intense heat of the day.

Perhaps the best-documented case of severe dehydration in the human is that of a Buddhist bishop (Soken Enami) who went for 7.6 days without food or water [19]. During bouts of severe thirst he rinsed his mouth with water; concurrent measurements suggest he did not swallow this water. Throughout this fast his body weight dropped from 51 to 43 kg (15%), virtually all of which loss was attributable to dehydration. It is also worth recalling that during hibernation bears are able to sleep for about 100 days without either food or water [20].

The inferences from a LSQ analysis of a *subset* of the *water intake* (by drinking) data taken from Nagy and Peterson [21] are reviewed in Table 10.10 and illustrated in

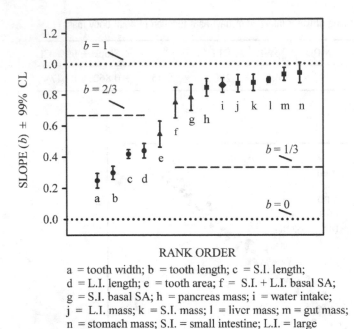

a = tooth width; b = tooth length; c = S.I. length;
d = L.I. length; e = tooth area; f = S.I. + L.I. basal SA;
g = S.I. basal SA; h = pancreas mass; i = water intake;
j = L.I. mass; k = S.I. mass; l = liver mass; m = gut mass;
n = stomach mass; S.I. = small intestine; L.I. = large
intestine; SA = surface area

Figure 10.5 Summary slopes for digestive system.

Figure 10.4. The data shown in Table 10.10 indicate that the slope of the best-fit LSQ line for daily water intake lies between 0.815 and 0.909, at the 99% CL. For a mean slope of 0.862, water intake increases by about 82% for a doubling of body mass $(100 \cdot (2^{0.862} - 1) = 82)$.

The two most deviant species in Figure 10.4b are the rock rat (*Petromus typicus*) and the rock mouse (*Aethomys namaquensis*).

Summary

In this chapter we reviewed the scaling of teeth (length, width, area), stomach mass, small and large intestinal volume, mass and surface areas, as well as the mass of gut, liver, and pancreas. No persuasive evidence was found that teeth scale in accord with geometrical similarity. The distribution of LSQ slopes for the digestive system, expressed in terms of body mass, plotted in rank order, is shown in Figure 10.5. Note that gut and stomach may scale in accord with relative invariance ($b = 1.0$), but the remaining slopes show little or no correlation with the reference slopes for either absolute or relative invariance. Neither do the slopes evince convincing evidence of scaling in accord with the dimensions of length ($b = 1/3$; the nearest to this are the circles in Figure 10.5) or area ($b = 2/3$) (triangles). The *functional* slope for water intake (see item i (diamond) in Figure 10.5) is not consistent with a slope of 1 (relative invariance).

Overall, the distribution of slopes for the digestive system (Figure 10.5) is quite different from the distributions seen in Chapters 8 (body composition) and 9 (circulatory system). On the other hand, broadly similar distributions will be seen for the respiratory (Chapter 15) and urinary (Chapter 16) systems.

References

1. Osborn, J.W. (1973). The evolution of dentitions. *American Scientist*, **61**:548–559.
2. Gould, S.J. (1975). On the scaling of tooth size in mammals. *American Zoologist*, **15**:351–362.
3. Pilbeam, D. and Gould, S.J. (1974). Size and scaling in human evolution. *Science*, **186**:892–901.
4. Fortelius, M. (1985). Ungulate cheek teeth: developmental, functional, and evolutionary considerations. *Acta Zoologica Fennica*, **180**:1–76.
5. Gingerich, P.D. and Smith, B.H. (1985). Allometric scaling in the dentition of primates and insectivores. In: Jungers, W.L. (ed.) *Size and Scaling in Primate Biology*. New York: Plenum Press, pp. 257–272.
6. Swindler, D.R. (1976). *Dentition of Living Primates*. London: Academic Press.
7. Chivers, D.J. and Hladik, C.M. (1980). Morphology of the gastrointestinal tract in primates: comparisons with other mammals in relation to diet. *Journal of Morphology*, **166**:337–386.
8. Schieck, J.O. and Millar, J.S. (1985). Alimentary tract measurements as indicators of diets of small mammals. *Mammalia*, **49**:93–104.
9. Calder, W.A. (1984). *Size, Function, and Life History*. Cambridge, MA: Harvard University Press. (Calder reported that gut lengths scale as the 0.83–0.85 power of body weight in cats. See p. 123. Using the same presumed dataset of five points for three species of cats, I find a slope of 0.415±0.390 (99% CL).)
10. Goetz, R.H. and Budtz-Olsen, O. (1955). Scientific safari: the circulation of the giraffe. *South African Medical Journal*, **29**:773–776.
11. Omura, H., Ohsumi, S., Nemoto, T., Nasu, K. and Kasuya, T. (1969). Black right whales in the North Pacific. *Scientific Reports Whales Research Institute*, **21**:1–78.
12. Brody, S. (1945). *Bioenergetics and Growth: With Special Reference to the Efficiency Complex in Domestic Animals*. New York: Reinhold.
13. Crile, G. and Quiring, D.P. (1940). A record of the body weight and certain organ and gland weights of 3,690 animals. *Ohio Journal of Science*, **40**:219–259.
14. Pitts, G.C. and Bullard, T.R. (1968). Some interspecific aspects of body composition in mammals. In: *Body Composition in Animals and Man*. Washington, DC: National Academy of Sciences, pp. 45–70. (Body weight data kindly provided via personal communication from the late W.A. Calder.)
15. Martin, R.D., Chivers, D.J., Maclarnon, A.M. and Hladik, C.M. (1985). Gastrointestinal allometry in primates and other mammals. In: Jungers, W.L. (ed.) *Size and Scaling in Primate Biology*. New York: Plenum Press, pp. 61–89.
16. Snipes, R.L. and Kriete, A. (1991). Quantitative investigation of the area and volume in different compartments of the intestine of 18 mammalian species. *Zeitschrift Säugetierkunde*, **56**:225–244.
17. Prothero, J.W. (1982). Organ scaling in mammals: the liver. *Comparative Biochemistry and Physiology*, **71A**:567–577.

18. Rindler, M.J., Xu, C-F., Gumper, I., Smith, N.N. and Neuber, T.A. (2007). Proteomic analysis of pancreatic zymogen granules: identification of new granule proteins. *Journal of Proteome Research*, **6**:2978–2992.

19. Yoshimura, H., Inoue, G., Yamamoto, M. *et al.* (1953). A contribution to the knowledge of dehydration of human body. *Journal of Biochemistry*, **40**:361–375.

20. Nelson, R.A., Wahner, H.W., Jones, J.D., Ellefson, R.D. and Zollman, P.E. (1973). Metabolism of bears before, during, and after winter sleep. *American Journal of Physiology*, **224**:491–496.

21. Nagy, K.A. and Peterson, C.C. (1988). Scaling of water flux rate in animals. *University of California Publications in Zoology*, **120**:1–172.

11 Integumentary system

When living forms emerged from primordial swamps, their external skin became specialized as a barrier between delicate living cells and a capricious, often unfriendly environment.

Rushmer *et al.* 1966

As the main interface between the "milieu intérieur" and the external environment, the vertebrate integument (L. "cover") or skin plays multiple roles in the body's economy. These include: barrier to infection, contributions to temperature regulation, prevention of dehydration, protection against ultraviolet light (via pigments found in melanocytes and red blood cells), resistance to mechanical stress, and, in some species, impacts on sexual attraction and camouflage (see above quote from Rushmer *et al.*).

Skin is a three-layered structure: the outer *epidermis* provides a permeability barrier (chiefly against dehydration) and resistance to stretch; the underlying *dermis* accounts for most of the bulk mechanical properties, conferred largely by collagen and elastin; the *hypodermis*, or subcutaneous layer, consists mainly of fat, and serves as a mechanical cushion, a thermal insulator, and an energy reserve. In addition, in mammals particularly, there are various skin derivatives, including glands (mammary, sebaceous, sweat), hair, horns, and nails (including claws and hooves). Baleen, supported from the upper jaw, is a type of filter composed of narrow plates of keratin found in toothless whales (Mysticeti). Scales (horny material) are seen on the tails of some rodents.

Hair at high density (as in fur) provides some mechanical protection against abrasion as well as thermal insulation and a barrier to evaporation. Hair also acts as a tactile receptor. However, hair is absent, or reduced in elephants, hippopotami, rhinoceros, and whales. Hair is present over most of the adult human body, but with a few exceptions is unobtrusive (hence the phrase "naked ape"). Skin contains mechanoreceptors for pressure, touch, and vibration as well as receptors for pain and temperature. Skin also plays an important role in generating vitamin D, which in turn participates in controlling levels of calcium and phosphate in the blood.

Skin has the largest mass (6–7% of body mass) of any wholly contiguous tissue in the adult human body (the fresh bones forming the skeleton comprise a larger fraction of adult body mass, but are only contiguous if one includes cartilage and ligaments). Adult human skin is composed of: 13% fat (by ether extraction), 22% crude protein, and 65% water [1]. Collagen represents some 86% of the protein in adult human skin [2].

This chapter considers four structural variables, three having the dimension of length (or thickness) and one the dimension of mass; there are five instances of thickness and three instances of skin mass. The chapter is based on the study of 569 records, all in tabular form. A summary plot of the slopes discussed in this chapter is provided in Figure 11.2.

Skin thickness

Skin thickness data were selected from the work of Sokolov [3] using primarily (85% of the present dataset) those results taken from the withers (portion of the back immediately below the neck). Sokolov [3] does not provide body size data; these came from the SBT (Chapter 4). Whether all the animals Sokolov [3] studied were adults is uncertain. The outcomes of LSQ analyses based on dermal, epidermal, and skin *thickness* as a function of body mass for presumptively adult mammals are furnished in Table 11.1 and Figure 11.1 (panels a, c).

Epidermal thickness (Table 11.1, row one) in this sample (95 species) ranges from 6 μm to 3 mm, with a median thickness of 28 μm. The slope for epidermal thickness drops substantially when cetaceans and sirenia are omitted (Table 11.1, row two). Dermal thickness (row four, 99 species) ranges from 88 to 44,700 μm, with a median of 900 μm. The mean slope for dermal thickness seems to be only modestly affected, if at all, by a few samples taken from different parts of the back (Table 11.1, rows three and four) (i.e., 0.391 − 0.366 = 0.025 is less than the 99% CL of either 0.053 or 0.042).

The best-fit LSQ line for overall skin thickness (see Table 11.1, row five, and Figure 11.1, panels a and c) has a slope of 0.408 ± 0.047 (99% CL) and a large MPD (62%). A slope of 0.408 implies a 33% increase in skin thickness for a doubling in body size ($2^{0.408} = 1.33$; $100 \cdot (1.33 - 1) = 33$) (Chapter 3). The two most deviant species in Figure 11.1 (panel c) are the goat (*Capra aegagrus*) and the goitered gazelle (*Gazella subgutturosa*).

Table 11.1 Scaling of dermal, epidermal, and skin thicknesses as a function of body mass in adult mammals

Row	N_{ord}	N_{sp}	MPD	pWR	DI (%)	r^2	Slope ± 99% CL	Thickness of:	Remarks
1	10	95	80	7.25	23.5	0.622	0.284 ± 0.060	Epidermis	-
2	8	82	39	5.3	13.3	0.587	0.171 ± 0.042	Epidermis	[a]
3	9	81	56	5.3	14.9	0.807	0.366 ± 0.053	Dermis	Withers only
4	10	99	56	7.2	23.7	0.861	0.391 ± 0.042	Dermis	Withers and elsewhere
5	10	100	62	7.2	23.7	0.841	0.408 ± 0.047	Skin	-

N_{ord} = number of orders; N_{sp} = number of species; MPD = mean percent deviation; pWR = log(BWmax/BWmin); DI = diversity index; r^2 = coefficient of determination; CL = confidence limit(s).
[a] Omitted cetaceans and sirenians from the dataset used in row one.

Table 11.2 Skin mass as a function of body mass in adult mammals

Row	N_{ord}	N_{sp}	MPD	pWR	DI (%)	r^2	Slope ± 99% CL	Source
1	2	5	-	2.6	0.6	0.99	0.942 ± 0.007 [a]	[4]
2	4	33	17.4	3.1	3.1	0.99	0.936 ± 0.045	[5]
3	12	74	23	6	22	0.99	0.954 ± 0.024	From Lit. [b]

See Table 11.1. [a] 95% CL; [b] this dataset includes all the entries used to construct row two; Lit. = literature.

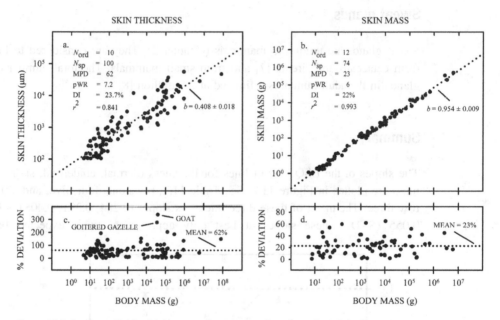

Figure 11.1 Scaling of skin thickness and mass as a function of adult body mass.

Skin mass

The outcomes of three LSQ analyses of skin mass as a function of body mass are given in Table 11.2 and Figure 11.1 (panels b and d). Note that these determinations of skin mass are to some degree problematic. The term "skin" could include hair (fur) and/or subcutaneous fat. In the present dataset, skin mass may include both ontogenetic and adult data. For example, Pace *et al.* [4] in their study of small mammals are not specific as to what is meant by skin mass (the authors provide no "raw" data), and it might be that some of their data is from subadult specimens (Table 11.2, row one). On the other hand, Pitts and Bullard [5] are explicit that hair was removed from the skin but subcutaneous fat was included (Table 11.2, row two). At least 60% of the records used to construct Table 11.2, row three, apparently include subcutaneous fat. In the newborn, human skin is said to constitute about 6% of body mass [6].

A slope of 0.954 ± 0.024 (99% CL) (see Table 11.2, row three; Figure 11.1, panel b) indicates that on average skin mass increases by 94% when body mass doubles ($2^{0.954}$ = 1.94; 100·(1.94 − 1) = 94) (Chapter 3). If one computes the slope for skin mass on body mass less skin mass (so as to remove the implicit correlation between the y- and x-variables), one then obtains a slope of 0.948 ± 0.027 (99% CL) (Table 11.2, row three). In this case the effect of any implicit correlation lies comfortably within the 99% CL. Human skin mass during ontogeny approaches the adult interspecific line from below [6] (data not shown).

Sweat glands

Sweat glands are unique to mammals (Chapter 2). They are considered to be absent from cetaceans and sirenia [7] and from small mammals. The total number of sweat glands in the adult human is estimated at 2.5 million [8].

Summary

The slopes of the LSQ best-fit lines for thickness (dermal, epidermal, skin) and skin mass are plotted in Figure 11.2. See Table 11.1, rows one, four, five, and Table 11.2, row three. The mean of these three thickness slopes (0.284, 0.391, 0.408) is 0.361 ± 0.055 (STD). These slopes at best only approximate those expected based on

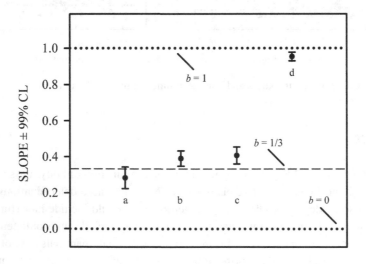

RANK ORDER

a = epidermal thickness; b = dermal thickness;
c = skin thickness; d = skin mass

Figure 11.2 Slope summary for integumentary system.

dimensionality (for length), with the three slopes for thickness (dermal, epidermal, skin) hovering around 1/3 (Chapter 3). In addition, the slope for skin mass falls roughly in the neighborhood of a slope of 1 (corresponding to relative invariance). The 99% CI for the scaling of skin mass is from 0.930 to 0.978.

Recall from Chapter 6 that body surface area scales as the 0.670 ± 0.027 (99% CL) power of body mass in adult mammals. From Table 11.1, row five, one finds that skin thickness scales as the 0.408 ± 0.047 (99% CL) power of body mass. Combining these two results leads to the expectation that skin mass should scale as the 1.078 power (b_{calc}) of body mass ($0.670 + 0.408 = 1.078$), far from the observed exponent of 0.954 ± 0.024 (99% CL) (Table 11.2, row three).

Even if we indulge in "special pleading" (Chapter 23), by selectively using just the negative 99% confidence limits (i.e., -0.047 for skin thickness and -0.027 for skin surface area), we find a predicted skin mass scaling as the 1.004 power of body mass $((1.408 - 0.047) + (0.670 - 0.027) = 1.004)$, rather than the observed slope of 0.954. Thus, a frankly biased analysis still yields an internal inconsistency ($b_{calc} > 1.0$). This may be due to the small sample size for body surface area ($N_{pnts} = 30$, $N_{sp} = 16$), or perhaps back skin scales differently from body skin as a whole.

References

1. Mitchell, H.H., Hamilton, T.S., Steggerda, F.R. and Bean, H.W. (1945). The chemical composition of the adult human body and its bearing on the biochemistry of growth. *Journal of Biological Chemistry*, **158**:625–637.
2. Widdowson, E.M. and Dickerson, J.W.T. (1964). Chemical composition of the body. In: Comar, C.L. and Bronner, F. (eds.) *Mineral Metabolism: An Advanced Treatise*. New York: Academic Press, pp. 1–217.
3. Sokolov, V.E. (1982). *Mammal Skin*. Berkeley, CA: University of California Press.
4. Pace, N., Rahlmann, D.F. and Smith, A.H. (1979). Scale effects in the musculoskeletal system, viscera and skin of small terrestrial mammals. *Physiologist*, **22**:S51–S52.
5. Pitts, G.C. and Bullard, T.R. (1968). Some interspecific aspects of body composition in mammals. In: *Body Composition in Animals and Man*. Washington, DC: National Academy of Sciences, pp. 45–70.
6. Wilmer, H.A. (1940). Quantitative growth of skin and subcutaneous tissue in relation to human surface area. *Proceedings of the Society for Experimental Biology and Medicine*, **43**:386–388.
7. Welsch, U. and Storch, V. (1976). *Comparative Animal Cytology and Histology*. Seattle, WA: University of Washington Press.
8. Adolph, E.F. (1969). *Physiology of Man in the Desert*. New York: Hafner Publishing Co.

12 Musculoskeletal system

MUSCLE

The most important conclusion... is that, with certain reservations, the ultimate chemical machinery underlying muscular contraction in every kind of animal is essentially the same.

Pantin 1956

Histologically speaking, there are two main types of muscle: striated and smooth. Striated muscle, named after the cross-striations seen microscopically, occurs in skeletal and cardiac muscle. Smooth or involuntary muscle lacks cross-striations. Purposeful movements in mammals, associated with abduction, adduction, bending, chewing, extension, flexion, and rotation, are all driven by skeletal muscle contraction. In the same way, bulk movements of air, blood, food, lymph, or waste products are due to muscular contraction. The internal movement of food is caused primarily by involuntary contraction (peristalsis) of smooth muscle. Blood circulation is attributable to cardiac muscle (Chapter 9), whereas respiration (Chapter 15) is actuated chiefly by the diaphragm (skeletal muscle). A remarkable feature of striated muscle is that maximal contractile force per unit cross-sectional area is, within a factor of two, a biological constant ($b \approx 0$, absolute invariance), applicable both to insects and vertebrates [1]. This invariance reflects the fact that the underlying biochemistry of muscle is essentially universal (see above quote by Pantin).

Skeletal muscle is quantitatively the dominant tissue in most adult mammals. But in some species of whales, at least seasonally, blubber may comprise a larger proportion of body mass than skeletal muscle [2]. There are roughly 650 distinguishable muscles in the adult human, varying in diameter from 10 to 100 μm and ranging in length [3] from 2 mm to 60 cm. It has been asserted that in larger vertebrates individual muscle fibers may reach 100 cm or more in length [4].

Modular construction

This theme of dissociation into separate modules... lies at the heart of any hope of resolving the natural development (ontogeny or phylogeny) of any complex system, biological or otherwise.

Gould 1992

Table 12.1 The modular design of skeletal muscle

Structure	Approximate diameter	Comment
Myosin filament	100 Å	Intracellular
Actin filament	60 Å	Intracellular
Myofibril	1–3 μm	Intracellular
Sarcomere	1–3 μm	Intracellular
Muscle fiber (myofiber)	50 μm	Single cell
Fascicles	0.5–2.0 mm	Bundle of fibers
Gross muscle	3 cm	Adult human biceps

$1 \text{ Å} = 10^{-8}$ cm.

Perhaps no other tissue better illustrates the principle of hierarchical modular construction than does skeletal muscle [5]. These modular components range in size from the molecular (chiefly actin and myosin) to the gross anatomical (see Table 12.1). The most striking feature of skeletal muscle at the cellular (myofiber or simply fiber) level of construction is the presence of longitudinal repeating units, or sarcomeres. Each sarcomere (about 2.2 μm long) constitutes a functional intracellular unit of contraction. At each hierarchical level, the components of a gross muscle exhibit an approximate cylindrical symmetry. This is most apparent at lower levels of resolution. The selection for cylindrical symmetry, over some six orders of magnitude in diameter (100 Å to 3 cm), may be a consequence of the fact that during contraction, overall muscle symmetry is only minimally disturbed. This modularity in construction of skeletal (and cardiac) muscle and the associated scaling relations (see below) are probably universal in adult mammals (see above quote by Gould). For a recent molecular model of muscle contraction see a paper by Duke [6].

This chapter presents results in tabular form for five y-variables, 12 instances, and 333 records. Figure 12.3 presents a striking summary plot for the major results discussed in this chapter.

Myofiber diameter and muscle mass

A LSQ analysis of a *subset* of the striated *myofiber diameter* data (for diaphragm muscle) reported by Gauthier and Padykula [7] gave a slope of 0.079 ± 0.090 (99% CL) ($N_{ord} = 2$, $N_{sp} = 7$, MPD = 20, pWR = 4.8, DI = 1.3%, $r^2 = 0.71$). These findings from a small dataset suggest that the best-fit slope for myofiber diameter on adult body mass in the diaphragm lies between −0.011 and 0.169. Note the wide confidence interval (CI) and low value of the coefficient of determination (r^2). If one (rather arbitrarily) assigns a slope of zero to the LSQ best-fit line for myofiber diameter, one obtains a mean intercept of 30 μm (Chapter 5).

Johnson and Beattie [8] determined myofiber diameter in three steers (age 630 days) and in six muscles. The smallest diameter fibers were in the biceps brachii muscle; they averaged 46 ± 2 (STD) μm in diameter, while the largest fibers were in the extensor digitorum communis muscle (located in the posterior forearm) and averaged 65 ± 5 (STD) µm in diameter. The thinnest striated muscle fibers in the adult human are the extrinsic muscles of the eye, averaging 9 to 17 µm in diameter [9]. Data on maximal myofiber length are scarce. Lockhart and Brandt [10] report finding myofibers in the human sartorius muscle (the longest muscle in the human body, located in the upper leg) at least 34 cm in length (see also Harris *et al.* [3]).

The fact that skeletal muscle cells are multinucleated may be an example of a *scale effect* (Chapter 3). That is, a given nucleus may only be competent to influence cellular metabolism in a defined small volume (relative to a myofiber length measured in cm). Neurons (some binucleate) may also have long processes (axons), up to a meter in length in the adult human and several meters in the neck of the giraffe; as a rule these long axons are likely to be one to two or more orders of magnitude smaller in diameter than comparably long myofibers.

Pollock and Shadwick [11] reported on the scaling of muscle *mass* as a function of body mass in four muscles (digital extensors, digital flexors, gastrocnemius, and plantaris) in 26 species drawn from five orders (pWR = 4.1, DI = 4.8%). They obtained LSQ slopes ranging from 0.93 to 1.03. These workers also reported that tendon *lengths* for the same muscles scale with body mass with slopes ranging from 0.34 to 0.38.

The outcomes of LSQ analyses of muscle mass on adult body mass using a subset of the data of Munro [12] (see Table 12.2, row one) as well as from a larger independent dataset drawn from the literature are reported in Table 12.2, row two, and Figure 12.1 (panels a, c). The extremes in muscle mass as a percent of body mass in the given sample (Table 12.2, row two) are from 23.4% for sheep to 62.5% for lion. The fact that the best-fit LSQ line has a slope of 1.001, with 99% CL of $\pm$ 0.012, implies that muscle mass is directly proportional to body mass ($b = 1.0$, relative invariance) on average, in the given sample of mammals. Note that the MPD is 13.4 for a pWR of 7.3. The precision of this dataset rivals those found in applied physics or engineering for comparable size ranges. For muscle mass, one finds after assigning a slope of 1.0 to the best-fit LSQ line (Table 12.2, row two) an intercept of 0.426, implying that on average mammals (in the given sample) consist of 42.6% skeletal muscle.

From Figure 12.1b, one sees that for both rat [13] (*Rattus norvegicus*) and human [14] (*Homo sapiens*) the *ontogenetic* lines approach the adult interspecific line from

Table 12.2 LSQ analyses of muscle mass as a function of adult body mass

N_{ord}	N_{sp}	MPD	pWR	DI (%)	r^2	Slope $\pm$ 99% CL	Source
6	10	12	5.3	5.2	0.998	1.008 ± 0.050	[12]
11	89	13.4	7.3	25.5	0.998	1.001 ± 0.012	From Lit.

N_{ord} = number of orders; N_{sp} = number of species; MPD = mean percent deviation; pWR = log(BWmax/BWmin); DI = diversity index; r^2 = coefficient of determination; CL = confidence limit(s); Lit. = literature.

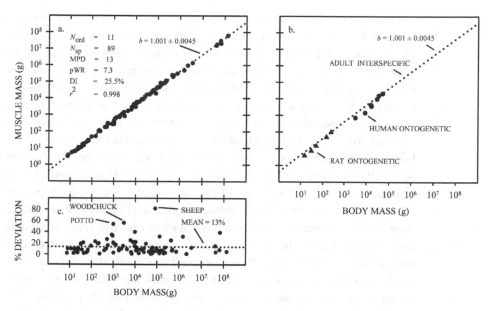

Figure 12.1 Scaling of muscle mass with body mass.

slightly below. In panel c, three especially deviant species are the potto (*Perodicticus potto*), sheep (*Ovis orientalis*), and woodchuck (*Marmota monax*).

If one computes the scaling of muscle mass on body mass less muscle mass (so as to remove the implicit dependence of body size on muscle mass), one obtains a slope of 0.999 ± 0.008 (99% CL). This finding again supports the concept of relative invariance ($b = 1.0$). It is worth noting that skeletal muscle mass, as calculated from the intercept of the best-fit LSQ line, constitutes on average 76% of body mass less skeletal muscle mass.

SKELETON

Vertebrates are practically unique among the Metazoa in their possession of a skeleton made from calcium phosphate rather than calcium carbonate.

Ruben and Bennett 1987

In vertebrates, many bones, linked by ligaments, form the skeleton which, in land animals, plays a central role in resisting the force of gravity. Bone also serves in a protective role, as in the skull and rib cage; as a reservoir, bone plays a key role in mineral homeostasis. It is estimated that there are 1,200 g of calcium in the adult human skeleton, another 500 mg in the intercellular fluid, and 280 mg in the blood plasma. Thus almost 100% of adult human body's calcium is in the skeleton. Currey [15] reported that calcium densities in the wet femur of cow (*Bos primigenius*), horse (*Equus ferus*), Roe deer (*Capreolus capreolus*), and wallaby (*Macropus rufogriseus*) vary between 252 and 283 mg/g, roughly consistent with relative invariance (see above quote from Ruben and Bennett).

Bone is divided into two basic histological types: cancellous and compact. Cancellous (or honeycomb) bone consists of branching trabeculae and is highly vascular. Compact bone, the more intensively studied form, is found notably in the shaft of long bones, such as the femur. Compact bone is comprised of primary osteons (in small mammals such as mouse and rat) and more highly organized secondary osteons in larger mammals. Secondary osteons are formed by resorption of small cavities inside pre-existing bone created by osteoclasts (thereby releasing minerals into the blood stream). The long axis of these cavities tends to be roughly parallel to the long axis of the bone. For an informative discussion of the evolution of bone see Ruben and Bennett [16]. The bone cells (osteocytes) come to occupy small lacunae within the bone, from which fine canaliculi extend into the interior of the osteon and to other osteocytes. According to Frost [17] there are about 20,000 osteocytes (lacunae) on average per mm^3 of lamellar bone in the adult human, corresponding to a mean center-to-center spacing of about 37 μm (take cube root of ($10^9/20,000$)). In the center of the osteon, in larger mammals, one finds a longitudinally running Haversian canal containing blood vessels and nerves. The whole secondary osteon is surrounded by a cement line. This process of resorption and laying down new secondary osteons goes on throughout life in response to changing mechanical and metabolic stresses. In addition, long bones enclose a marrow cavity containing fat cells. The total quantity of marrow in the adult human ranges from 1,500 to 4,000 g. White marrow (mostly fat) is found in the long bones of the adult; red marrow (red-blood-cell producing) is found in "flat" bones such as the sternum and bones of the skull. (A large marrow cavity is said to be absent in the long bones of the elephant. Bats have notably small marrow cavities.)

Bone blood flow

Copp and Shim [18] estimated that total bone blood flow in dog and rabbit constitutes 5–7% of cardiac output. This figure was considered to be "far exceeding the metabolic needs of bone." Ganong [19] places total bone blood flow in the adult human at 3.6 to 7.3% of cardiac output. Because of its rigidity and complex architecture, the measurement of blood flow and oxygen consumption in bone is a challenging problem. Data on more species using varied methods are needed before reliable scaling inferences can be drawn.

Bone compressive strength

The products of a LSQ analysis of the compressive strength of the femur based on data provided by Yamada [20] are shown in Table 12.3 (body size data taken from the SBT, Chapter 4).

The implication of Table 12.3, for this small sample, is that compressive strength exhibits (approximately) absolute invariance ($b \approx 0$) for pWR = 3.3.

Table 12.3 Compressive strength (kg/mm^2) of wet femur at mid-shaft as a function of body mass

N_{ord}	N_{sp}	MPD	pWR	DI (%)	r^2	Slope $\pm$ 99% CL
5	8	10	3.3	2.4	0.524	0.044 $\pm$ 0.064

See Table 12.2.

Table 12.4 LSQ analysis of bone density (g/cm^3) as a function of adult body mass [22] [a]

N_{ord}	N_{sp}	MPD	pWR	DI (%)	r^2	Slope $\pm$ 99% CL	Entity
6	27	6	3.3	4.6	0.108	0.0119 $\pm$ 0.019	Femur in land mammals
4	21	8	3.1	2.7	0.029	−0.010 $\pm$ 0.037	Femur in aquatic mammals
6	26	5	3.3	4.6	0.154	0.0178 $\pm$ 0.024	Humerus in land mammals
5	22	8	3.1	3.4	0.175	−0.029 $\pm$ 0.040	Humerus in aquatic mammals

See Table 12.2. [a] Body mass from SBT; the author states that densities were determined from "thoroughly soaked" museum specimens.

Bone density

Dry bone is about twice as dense as most other body tissues. As bones become larger and heavier, the muscles must do more work to move them. Hence there is likely to be a strong selective pressure to keep bones as light as possible, commensurate with their mechanical function [21]. The results of a LSQ analysis of bone density in adult mammals based on data from Wall [22] are displayed in Table 12.4. Body weights are taken from the SBT (Chapter 4).

The results presented in Table 12.4 are consistent with the inference that bone density, in the femur and humerus, in these modest samples, is independent of body size ($b \approx 0$, absolute invariance).

Joint surface area

Godfrey *et al.* [23] established that the surface area of the head of the humerus scales as the 0.677 power of body mass ($N_{ord} = 6$, $N_{sp} = 73$, pWR = 2.4, DI = 10%, $r^2 = 0.83$) ("raw" data not provided). They interpret their findings as being consistent with geometric similarity.

Skeletal mass

The most widely cited study of the scaling of *skeletal mass* in mammals is perhaps that of Prange *et al.* [24]. Their study of *dry* skeletal mass was based on 48 data points, and

Table 12.5 Scaling of wet (fresh) skeletal mass as a function of body mass

N_{ord}	N_{sp}	MPD	pWR	DI (%)	r^2	Slope $\pm$ 99% CL	Entity	Source
1	10	16	3.2	0.5	0.996	1.050 ± 0.082	Cetaceans	[31, 32] [a]
10	60	24	7.2	21	0.996	1.019 ± 0.022	Mammals at large	From Lit.

See Table 12.2; row two includes the data of row one. [a] Some data kindly provided by J.G. Mead, Marine Mammal Program, Smithsonian Institute.

some 27 species drawn from eight taxonomic orders, spanning six orders of magnitude in body mass. In most cases only genus names were given. They reported a slope of 1.09 (no CL given). Their dataset included skeletal mass (1,782 kg or 27% of body mass) for an adult African elephant (*Loxodonta africana*) having a body mass of 6,600 kg. This point exceeds the prediction from the best-fit LSQ line by a factor of 2. The next largest entry was for the human, weighing 67 kg. Thus the point for the elephant is a substantial outlier on both the *x*- and *y*-axes.

Moreover, the value for skeletal mass in the elephant used by these workers and others is of doubtful validity. The source cited for the elephant data is Crile [25], but data for the skeletal mass of an elephant were not found in this work. A more likely source is Quiring [26] and later Martin and Fuhrman [27] (see my comment in references). When the point for the elephant is omitted from this dataset [24] the LSQ slope drops to 1.040 ± 0.062 (99% CL). This apparently mistaken value for skeletal mass in the elephant then propagated through the literature [28, 29]. Anderson *et al.* [30] reported slopes of 1.090 and 1.118 for *dry* skeletal mass on body mass in land mammals (pWR = 6.0) and cetaceans (pWR = 1.2), respectively ("raw" data not provided).

The outcomes of the present LSQ analyses of *fresh* skeletal mass in cetaceans [31 – 33] and in mammals at large, including cetaceans, are brought together in Table 12.5, rows one and two, and Figure 12.2. (It is not always stated clearly that skeletal mass is "fresh," but an attempt was made to select those data where this seemed likely to be the case.) The slopes for the two groups are not significantly different. For cetaceans, skeletal mass comprised on average 11% of body mass, with a range from 6.5 to 15%. For mammals at large, including cetaceans, skeletal mass averages 11% of body mass, with a range of 6.3 to 17.5%. These findings agree with the inference of Anderson *et al.* [30] that skeletal mass does not exceed 20% of adult body mass in mammals generally.

The slopes for cetaceans and for mammals are not significantly different (Table 12.5). The best-fit slope (1.019 ± 0.022) (99% CL) for fresh skeletal mass on body mass in mammals at large (Table 12.5, row two) lies between 0.998 and 1.041. If one assigns a slope of 1 (rather than 1.019) to the best-fit line, an intercept of 0.095 is found, implying that to a first approximation, skeletal mass makes up 9.5% of body mass in the given sample. If we omit the data for cetaceans from Table 12.5, row two, we obtain a slope of 1.015 ± 0.031 (99% CL) for 50 species of mammals.

When one computes a slope for skeletal mass on body mass *less* skeletal mass (so as to remove the implicit correlation between body mass and skeletal mass) one finds that $b = 1.021 \pm 0.024$ (99% CL), not significantly different from the slope of 1.019 ± 0.022 (99% CL) (Table 12.5, bottom row).

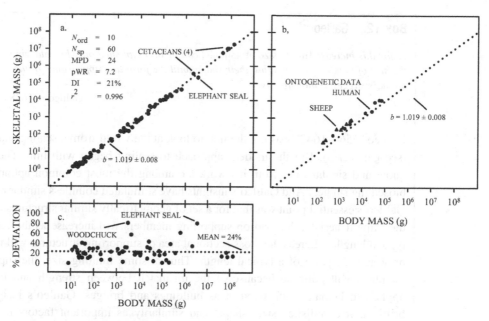

Figure 12.2 Scaling of wet (fresh) skeletal mass as a function of body mass.

The data for the scaling of wet skeletal mass in mammals generally are also depicted in Figure 12.2. It is striking, and probably significant, that the range of slopes for skeletal mass (0.99–1.041) overlaps the range of slopes (0.989–1.013) for the scaling of skeletal muscle (see Table 12.2). Recall that in Chapter 8, it was found that whole body ash scales as the 1.020 ± 0.040 (99% CL) power of body mass. Thus the exponents for muscle mass, skeletal mass, and total ash content are each reasonably consistent with relative invariance ($b = 1.0$).

The human and sheep (*Ovis orientalis*) ontogenetic data (Figure 12.2, panel b) are from Krogman [33] and Barcroft [34], respectively. Both ontogenetic datasets appear to approach the adult interspecific line from above. For a prior study of bone scaling in adult mammals, see Prothero [35].

Similarity

Galileo (1564–1642) argued that bones in land mammals could not scale in agreement with geometric similarity [36] (see Box 12.1 and Chapter 3). In 1975, McMahon [37] made the provocative proposal that the dimensions of long bones scale in accord with elastic similarity. He derived a series of exponents for the dependence of length on diameter, and both on body mass (somatic exponents), explicitly assuming, with limited evidence, that bone volume scales directly with body mass. This assumption was not in accord with what was then known of skeletal scaling (although it accords better with present evidence; see Table 12.5). For example, Schmidt-Nielsen, in a well-known paper published in 1970, cited a slope of 1.13 (far from 1.0) for dry skeletal scaling as a function of body mass [38].

Box 12.1 Galileo

. . . for this increase [in the size of animals] *can be accomplished only by enlarging the size of bones, thus changing their shape until the form and appearance of the animals suggest a monstrosity. . .*

Galileo *c.* 1638

Galileo (1564–1642) was not the first to look at the world from a scaling standpoint (see Chapter 3). But the modern approach to scaling began with him. The above quote and similar ones from his work are among the most common aphorisms in bioscaling studies. And Galileo's insights have stimulated countless similar remarks. Galileo's essential point was that for a set of geometrically similar structures made of the same materials, the load on supporting members will increase as the cube of a typical length, whereas the cross-sectional area of supporting structures will increase only as the square of a typical length. Thus at some limiting size, the supporting members will collapse because of the mismatch between strength and load. As applied to human artefacts, such as buildings and bridges, Galileo's insight was brilliant. It established size, shape, and similarity as important factors in scaling studies. When Newton spoke of standing on the shoulders of giants, he no doubt had Galileo, among others, in mind.

In effect, Galileo argued that if the skeletons of land mammals scaled in accord with geometric similarity, large mammals would collapse under their own weight. But large land mammals, such as the elephant, do not collapse under their own weight. Ergo, the assumption of geometric scaling must be false. This style of *reductio ad absurdum* argument is not uncommon in mathematics but is rare in science. (It implicitly assumes a double-valued logic: statements are either true or false.) Although quotes such as the above still appear in scaling studies, we now know that the extrapolations Galileo made to the scaling of the skeleton in land and marine mammals were wide of the mark. This may be regarded as an early example of a failure in reductionism as applied to biology (see Box 9.1, Chapter 9).

Galileo implicitly assumed that static compressional forces (here due to body weight) were the central factor limiting the size of structures under a gravitational load. But in mammals it is locomotion that generates peak stresses, not static load. Galileo failed to see that changes in posture and behavior could alter these peak stresses (gazelles race across the veldt, elephants shuffle). Understandably, he could not have known that natural selection would tend to minimize the burden of skeletal weight or mass as far as practicable. As we have seen above, current evidence is that skeletal weight increases only slightly, if at all, faster than body weight in a broad sample of mammals. And there is as yet no convincing evidence that skeletal weight in cetacean scales very differently from that in land mammals. However, it should be said that these qualifications to Galileo's insights as applied to scaling in mammals do not detract from the overall significance of his work, which, on scaling and much besides, was monumental.

Table 12.6 Similarity analysis for three long bones [39]

Exponents (± 95% CL)		Derived exponents		Entity
			Geometric similarity	
Diameter	Length	Bone volume (95% CL)	(for lengths)	
0.36 ± 0.01	0.36 ± 0.02	1.08 ± 0.04	0.36	Femur
0.38 ± 0.01	0.36 ± 0.02	1.12 ± 0.04	0.37	Humerus
0.36 ± 0.02	0.32 ± 0.02	1.04 ± 0.06	0.35	Tibia

McMahon's assumption of direct proportionality between bone volume and body mass was empirically valid for the family of *artiodactyls* he studied, namely the bovidae. For this group of mammals ($N_{ord} = 1$, $N_{sp} = 83$, pWR = 3.4, DI = 1.1%) the somatic exponents McMahon derived in theory for elastic similarity (0.25 for lengths, 0.375 for diameters) were in reasonable agreement with the empirical exponents he derived for the femur, humerus, tibia, and ulna (no CL were given for the empirical slopes).

In 1979 Alexander and his colleagues published a pivotal paper showing that McMahon's empirical findings were true only for a special case (the bovidae), and were not typical of mammals in general [39]. (Alexander, a pioneer in contemporary biomechanics, has a broad background in zoology and a firm grasp of engineering concepts and methods.) The findings these authors published for three long bones in mammals at large are shown in Table 12.6 ($N_{ord} = 8$, $N_{sp} = 36$, $N_{pnts} = 41$, pWR = 5.9, DI = 12.1%).

One may think of long bones as roughly cylindrical; hence bone volume is proportional to the product of length and diameter squared. Thus the somatic exponents for bone *volume* (Table 12.6, column three, row one) are calculated as the sum of the exponent for length plus twice the exponent for diameter (e.g., $0.36 + 2 \cdot 0.36 = 1.08$). The 95% CL are those given by the authors. The exponents in column four for diameters and lengths are computed as 1/3 of the entries in column three (e.g., 1.08/3 = 0.36) (Chapter 3). Since the datasets for diameter and length are the same, one is entitled to compute the CL in bone volume in the same way as one computes volume (e.g., for the first row $2 \cdot 0.01 + 0.02 = 0.04$) (Table 12.6, row one). Two relevant criteria for judging geometric similarity may be invoked. The first is equality of the somatic exponents for diameter and length. It is seen in Table 12.6, columns one and two, that the somatic exponents agree with each other to within the given 95% CL The second criterion is that the somatic exponents (columns one and two) agree with the amended exponents for geometric similarity (Table 12.6, column four) (Chapters 3, 24). This is also the case within the given 95% CL. It is concluded that the scaling of long bones in the given sample is consistent with geometric similarity.

In 1990 Bertram and Biewener [40] published an extensive dataset for the scaling of bone diameter and length in carnivores. A LSQ analysis of a *subset* of their data was carried out. For these data, with two exceptions, $N_{ord} = 1$, $N_{sp} = 106$, pWR = 3.6, DI = 1.2 (i.e., for the diameter and length of the radius, the $N_{sp} = 105$). The outcomes of this analysis are seen in Table 12.7 (which is structured in the same way as Table 12.6). In

Table 12.7 Similarity analysis for four long bones [40]

Exponents (± 95% CL)		Derived exponents		Entity
Diameter	Length	Bone volume	Geometric similarity (for lengths)	
0.358 ± 0.024	0.346 ± 0.030	1.062 ± 0.078	0.354	Femur
0.388 ± 0.024	0.336 ± 0.028	1.112 ± 0.076	0.371	Humerus
0.395 ± 0.031	0.362 ± 0.040	1.152 ± 0.102	0.384	Radius
0.368 ± 0.025	0.2985 ± 0.035	1.035 ± 0.084	0.345	Tibia

Table 12.8 LSQ analysis of the ratio of internal to external diameter in three long bones

N_{ord}	N_{sp}	MPD	pWR	DI (%)	r^2	Slope ± 99% CL	Entity
6	20	8	5.6	7.2	0.381	−0.032 ± 0.027	Femur
4	19	11	2.9	2.4	0.02	−0.011 ± 0.057	Humerus
6	21	14	5.6	7.3	0.179	−0.030 ± 0.042	Tibia

See Table 12.2.

Table 12.7 one finds that the femur and the radius meet the criteria for equality of exponents and agreement with the amended exponents for geometric similarity (Table 12.7, column four), both within the given 95% CL. The exponents for humerus and tibia do not meet these criteria. Bertram and Biewener [40], in a more detailed analysis than given here, concluded that these two bones (humerus and tibia) exhibit differential scaling as between small and large carnivorous mammals. That is, a single straight line is an inadequate representation of these data.

The authors suggest that small mammals tend to geometric scaling of the long bones, but larger mammals may not. For more recent reviews of the literature on bone scaling see Garcia and Silva [41] and Biewener [42].

The ratio of internal to external bone diameter

In 1985 Currey and Alexander published the outcomes of measurements of the internal and external diameters in different long bones of a number of species of mammals [43]. A synopsis of a LSQ analysis of a subset of their data is shown in Table 12.8.

A reasonable inference from Table 12.8 is that the ratio of internal to external diameter in the given samples is independent of body size ($b \approx 0$, absolute invariance).

Summary

A synopsis of the major LSQ slopes cited above, in rank order, is presented in Figure 12.3. Looking at items a to h in Figure 12.3, we find that the mean slope for

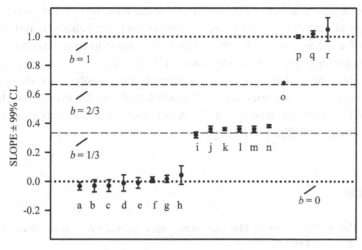

RANK ORDER

a = r.i.e.d.*(femur); b = r.i.e.d. (tibia); c = bone density in aquatic
mammals (humerus); d = r.i.e.d. (humerus); e = bone density in aquatic
mammals (femur); f = bone density in land mammals (femur); g =
bone density in land mammals (humerus); h = compressive strength
of femur mid-shaft; i = tibia length; j = tibia diameter; k = femur
diameter; l = femur length; m = humerus length; n = humerus
diameter; o = humerus head surface area (99 % CL not available);
p = muscle mass; q = wet skeletal mass in mammals generally;
r = wet skeletal mass in cetaceans

* r.i.e.d. = ratio of internal to external diameter

Figure 12.3 Slope summary for musculoskeletal system.

these eight entries is –0.0048 ± 0.0255 (STD). This result is in reasonable agreement
with absolute invariance (b = 0.0). When we consider the next six items (i to n), we
find a mean slope 0.357 ± 0.018 (STD). The mean slope is in fair but not excellent
agreement with scaling as the one-third power of body mass (i.e., for constant
shape). Item o has a slope of 0.677 ± 0.001, not far from the reference slope of
2/3. Items p to r have a mean slope of 1.023 ± 0.020. Together these three items are
in reasonable agreement with relative invariance (b = 1.0). It is true, however, when
we consider the 99% CL for item r, its slope could be far from 1.0 (i.e., 1.05 + 0.082
= 1.132).

If one fits a LSQ line to a plot (not shown) of the 18 *empirical* slopes (see Figure 12.3)
against their respective *reference* slopes we obtain a line with a slope of 1.028, a
coefficient of determination of 0.996, and a mean absolute deviation between the
empirical slopes and the reference slopes of 0.021 ± 0.010 (STD).

Considering the whole picture, one may say that the mean of eight slopes is in good
agreement with absolute invariance (b = 0.0), the mean of another six slopes is in fair
agreement with a slope of 1/3, and one slope (0.677) is in reasonable agreement with
a slope of 2/3. Two of three slopes are in reasonable agreement with relative invariance
(b = 1.0) but one slope may not be. That is, we have 17 out of 18 instances where the
empirical slopes are reasonably consistent with the reference slopes based on either

invariance or dimensionality. The scaling of muscle mass ($b = 1.001$; Table 12.2, row two) suggests that the non-dimensional ratio (muscle mass/body mass) may be an example of a quantitative similarity criterion, analogous to those criteria such as the Reynolds number in physical scaling studies (Chapter 3).

This chapter shows that by drawing together a number of slopes and displaying them in relation to reference slopes, one may obtain potentially useful insights into the overall scaling of the musculoskeletal system. Such insights are much less apparent when, as is customary, one studies slopes for a given system one or two at a time.

References

1. Alexander, R.McN. (1985). The maximum forces exerted by animals. *Journal of Experimental Biology*, **115**:231–238.
2. Omura, H., Ohsumi, S., Nemoto, T., Nasu, K. and Kasuya, T. (1969). Black right whales in the North Pacific. *Scientific Reports of Whales Research Institute*, **21**:1–78.
3. Harris, A.J., Duxson, M.J., Butler, J.E. *et al.* (2005). Muscle fiber and motor unit behaviour in the longest human skeletal muscle. *Journal of Neuroscience*, **25**:8528–8533.
4. Ackerman, E. (1962). *Biophysical Science*. Englewood Cliffs, NJ: Prentice-Hall.
5. Huxley, H.E. (1958). The contraction of muscle. *Scientific American*, **199**:66–82.
6. Duke, T.A.J. (1999). Molecular model of muscle contraction. *Proceedings of the National Academy of Sciences*, **96**:2770–2775.
7. Gauthier, G.F. and Padykula, H.A. (1966). Cytological studies of fiber types in skeletal muscle. A comparative study of the mammalian diaphragm. *Journal of Cell Biology*, **28**:333–354.
8. Johnson, E.R. and Beattie, A.W. (1973). Variation in muscle fibre diameter among sections and intrasections and between contralateral muscles in seven bovine muscles. *Journal of Agricultural Science*, **81**:9–14.
9. Montagnani, S. and De Rosa, P. (1989). Morphofunctional features of human extrinsic ocular muscles. *Documenta Ophthalmologica*, **72**:119–128.
10. Lockhart, R.D. and Brandt, W. (1938). Length of striated muscle-fibers. *Journal of Anatomy*, **70**:470.
11. Pollock, C.M. and Shadwick, R.E. (1994). Allometry of muscle, tendon, and elastic energy storage capacity in mammals. *American Journal of Physiology*, **266**:R1022–R1031.
12. Munro, H.N. (1969). Evolution of protein metabolism in mammals. In: Munro, H.N. (ed.) *Mammalian Protein Metabolism*. New York: Academic Press, pp. 133–182.
13. Cheek, D.B., Powell, G.K., Scott, R.E. *et al.* (1965). Growth of muscle mass and skeletal collagen in the rat. I. Normal growth. *Bulletin Johns Hopkins Hospital*, **116**:378–387.
14. Blackfan, K.D. (ed.) (1933). *White House Conference on Child Health and Protection: Part II. Anatomy and Physiology*. New York: Century Company.
15 Currey, J.D. (1987). The evolution of the mechanical properties of amniote bone. *Journal of Biomechanics*, **20**:1035–1044.
16. Ruben, J.A. and Bennett, A.A. (1987). The evolution of bone. *Evolution*, **4**:1187–1197.
17. Frost, H.M. (1961). Measurements of the diffusion pathway between osteocyte lacuna and blood. *Henry Ford Hospital Medical Bulletin*, **9**:137–144.

18. Copp, D.H. and Shim, S.S. (1965). Extraction ratio and bone clearance of Sr^{85} as a measure of effective bone blood flow. *Circulation Research*, **16**:461–467.

19. Ganong, W.F. (1965). *Review of Medical Physiology*. Los Altos, CA: Lange Medical Publications.

20. Yamada, H. (1970). *Strength of Biological Materials*. Baltimore, MD: Williams & Wilkins Co.

21. Martin, R.B. (2007). The importance of mechanical loading in bone biology and medicine. *Journal of Musculoskeletal Neuronal Interactions*, **7**:48–53.

22. Wall, W.P. (1983). The correlation between high limb-bone density and aquatic habits in recent mammals. *Journal of Paleontology*, **57**:197–207.

23. Godfrey, L., Sutherland, M., Boy, D. and Gomberg, N. (1991). Scaling of limb joint surface areas in anthropoid primates and other mammals. *Journal of Zoology*, **223**:603–625.

24. Prange, H.D., Anderson, J.F. and Rahn, H. (1979). Scaling of skeletal mass to body mass in birds and mammals. *American Naturalist*, **113**:103–122.

25. Crile, G. (1941). *Intelligence, Power and Personality*. New York: McGraw-Hill. (Data for the skeletal mass of an elephant were not found in this work.)

26. Quiring, D.P. (1939). Notes on an African elephant (*Loxodonta africana*). *Growth*, **3**:9–13. (The mass of the skeleton (1,238 lbs) given in this work is only 8% of body mass (14,640 lbs). The skeleton had been left to "weather" for 10 days in Tanganyika before weighing.)

27. Martin, A.W. and Fuhrman, F.A. (1955). The relationship between summated tissue respiration and metabolic rate in the mouse and dog. *Physiological Zoology*, **28**:18–34. (Skeletal mass for an African elephant is given as 27% of body mass. The specific values for body, kidney, and liver masses suggest this is the same elephant as discussed by Quiring (see Reference [26]), except that 8% for skeletal mass has become 27%. Perhaps someone inadvertently took the body mass of this elephant to be 4,640 lbs rather than 14,640 lbs (100*1238/4640 = 27)?)

28. Kayser, C. and Heusner, A.A. (1964). Étude comparative du métabolism énergétique dans la série animale. *Journal Physiologie*, Paris **56**:489–524.

29. Schmidt-Nielsen, K. (1975). Scaling in biology: the consequences of size. *Journal of Experimental Zoology*, **194**:287–307.

30. Anderson, J.F., Range, H. and Prange, H.D (1979). Scaling of supportive tissue mass. *Quarterly Review Biology*, **54**:139–148.

31. Quiring, D.P. (1943). Weight data on five whales. *Journal of Mammalogy*, **24**:39–45.

32. Winston, W.C. (1950). The largest whale ever weighed. *Natural History*, **59**:392–398.

33. Krogman, W.M. (1941). Growth of Man. *Tabulae Biologicae*, **20**:1–963.

34. Barcroft, J. (1947). *Researches on Pre-Natal Life*. Springfield, IL: Charles C. Thomas.

35. Prothero, J. (1995). Bone and fat as a function of body weight in adult mammals.*Comparative Biochemistry and Physiology*, **111A**:633–639.

36. Galilei, G. (1954). *Dialogues Concerning Two New Sciences*. New York: Dover Publications.

37. McMahon, T.A. (1975). Using body size to understand the structural design of animals: quadrupedal locomotion. *Journal of Applied Physiology*, **39**:619–627.

38. Schmidt-Nielsen, K. (1970). Energy metabolism, body size, and problems of scaling. *Federation Proceedings*, **29**:1524–1532.

39. Alexander, R.McN., Jayes, A.S., Maloiy, G.M.O. and Wathuta, E.M. (1979). Allometry of the limb bones of mammals from shrews (*Sorex*) to elephant (*Loxodonta*). *Journal of Zoology*, **189**:305–314.

40. Bertram, J.E.A. and Biewener, A.A. (1990). Differential scaling of the long bones in the terrestrial carnivora and other mammals. *Journal of Morphology*, **204**:157–169.

41. Garcia, G.J.M. and da Silva, J.K.L. (2006). Interspecific allometry of bone dimensions: a review of theoretical models. *Physics of Life Reviews*, **3**:188–209.

42. Biewener, A.A. (2005). Biomechanical consequences of scaling. *Journal of Experimental Biology*, **208**:1665–1676.

43. Currey, J.D. and Alexander, R.McN. (1985). The thickness of the walls of tubular bones. *Journal of Zoology*, **206A**:453–468.

13   Neuroendocrine system

*The traditional distinctions between the endocrine and nervous systems appear
to be largely artificial. The two systems function together to transform
a loosely bound collection of cells, tissues, and organs into a cooperative
enterprise...*

Turner 1966

Because the nervous and endocrine systems interact closely with one another (see above quote from Turner), it is convenient and appropriate to consider both in the same chapter. We begin with a discussion of the nervous system.

NERVOUS SYSTEM

*No one, I presume, doubts that the large proportion which the size of man's brain bears
to his body, compared with the same proportion in the gorilla or orang, is closely
connected with his higher mental powers.*

Darwin 1871

*However, when the effects of confounding variables such as body size and
socioeconomic status are excluded, no correlation is found between IQ and brain size
among modern humans.*

Martin 1996

Of the various systems and subsystems into which mammals may be apportioned, the one of most scientific and philosophic interest is the nervous system. In particular, the human brain is the most complex structure known. We look to science to answer questions of how, when, where, and why. However, when we look for an explanation of human associative recall, consciousness, creativity, empathy, humour, imagination, insight, intelligence, intuition, language, logic, music, and pattern recognition we look largely in vain. The brain's basic workings remain enigmatic. Scaling studies stand to provide possible insights into the compromises and trade-offs inherent in the evolution of the nervous system, and further, to reveal possible *scale effects* that may limit the useful size of the mammalian brain (see below). This chapter is based on a study of 13 y-variables, for which there are 20 instances, involving a total of 2,101 records, all in tabular form. For a summary plot of the major slopes derived in this chapter see Figure 13.6.

Length of "wiring" in human brain

The total length of "wiring" or neuronal connections in the adult human brain is estimated at 10^5 to 10^7 km [1]. This range is equivalent to 2.5 to 250 times the equatorial circumference of the earth (40,000 km).

Brain size

Modules are the. . . processing units that have allowed large-brained animals to evolve.

Leise 1990

The first published quantitative scaling study of brain size in mammals is apparently that of Snell [2] in 1891. He reported, for 23 species, a slope of 0.68 (not obtained by LSQ). Using a subset of his data (with one point per species), one finds a best-fit LSQ slope of 0.652 ± 0.124 (99% CL) (see Table 13.1, row one). Unfortunately the 99% CL are too wide to make this more than a suggestive outcome. Since then many other studies have been reported. Jerison [3] argued that brain size *should* scale as the two-thirds power of body size, in accord with the cortical representation of the body surface. But the cortical representation of the body surface in cerebral cortex is not a conspicuous feature of the mammalian brain; it is perhaps unlikely to be an important determinant of brain size in mammals. The results of a LSQ analysis using the reduced major axis method in 309 species of placental mammals were published by Martin in 1981 [4] (see Table 13.1, row two). He obtained a slope of 0.755 ± 0.025 (95% CL). For a brief discussion of the major axis method see Chapter 23.

The main findings of the present LSQ analysis of brain size as a function of body mass in mammals are depicted in Table 13.2 and Figure 13.1; data from the literature (111 records were by personal communication from J.F. Eisenberg, another 93 records from R.D. Martin). Note that when workers report brain size they usually do not indicate whether the cerebellum is included. Observe that the entries in Table 13.2, row two, are those of row three less the data for primates.

When these data (Table 13.2, row three) are tested for non-linearity by binning (Chapters 3, 17), there is some evidence (not shown) of a slight overall concavity downwards (see also Figure 13.2).

Table 13.1 Two prior scaling studies of brain mass as a function of adult body mass

N_{ord}	N_{sp}	MPD	pWR	DI (%)	r^2	Slope $\pm$ CL	CL	Source
8	23	109	7	13	0.913	0.652 ± 0.124	99%	[2] [a]
13	309	-	7	37	0.92	0.755 ± 0.025	95%	[4]

N_{ord} = number of orders; N_{sp} = number of species; MPD = mean percent deviation; pWR = log (BW_{max}/BW_{min}); DI = diversity index; r^2 = coefficient of determination; CL = confidence limit(s); [a] body size data from SBT, Chapter 4.

Table 13.2 Present LSQ analysis of brain mass as a function of body mass

Row	N_{ord}	N_{sp}	MPD	pWR	DI (%)	r^2	Slope ± 99% CL	Entity
1	1	88	27.7	3.3	1.0	0.922	0.780 ± 0.064	Primates only
2	18	710	33	7.4	62.5	0.975	0.741 ± 0.012	Omit primates
3	19	798	38	7.4	66.8	0.967	0.759 ± 0.013	FDS

See Table 13.1.

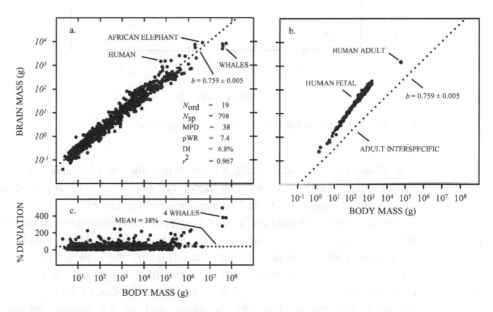

Figure 13.1 Scaling of brain size as a function of body size in mammals.

The FDS (Table 13.2, row three) is the second largest dataset in this study (see Chapter 6); it is the largest dataset for any single organ in this work, testifying to the great interest in the brain. The diversity index (DI) for the FDS approaches 67%. The best-fit LSQ slope obtained here (Table 13.2, row three) is virtually identical to that reported by Martin [4] (Table 13.1, row two). The slope for primates (Table 13.2, row one) is not significantly different from that for the FDS (row three).

In regard to the above quote from Darwin, recall that those human beings with a brain size of about 800 g are considered to be of normal "mentality" (e.g., can learn a language) [5]. This brain size (800 g) (relative to an adult human average of, say, 1,500 g) may be compared to that of a gorilla (*Gorilla gorilla*) (500 g) or a Pacific white-sided dolphin (*Lagenorhynchus obliquidens*) (1,500 g). Evidently a larger brain need not imply greater intelligence, and any attempt to relate brain size to intelligence over a factor of only 2 or so in brain size may be problematic; see above quote from Martin. Darwin perhaps would not have disputed this, as he goes on to say: "Under this point of view, the brain of an ant [with perhaps a million neurons] is one of the most

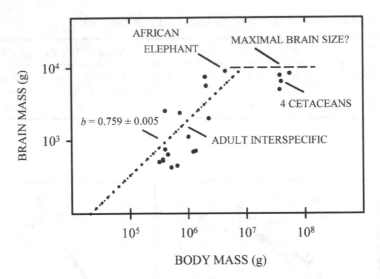

Figure 13.2 Apparent maximal brain mass in adult mammals.

marvellous atoms of matter, perhaps more so than the brain of a man." It seems likely that modularity (as for muscle; see Chapter 12) will prove to be of importance in the scaling of the mammalian brain (see above quote from Leise).

The human ontogenetic data (T.H. Shepard, personal communication) for brain mass are compared with the adult interspecific LSQ line in Figure 13.1b. This illustrates the most striking *structural* (and perhaps fateful) distinction between humans and other mammals, namely the protracted period of virtually linear growth (on a log–log plot) of brain size with body size before birth. Everywhere the human fetal line lies well above the adult interspecific line. At the upper end of the human ontogenetic line (Figure 13.1b, fetal body weight about 1,490 g) we find that brain weight (177 g) is about 13 times that predicted by the adult interspecific line (i.e., 13.7 g) (177/13.7 = 13). At birth the human brain has about 25% of its adult mass [6]. By the end of one year after birth it has increased to about 1,000 g.

A second important observation in Figure 13.1a is that the data for four species of whales (*Balaenoptera borealis, B. physalus, Megaptera novaeangliae, Physeter catodon*) all fall well below the best-fit adult interspecific line. This finding is shown more clearly in Figure 13.2, where it is suggested that there may be an upper limit to brain size in mammals at around 10 kg [7]. The brain mass data for four cetaceans (see above) and an African elephant [8] (*Loxodonta africana*) all lie below this hypothetical maximum (compare with Figure 13.1a).

There is a minimum diameter (of less than 1 μm) to axons (myelinated or unmyelinated), implying that the number of axons crossing from one hemisphere to another via the corpus callosum cannot scale as rapidly as does cortical volume (see Table 13.6, row four, $b = 1.04$). Here one assumes that the cross-sectional area of the corpus callosum scales roughly as the two-thirds power of brain volume. Because of such constraints, there may be a threshold brain size (perhaps around 10 kg) at which increasing brain size produces few or no net benefits. This may be viewed as a biological *scale effect* (Chapter 3). Finally, while it is often said that the sperm whale has the largest brain (at 7.8 kg) in the animal kingdom, there are data in the literature [8] which put the brain of a 4,000 kg male African elephant (*Loxodonta africana*) significantly higher (at 9.0 kg).

Cerebellar size

Despite continuing work on the structure and function of the cerebellum, there is still no consensus as to what it does or how it does it.

Thach *et al.* 1992

The cerebellum (L. "small brain") is involved in the regulation of physical equilibrium, coordination of movement, gait, and posture. It is more highly developed in active animals with complex movement patterns (as for many species of fish, as well as birds and mammals) and less developed in less active animals (e.g., reptiles and amphibians) [9]. As with many other parts of the nervous system, the first conceptions of cerebellar function were inferred from clinical observations. Symptoms of cerebellar dysfunction include asthenia (muscle weakness), ataxia (incoordination of movement), hypotonia (loss of muscle tone), and tremor. Unlike motor cortex, where contralateral muscles are stimulated, the muscles regulated by the cerebellum are ipsilateral. The actions of the cerebellum are not known to impinge on consciousness. Overall, like the cerebrum, the functioning of the cerebellum is not yet well understood (see above quote by Thach *et al.*).

The outcomes of a LSQ analysis of cerebellar size based on data taken from the literature are shown in Table 13.3 and Figure 13.3. According to these findings ($b = 0.785 \pm 0.055$) (99% CL) the cerebellum apparently scales in much the same way with body size as does the brain as a whole ($b = 0.759$) (see Table 13.2, row three). From Figure 13.3b, one sees that most of the human ontogenetic cerebellar points [10] lie above the adult interspecific line. In Figure 13.3c, the four most deviant points in the upper left-hand corner are for three species of tenrecs (*Echinops telfairi*, *Tenrec ecaudatus*, *Setifer setosus*), and for the European hedgehog (*Erinaceus europaeus*). The point for the cow (*Bos primigenius*) is also highly deviant. These five points all lie below the best-fit LSQ line. Keep in mind that the scatter around the best-fit line is considerable (MPD = 50%) (Table 13.3; Figure 13.3c).

Table 13.3 LSQ analysis of cerebellar mass as a function of adult body mass [a]

N_{ord}	N_{sp}	MPD	pWR	DI (%)	r^2	Slope $\pm$ 99% CL
12	100	49.5	5	20	0.935	0.785 ± 0.055

See Table 13.1. [a] Data from the literature.

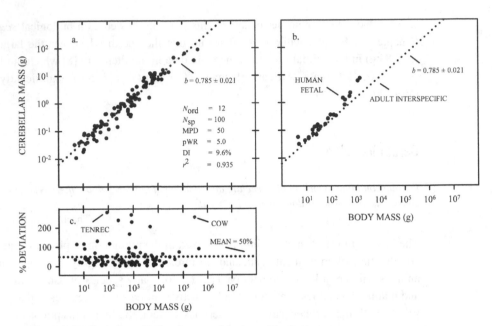

Figure 13.3 Scaling of cerebellar size as a function of body size.

Eye size

The eye, in short, can never be very small and need never be very big; it has its own limitations and conditions apart from the size of the animal.

D'Arcy Thompson 1943

For the eye has every possible defect that can be found in an optical instrument.

Helmholtz 1873

The eyeball is roughly spherical and in the adult human is about 24 mm in diameter. The major optic surface is the transparent cornea, which accounts for about two-thirds of the total dioptric (refractive) power of the relaxed young adult human eye. The cornea is roughly 1 mm thick and 12 mm in diameter. It is continuous posteriorly with the outermost layer of the eye, the sclera. Internally, the lens (about 10 mm in diameter) and its attachments partition the eyeball into two compartments. The anterior compartment is filled with watery aqueous humor, the posterior with the gel-like vitreous humor. The amount of light entering the eye through the pupil is controlled by the iris. In the young adult human, contractions of the iris can reduce the light intensity impinging on the retina by a factor of about 16.

The outcomes of the present analysis of data from the literature pertaining to the scaling of eyeball size are shown in Table 13.4. In row one of Table 13.4, 13 species (out of 61) deviate from the best-fit LSQ line by more than 100%. Among these 13 deviant species, two are cetaceans, three are semiaquatic, and one is fossorial. Seven species are considered to be nocturnal, mostly nocturnal or crepuscular. Ontogenetic data for eye size in the rat (not shown) are available from Hatai [11]. LSQ findings for a reduced dataset (RDS) are given in Table 13.4, row two.

Table 13.4 LSQ analysis of eyeball size (mass or volume) as a function of adult body mass

N_{ord}	N_{sp}	MPD	pWR	DI (%)	r^2	Slope ± 99% CL	Sample
10	61	107	7	21	0.893	0.686 ± 0.082	FDS
8	48	47	5.9	13	0.919	0.647 ± 0.076	RDS [a]

See Table 13.1. [a] 13 records for which the deviation from the best-fit LSQ FDS line was greater than 100% were removed from the FDS, thereby generating a reduced dataset (RDS) (bottom row).

From the data of Howland *et al.* [12] for 96 species (one point per species) I find that eye diameter scales as the 0.184 ± 0.028 (99% CL) power of body mass in adult mammals. Assuming a spherical eyeball, this result implies that eye volume scales as the 0.552 power of body mass (3·0.184 = 0.552). The data for eye size are unusual, but not unique, in having a considerable number of species that systematically fall well below the best-fit LSQ line (not shown). Even acknowledging this scatter, the above data do not support the notion that eye size scales independently of body size (see above quote from Thompson and Box 13.1). The eye has often been cited as an object whose alleged perfection must result from "intelligent design." In fact, like other organs, the eye has defects that are an inevitable result of an opportunistic response by natural selection to problems, not of the unknowable long-term, but of the mere moment (see above quote from Helmholtz).

Spinal cord

...the spinal cord is organized in essentially the same way in all living vertebrates.
Jerison 1973

From the standpoint of motor control and sensory input, the nervous system exhibits two major divisions: somatic and autonomic. The somatic motor system relates to the voluntary (or skeletal) muscles and somatic sensation (chiefly due to stimulation from the skin). The autonomic portion of the nervous system is responsible for the motor innervation of blood vessels, glands, heart, and smooth muscle. Unlike the cerebrum, where most of the gray matter is on the surface (forming the cortex) and the white matter lies interiorly, the gray matter of the spinal cord is situated centrally, forming vertical columns, in an H-shaped configuration, with the white matter lying on the outside. The ventral arms of the H are motor, the dorsal arms sensory. Moreover, somatic motor nerves are more ventral and somatic sensory more dorsal than are the visceral motor and sensory nerves. Thus the organization of the spinal cord (and cerebellum) is more stereotyped than is the case for the cerebrum (see above quote from Jerison).

The cerebral control of spinal cord centers appears to be greater in primates (monkey, man) than in carnivores (dog). A cervical enlargement related to the forelimbs is present

Box 13.1 D'Arcy Thompson

D'Arcy Thompson combined within a single individual the capabilities of a classicist, a mathematician, and a naturalist [B1]. He was distinguished in each of these fields. His essay *On Growth and Form* [B2] is widely regarded as one of the finest scientific works in the English language. Chapter II, entitled "On Magnitude," still provides a useful introduction to bioscaling. Like Galileo (see Box 12.1) D'Arcy brought together the diverse issues of scaling and form (or pattern formation) between the covers of one remarkable opus.

Throughout his book [B2], D'Arcy Thompson makes the implicit assumption that growth and form in living organisms are physically caused. Only occasionally does he make this assumption explicit, as when he says: "Nor is it otherwise with the material form of living things... Their problems of form are in the first instance mathematical problems, their problems of growth are essentially physical problems..." (p. 10). Towards the end of his great work he is even more explicit about the implications of his underlying assumption, when he says: "It is hard indeed (to my mind) to see in such a case as this [of spiral forms in nature] where Natural Selection necessarily enters in, or to admit that it has had any share whatsoever in the production of these varied conformations" (p. 849).

Few would now concede that physics plays the direct role in the determination of biological form that D'Arcy Thompson seems to have envisaged. It is too much to ask of physics. Thompson failed to recognize the possibility that there might be only a limited, even if large, number of ways in which repeating units, such as cells, can be packed efficiently in 3-D space. But we should not dismiss his views out-of-hand. Akin to Thompson, my late friend and colleague Brian Goodwin, a well-known theoretical biologist, took strong exception to the widespread reductionist view that biology is dominated by genes [B3]. So far the gene-centered view of biology, by itself, has largely failed to account for the remarkable processes that occur during the development of multicellular organisms [B4]. Thus, as Brian argued at length, there may yet prove to be an important role for self-organization in biology that is not now widely appreciated. While D'Arcy Thompson was no doubt mistaken in his belief that there is a direct immediate role for physics in the production of biological form, there is a wider perspective, a non-reductionist one, that takes into account the possibly very special properties of self-organizing systems, where his underlying intuition may eventually prove to have been essentially correct.

B1. Medawar, P.B. (1958). Postscript: D'Arcy Thompson and Growth and Form. In: D'Arcy Thompson, R. *D'Arcy Wentworth Thompson: The Scholar-Naturalist (1860–1948)*. New York: Oxford University Press, pp. 219–233.

B2. Thompson, D.W. (1943). *On Growth and Form*. New York: Cambridge University Press.

B3. Goodwin, B. (1994). *How the Leopard Changed its Spots*. New York: Charles Scribner's Sons.

B4. Gilbert, S.F. and Epel, D. (2009). *Ecological Developmental Biology. Integrating Epigenetics, Medicine and Evolution*. Sunderland, MA: Sinauer Associates.

Table 13.5 Spinal cord mass as a function of adult body mass

N_{ord}	N_{sp}	MPD	pWR	DI (%)	r^2	Slope ± 99% CL	Source
4	21	15	3.7	3.2	0.984	0.681 ± 0.057	[14]
9	31	23.5	5.1	11.3	0.979	0.715 ± 0.053 [a]	From Lit.

See Table 13.1. [a] About two-thirds of the records in this dataset are from [14]; Lit. = literature.

Table 13.6 Scaling of various cerebral parameters as a function of brain mass or volume

Row	N_{ord}	N_{sp}	MPD	r^2	Slope ± 95% CL	Entity	Units	Source
1	7	15	14	0.62	0.08 ± 0.02	Cortical thickness	cm	[16]
2	4	15	16	0.98	0.74 ± 0.04	Total gyral length	cm	[16]
3	13	44	16	1.0	0.91 ± 0.02	Cortical SA [a]	cm^2	[16]
4	10	96	19	1.0	1.04 ± 0.02	Cortical volume	cm^3	[16]
5	4	18	24	0.76	−0.312 ± 0.093	Cortical ND [b]	neurons/mm^3	[18]

See Table 13.1. [a] SA = surface area; [b] ND = neuron density.

in all mammals, and an analogous enlargement in the lumbar region for the hind limbs is found in most mammals [13].

The inferences from LSQ analyses of spinal cord size in adult mammals are brought together in Table 13.5. Row one gives the outcome of a LSQ analysis of a *subset* of the data of MacLarnon [14]. The best-fit LSQ line has a slope of 0.681 ± 0.057 (99% CL). The present LSQ analysis of data taken from the literature, for 31 species, including 15 species of primates, is shown in Table 13.5, bottom row. The LSQ slope is 0.715 ± 0.053 (99% CL). The two slopes given in Table 13.5 are not significantly different. Human ontogenetic data (not shown) are found in Jackson [15].

Cerebral parameters

My friend and colleague John Sundsten and I carried out a study of cerebral scaling in adult mammals [16]. This study drew on work carried out in a small number of carefully selected laboratories. A natural aim in scaling studies, to have as large a sample size as possible, must be offset, especially in morphometric studies at the histological level, by the recognition that tissue shrinkage, particularly in older studies, may introduce large and unpredictable distortions. It is possible to correct for artefacts due to shrinkage [17], but this has seldom been done. The results in Table 13.6 bring together a number of our published findings as to how various brain parameters may scale with *brain size* (see also Prothero [18]). Unfortunately the original "raw" data for this study are no longer available; as a result pWR and DI are not calculable.

ENDOCRINE SYSTEM

Although often referred to as a "system," there is no continuity of structure, function,
or location of endocrine organs to justify such a reference in the strict sense.

Gorbman and Bern 1962

Consider now what is loosely termed the endocrine "system" [19]. The nervous and endocrine systems work on quite different time scales: the endocrine system in seconds or longer, the nervous system in milliseconds. The endocrine glands secrete their products into the blood. For a discussion of the scaling of the pancreas, only a small portion of which is endocrine, see Chapter 10 [20]. Discussion of the scaling of the gonads is deferred to the Reproductive System (Chapter 14). Here our attention is confined to the adrenal, parathyroid, pituitary, and thyroid glands. Although referred to as a system, the term is of limited cogency (see above quote from Gorbman and Bern).

Adrenal gland

The adrenal gland of mammals consists of two parts of entirely different origin,
function and structure: adrenal cortex and adrenal medulla.

Welsch and Storch 1976

Adrenal tissue is present in all vertebrates; only in mammals is the adrenal gland divided into two distinct components [21]. The first component, the *adrenal medulla*, secretes epinephrine and norepinephrine; in effect it functions as a sympathetic ganglion (which is involved in the fight-or-flight response). Epinephrine raises the blood sugar level and in moderate amounts increases metabolic rate. The adrenal medulla is not considered essential to life. The second component of the adrenal gland, namely the *adrenal cortex*, secretes glucocorticoids (steroids), aldosterone, and androgenic steroids, with diverse effects on carbohydrate and protein metabolism. Loss of glucocorticoids, in the absence of replacement therapy, is fatal. The two components of the adrenal gland function independently of one another (see above quote from Welsch).

The findings from LSQ analyses of adrenal size on body size are displayed in Table 13.7 and Figure 13.4 (panels a and c). The data used to construct the first row of Table 13.7 are from Crile and Quiring [22].

Brody [23] reported a slope of 0.798 ± 0.014 (s.e.) for the scaling of adrenal gland weight on body weight. From Table 13.7, row two, one obtains a mean slope of 0.801 ± 0.053 (99% CL). This result implies that on average adrenal size increases by 74% for a

Table 13.7 Scaling of adrenal gland mass as a function of body mass in adult mammals

N_{ord}	N_{sp}	MPD	pWR	DI (%)	r^2	Slope $\pm$ 99% CL	Source
13	75	49	6.5	25.9	0.928	0.802 ± 0.069	[22]
14	110	51.5	6.9	32.5	0.935	0.801 ± 0.053	From Lit.

See Table 13.1. Lit. = literature.

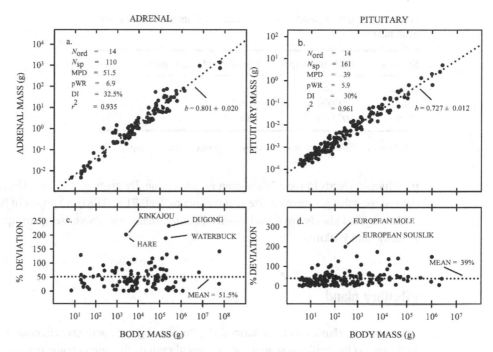

Figure 13.4 Scaling of adrenal and pituitary gland size in adult mammals.

doubling in adult body size $(100 \cdot (2^{0.801} - 1) = 74)$, or alternatively, the gland is 39 times smaller in a mature blue whale than it would be if it scaled in accord with direct proportionality to body size $(\Delta b = 1 - 0.801 = 0.199; N_d = 26.6 \cdot 0.199 = 5.3; f = 2^{5.3} = 39)$ (Chapter 3). In Figure 13.4c, two deviant species are the kinkajou (*Potus flavus*) and hare (*Lepus arcticus*) which have, coincidentally, nearly identical body sizes and percent deviations. The other two deviant species in panel c are the dugong (*Dugong dugon*) and the waterbuck (*Kobus ellipsiprymnus*). Rat ontogenetic data for the adrenal gland are available from Hatai [11].

Parathyroid gland

...parathyroids are the only example of an endocrine gland not found in all classes of vertebrates.

Andrew and Hickman 1974

In 1925 Collip [24] reported the discovery of an active substance secreted by the parathyroid glands, which he termed "parathormone." This hormone, among other functions, regulates calcium (and phosphate) levels in the blood. Decreased levels of blood calcium lead to increased secretion of parathormone into the blood, which in turn indirectly promotes calcium resorption from bone. Commonly, there are one or two pairs of parathyroid glands, usually embedded in the dorsal surface of the thyroid gland. Loss of the parathyroid glands leads to muscle spasm and death. Few measurements of parathyroid size and body size were found; these are shown in Table 13.8. The size of the

Table 13.8 Scaling of parathyroid gland mass as a function of body mass in adult mammals

Species	Body mass (g)	Parathyroid mass (mg)	Source
Rabbit	2,240	12.6	[25]
Human	51,000	166 [a]	[26]
Cow	607,000	645	[27]

[a] This datum is derived from post-mortem psychotic patients.

parathyroid glands in the adult human is conventionally given as 140 mg. These meager data suggest that parathyroid size scales about as the 0.71 ± 0.08 (s.e.) power of body size.

Parathyroid glands are found chiefly in amphibians and above (see prior quote from Andrew and Hickman).

Pituitary gland

The pituitary gland was once termed the "master" gland, perhaps reflecting the perennial (and so far fruitless) search for biological entities that are in some overall sense "in charge". Nevertheless, the pituitary gland has such an impressive range of important functions that it warrants assignment as the most important of the endocrine glands. The pituitary gland secretes two classes of hormones: those which directly affect "target" tissues, such as cartilage and the growth of the epiphysial region of bone; and trophic hormones, which influence secretion by other endocrine glands.

The pituitary is usually divided into anterior, intermediate, and posterior portions. The intermediate portion is reduced or absent in some species, including the adult human. The anterior lobe secretes six hormones (known by the acronyms TSH, ACTH, GH, FSH, LH, and PRL (formerly LTH)), which influence the thyroid and adrenal glands, body growth, ovarian follicular growth, ovulation, and mammary glands, respectively. These hormones are thought to be derived from a common ancestor. The posterior lobe secretes ADH, which fosters water retention by the kidneys, and oxytocin, which facilitates milk release during lactation. In the adult human the pituitary gland weighs about half a gram. Brody [23] reported a slope of 0.762 ± 0.018 (s.e.); Stahl [28] gave a slope of 0.56 ± 0.02 (s.e.) for pituitary size on body size in mammals.

The present analysis of the scaling of pituitary gland size in terms of adult body size, based on a subset of the data of Bauchot and Legait [29], is summarized in Table 13.9 and Figure 13.4, panels b and d.

The slope of 0.727 ± 0.030 (99% CL) (if applicable to mammals generally) implies that pituitary size in a blue whale will be smaller by 7.3 doublings ($\Delta b = 1 - 0.727 = 0.273$; $N_d = 26.6 \cdot 0.273 = 7.3$), or by a factor ($f$) of 158 ($f = 2^{7.3} = 158$), than for direct proportionality ($b = 1.0$). The two deviant species in Figure 13.4d are the European mole (*Talpa europaea*) and the European souslik (*Spermophilus citellus*).

Table 13.9 Scaling of pituitary mass as a function of adult body mass

N_{ord}	N_{sp}	MPD	pWR	DI (%)	r^2	Slope $\pm$ 99% CL	Source
14	161	39	5.9	30	0.961	0.727 $\pm$ 0.030	[29]

See Table 13.1.

Table 13.10 Scaling of thyroid mass as a function of adult body mass

N_{ord}	N_{sp}	MPD	pWR	DI (%)	r^2	Slope $\pm$ 99% CL	Source
12	75	55	6.6	24.3	0.947	0.961 $\pm$ 0.070 [a]	[22]
14	103	59	6.6	30.3	0.941	0.945 $\pm$ 0.062	From Lit.

See Table 13.1. Lit. = literature; [a] data for peccary omitted (specimen reported to have goiter).

Thyroid gland

Thyroid tissue is found in all vertebrates. In adult mammals the thyroid gland secretes the hormones thyroxine (T_4) and triiodothyronine (T_3) which play an important role in the regulation of cellular oxygen consumption, as well as in lipid and carbohydrate metabolism. The thyroid gland also plays an important role in growth. The parafollicular cells of the thyroid secrete calcitonin, a regulator of calcium levels in the body. The thyroid is unique among endocrine glands in storing its hormones extracellularly, in follicles. In a normal adult human the thyroid gland weighs about 30 g. Blood flow through the thyroid, per gram tissue, is higher than in either brain or kidney. The most active form of thyroid hormone (T_3) is covalently bound to three atoms of iodine. Iodine, a rare element, is present in the earth's crust in about 0.5 parts per million and in the human body in about 0.4 parts per million. The thyroid gland concentrates iodine present in arterial blood by a factor of several hundred. Brody [23] reported that the size of the thyroid gland scales on body size with a slope of 0.924 $\pm$ 0.017 (s.e.). The outcomes of an analysis of the thyroid data of Crile and Quiring [22] and from the literature generally are synopsized in Table 13.10 and Figure 13.5.

Note that the mean slope of 0.945 (see Table 13.10, row two) corresponds to an increase of about 92.5% in thyroid size for a doubling in body size ($100 \cdot (2^{0.945} - 1) = 92.5$), or 1.5 fewer doublings in a blue whale (assumed pWR = 8.0) than would be the case if the thyroid gland scaled with a slope of one ($26.6 \cdot (1 - 0.945) = 1.5$). The human ontogenetic data in Figure 13.5b are from T. Shepard (personal communication). These data lie almost entirely above the adult interspecific line. The rat ontogenetic data for the thyroid (not shown) also approach the adult interspecific line from above (see Hatai [11]). Three deviant species identified in Figure 13.5, panel c, are the arctic hare (*Lepus arcticus*), the red fox (*Vulpes vulpes*), and the Weddell seal (*Leptonychotes weddelli*).

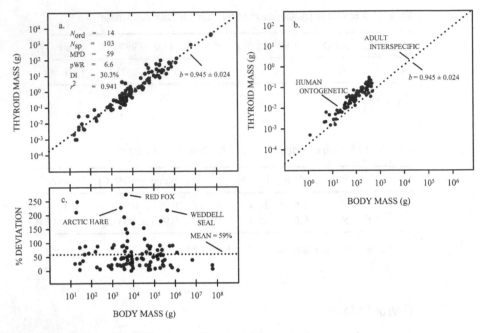

Figure 13.5 Scaling of thyroid size as a function of body size in mammals.

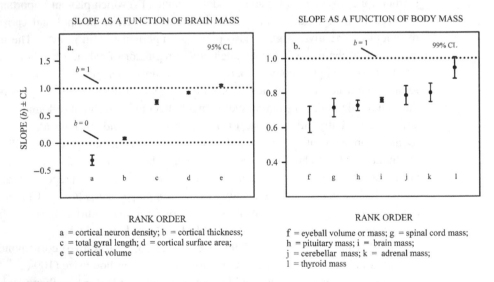

SLOPE AS A FUNCTION OF BRAIN MASS

SLOPE AS A FUNCTION OF BODY MASS

a = cortical neuron density; b = cortical thickness;
c = total gyral length; d = cortical surface area;
e = cortical volume

f = eyeball volume or mass; g = spinal cord mass;
h = pituitary mass; i = brain mass;
j = cerebellar mass; k = adrenal mass;
l = thyroid mass

Figure 13.6 Slope summary for the neuroendocrine system.

Summary

The major slopes discussed above are displayed in Figure 13.6. Note that the results are expressed as a function of either *brain volume* or mass (panel a) or *body mass* (panel b). The CL for panel a are 95% (original data no longer available) and 99% for panel b. The slopes in panel a for cortical thickness (Table 13.6, row one, $b = 0.08 \pm 0.02$), for cortical surface area (Table 13.6, row three, $b = 0.91 \pm 0.02$), and for cortical volume (Table 13.6, row four, $b = 1.04 \pm 0.02$) are approximately in agreement with slopes of 0.11, 0.89, and 1.00, respectively, as predicted by a repeating units model (see Chapters 3, 24).

It is notable in panel b that only one of the seven organ slopes (that for thyroid) is even close to 1. In panel b, with the possible exception of the thyroid, none of the slopes agree with the reference slopes ($b = 0.0$ or 1.0) for invariance. The deviation from the reference slope of 1 is greater for the brain and cerebellum than for any other major organs (see also Table 13.4). Figure 13.6b resembles Figure 8.4, except that here the dispersion in slopes is much wider. Generally speaking, each system has, with a few exceptions, and to a first approximation, its own distinctive pattern of slopes.

References

1. Murre, J.M.J. and Sturdy, D.P.F. (1995). The connectivity of the brain: multi-level quantitative analysis. *Biological Cybernetics*, **73**:529–545.
2. Snell, O. (1891). Das Gewicht des Gehirnes und des Hirnmantels der Säugethiere in Beziehung zu deren geistigen Fähigkeiten. *Sitzungsberichte der Gesellschaft für Morphologie und Physiologie*, **7**:90–94. (Even at this early stage the author was interested in the possible relation between brain size and intellectual abilities.)
3. Jerison, H.J. (1973). *Evolution of the Brain and Intelligence*. New York: Academic Press.
4. Martin, R.D. (1981). Relative brain size and metabolic rate in terrestrial vertebrates. *Nature*, **293**:57–60.
5. Cobb, S. (1965). Brain size. *Archives Neurology*, **12**:555–561.
6. Tanner, J.M. (1990). *Foetus into Man: Physical Growth from Conception to Maturity*. Cambridge, MA: Harvard University Press.
7. Pilleri, G. and Gihr, M. (1970). The central nervous system of the Mysticete and Odontocete whales. *Investigations Cetacea*, **2**:89–127.
8. Robertson-Bullock, W. (1962). The weight of the African Elephant, *Loxodonta africana*. *Proceedings of the Zoological Society London*, **138**:133–135.
9. Romer, A.S. (1970). *The Vertebrate Body*, 4th edn. Philadelphia, PA: W.B. Saunders.
10. Howard, E., Granoff, D.M. and Bujnovszky, P. (1969). DNA, RNA, and cholesterol increases in cerebrum and cerebellum during development of human fetus. *Brain Research*, **14**:697–706.
11. Hatai, S. (1913). On the weights of the abdominal and the thoracic viscera, the sex glands, ductless glands and the eyeballs of the Albino rat (*Mus norvegicus albinus*) according to body weight. *American Journal of Anatomy*, **15**:87–119.
12. Howland, H.C., Merola, S. and Basarab, J.R. (2004). The allometry and scaling of the size of vertebrate eyes. *Vision Research*, **44**:2043–2065.
13. Andrew, W. and Hickman, C.P. (1974). *Histology of the Vertebrates. A Comparative Text*. Saint Louis, MO: C.V. Mosby.
14. MacLarnon, A. (1996). The scaling of gross dimensions of the spinal cord in primates and other species. *Journal of Human Evolution*, **30**:71–87.
15. Jackson, C.M. (1909). On the prenatal growth of the human body and the relative growth of the various organs and parts. *American Journal of Anatomy*, **9**:119–165.

16. Prothero, J.W. and Sundsten, J.W. (1984). Folding of the cerebral cortex in mammals: a scaling model. *Brain, Behavior and Evolution*, **24**:152–167.

17. McLean, M. and Prothero, J.W. (1991). Three-dimensional reconstruction from serial sections. V. Calibration of dimensional changes during tissue preparations and data processing. *Analytical Quantitative Cytology and Histology*, **13**:69–278.

18. Prothero, J. (1997). Scaling of cortical neuron density and white matter volume in mammals. *Journal of Brain Research*, **38**:513–524.

19. Gorbman, A. and Bern, H.A. (1962). *A Textbook of Comparative Endocrinology*. New York: John Wiley & Sons.

20. Parsons, J.A., Bartke, A. and Sorenson, R.L. (1995). Number and size of Islets of Langerhans in pregnant, human growth hormone-expressing transgenic, and pituitary dwarf mice: effect of lactogenic hormones. *Endocrinology*, **136**:2013–2021. (In mice these workers found that the Islets comprise less than 0.5% of pancreatic weight.)

21. Turner, C.D. (1966). *General Endocrinology*. Philadelphia, PA: W.B. Saunders Co.

22. Crile, G. and Quiring, D.P. (1940). A record of the body weight and certain organ and gland weights of 3,690 animals. *Ohio Journal of Science*, **40**:219–259.

23. Brody, S. (1945). *Bioenergetics and Growth*. New York: Reinhold.

24. Collip, J.B. (1925). The extraction of a parathyroid hormone. *Journal of Biological Chemistry*, **63**:395–438.

25. Brown, W.H., Pearce, L. and Van Allen, C.M. (1926). Organ weights of normal rabbits. *Journal of Experimental Biology*, **43**:733–741.

26. Freeman, W. (1934). The weight of the endocrine glands: biometrical studies in psychiatry. *Human Biology*, **6**:489–523.

27. Stott, G.H. and Smith, V.R. (1964). Histology, cytology, and size of the parathyroid in bovine related to age and function. *Journal of Dairy Science*, **47**:426–432.

28. Stahl, W.R. (1965). Organ weights in primates and other mammals. *Science*, **150**:1039–1041.

29. Bauchot, R. and Legait, H. (1978). Le volume de l'hypohyse et des lobes hypophysaires chez mammifères: corrélations et allométries. *Mammalia*, **42**:178–253.

14 Reproductive system

The biological problem of the perpetuation of the species has been solved in a
bewildering variety of ways in nature.

Nalbandov 1976

Reproduction is the intricate process by which new individuals arise as a result of the
production by adult males and females of (haploid) gametes (sperm and ova) which fuse
during fertilization, giving rise to a diploid zygote. Subsequent divisions of the zygote,
through an astonishingly complex process of development, give rise to a new adult,
similar, yet not identical, to other adults of the same species. It is reproduction, usually
with slight variations, that makes natural selection possible. Each new adult consists of
germ cells (gametes), which in a sense are potentially immortal, and somatic cells,
which die when the proprietor dies. It is estimated that the adult human consists of some
10^{14} cells, implying that the human zygote undergoes at least 46 doublings ($2^{46} \approx 10^{14}$).
It is sobering to reflect that each of us is the end product of an extraordinarily robust
scale-up process (see above quote by Nalbandov).

This chapter contains reviews for eight y-variables expressed in 20 instances, for a
total of 3,566 records, all in tabular form. Summary plots of the main slopes derived in
this chapter are provided in Figure 14.3. For a scaling review of growth and reproduction, partly in mammals, see Reiss [1].

Gestation period

In the monotremes there is no gestation as such...

Eisenberg 1981

Marsupials are characterized by very short gestations and long lactations.

Loudon 1987

The long gestation period of placental mammals is possible due to the evolution of
the... trophoblast, which is responsible for preventing rejection of the embryo by the
maternal immune system.

Carroll 1988

The details of reproduction vary significantly across the three major groups of mammals:
monotremes, marsupials, and eutherians. In monotremes, a fertilized egg is hatched either

Table 14.1 Selected prior LSQ analyses of gestation period as a function of adult body mass

Row	N_{ord}	N_{sp}	MPD	pWR	DI (%)	r^2	Slope ± 99% CL	Year	Source
1	9	42	46	5.4	13	0.723	0.259 ± 0.069	1931	[3]
2	13	75	38	5.1	20.2	0.748	0.267 ± 0.048	1974	[4]
3	6	43	53	4.2	6.75	0.386	0.178 ± 0.095	1979	[5]
4	17	451	-	6.3	46.5	0.692	0.241 [a]	1980	[6]
5	11	197	41	6.2	25.5	0.725	0.243 ± 0.028	1981	[7]
6	16	287	56.4	7.65	49.2	0.591	0.209 ± 0.027	1981	[8]

N_{ord} = number of orders; N_{sp} = number of species; MPD = mean percent deviation; pWR = log (BW_{max}/BW_{min}); DI = diversity index; r^2 = coefficient of determination; CL = confidence limit(s). [a] Cetaceans not included in sample, no "raw" data provided.

in a nest (platypus) or in a pouch (echidna) (see above quote from Eisenberg). The young of marsupials leave the uterus after two to three weeks and, while still very immature (late embryonic to early fetal), make their way to the pouch where they suckle (see quote from Loudon) for roughly six months [2]. During this period the mother may conceive; however, implantation is then delayed. In eutherian mammals, the mother develops a placenta (from a precursor trophoblast) which allows material exchange between the mother and developing embryo/fetus. The existence of a placenta permits a much longer period of gestation than in marsupials (see quote from Carroll). Many workers have taken up the LSQ analysis of gestation time. A synopsis of some of this earlier work is given in Table 14.1.

The mean of the six slopes listed in Table 14.1 is 0.233 ± 0.031 (STD). Omitting the most deviant slope (0.178, row three), the mean slope is 0.244 ± 0.020 (STD). None of the six slopes differs significantly from this mean.

The results of the present LSQ analysis of gestation period are presented in Table 14.2 and Figure 14.1 (panels a and c); data for Table 14.2 (row one) are from Eisenberg [8] (287 records), Doyle [9] (20), and Gittleman [10] (38). Row two shows the FDS (row one) less marsupials and the third row the FDS less cetaceans. Where possible maternal body weights were used; but most body weights are species averages. The most representative slope (0.204 ± 0.026) (99% CL) is that in row one (Table 14.2). This slope departs significantly from a hypothetical reference slope of 1/4 (Chapter 3, Table 3.5).

Observe that five species of marsupials identified in panel c of Figure 14.1 deviate widely in gestation period from the MPD (56%) for the FDS (see Table 14.2, row one). In order of decreasing deviation these are: quoll (*Dasyurus viverrinus*), bandicoot (*Perameles gunni*), southern opossum (*Didelphis marsupialis*), brindled bandicoot (*Isoodon macrourus*), and brush-tailed possum (*Trichosurus vulpecula*). Still, when either marsupials or cetaceans are omitted the effect on the slope is not significant (see Table 14.2, rows two, three).

The longest known gestation periods (up to 22 months) are associated with elephants, whereas the gestation period of the much larger blue whale is 11 months [11]. The gestation period in humans is about 280 days, a fact known to Hippocrates (*c*. 460–377 BC). For an extensive study of gestation period that takes into account marked differences among altricial (largely helpless at birth) and precocial (largely independent at birth) species see Martin and MacLarnon [12].

Table 14.2 Outcomes of the present LSQ analysis of gestation period versus adult body mass

N_{ord}	N_{sp}	MPD	pWR	DI (%)	r^2	Slope $\pm$ 99% CL	Sample
16	345	56	7.65	50.8	0.549	0.204 ± 0.026	FDS [a]
15	333	50.4	7.65	47.4	0.587	0.202 ± 0.024	Omit marsupials
15	338	55.9	6.1	38.1	0.527	0.209 ± 0.028	Omit cetaceans

See Table 14.1. [a] The mean deviation from the best-fit LSQ line for the 12 species of marsupials represented in this sample is 281%.

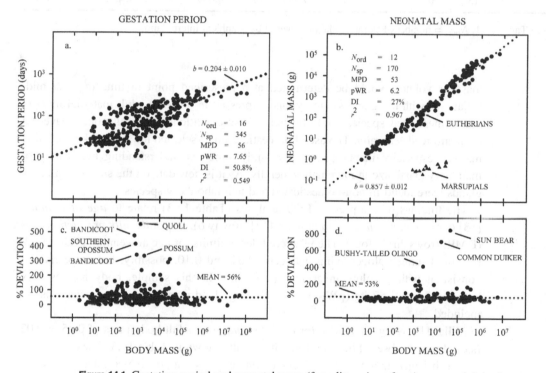

Figure 14.1 Gestation period and neonatal mass (for a litter size of one) versus adult body mass.

Lactation

The universal and almost unique feature of mammalian reproduction is the production of secretions by the mother which nourish the offspring after birth... Lactation provides the young with a buffer against the hazards of fluctuations in quantity, quality and toxicity of the normal food supply... and has allowed for a much higher degree of specialization in feeding mechanisms and food sources of the adults.

Pond 1977

As remarked in Chapter 2, lactation has profound implications for social behavior in mammals (see also above quote from Pond). In many species, mammary glands undergo a cycle of growth and regression coinciding with periods of pregnancy and lactation. The most obvious exception is the human female. Ideally mammary gland

Table 14.3 LSQ analyses of age at weaning, mammary gland mass, and daily milk production (DMP) as a function of adult body mass

N_{ord}	N_{sp}	N_{pnts}	MPD	pWR	DI (%)	r^2	Slope ± CL	CL	Entity	Source
6	71	71	30	3.9	7.1	0.02	0.0315 ± 0.071	99%	Age at weaning	[13]
6	10	19	-	~4	~3.9	0.98	0.87 ± 0.03	s.e.	Mammary gland mass	[14]
4	11	14	-	~4	~2.7	0.98	0.73 ± 0.03	s.e.	DMP	[14]
6	13	13	20	4.3	4.7	0.989	0.777 ± 0.077	99%	DMP	[14] [a]

See Table 14.1. N_{pnts} = number of points; [a] data recovered by graphical interpolation.

mass or volume would be determined at a prescribed point in time (e.g., at mid-lactation), with the gland(s) empty, and expressed in terms of adult maternal mass in non-domesticated species. Unfortunately, these conditions, or similar ones, have not been enforced generally. The use of domestic animals selected for high milk production may be especially inappropriate in a study aimed at understanding the scaling of mammary gland size in mammals generally. As it is, few data on the size of mammary glands were found in non-domesticated and non-laboratory species.

The findings of the present LSQ analyses (Table 14.3) relate to *age at weaning* [13] (row one), *mammary gland mass* [14] (row two), and *daily milk production* [14] (DMP) (rows three, four). The 99% confidence limits for age at weaning (Table 14.3, row one) furnish slopes lying between –0.04 and 0.10, possibly, but not certainly, consistent with absolute invariance ($b = 0.0$). In this sample, body masses vary between 7 g and 57 kg (pWR = 3.9). No large land mammals and no cetaceans are included.

Millar [13] reported a slope for *age at weaning* on adult body mass of 0.05 ± 0.02 (s.e.). From a subset of his data I find the results shown in Table 14.3, row one. It may be that if larger mammals were included one might see a positive correlation between age at weaning and adult body mass. Hanwell and Peaker [15] reported slopes of 0.82 ± 0.026 (s.e.) and 0.77 ± 0.022 (s.e.) for the scaling of *mammary gland mass* and daily *milk production*, respectively. The results shown in Table 14.3, rows two and three, for mammary gland mass and daily milk production are taken directly from Linzell [14]. The results shown in Table 14.3, row four, for daily milk production are derived from the data of Linzell [14] by graphical interpolation. For reviews of the evolution of the mammary gland see Asdell [2] and Blackburn [16].

Neonatal size

Relative to other mammals, all bats are enormous at birth. . . They weigh approximately 15–30% of the mass of the postpartum mother.

Tuttle and Stevenson 1982

Table 14.4 Neonatal mass as a function of adult body mass for a litter size of one [a]

N_{ord}	N_{sp}	MPD	pWR	DI (%)	r^2	Slope $\pm$ 99% CL	Sample	Source
7	20	32.4	5	7.4	0.988	0.849 $\pm$ 0.064	-	[17]
12	170	53	6.2	27	0.967	0.857 $\pm$ 0.032	FDS	From Lit.

See Table 14.1. Lit. = literature; [a] monotremes and marsupials not included in either of the LSQ calculations.

Table 14.5 Total neonatal litter mass as a function of adult body mass for varying litter sizes (greater than or equal to one)

N_{ord}	N_{sp}	MPD	pWR	DI (%)	r^2	Slope $\pm$ 99% CL	Sample	Year	Source
9	50	57.6	5	12.5	0.954	0.783 $\pm$ 0.067	-	1959	[17]
15	83	68	5.5	26	0.926	0.796 $\pm$ 0.066	-	1974	[4]
16	416	61	6.5	44.6	0.952	0.800 $\pm$ 0.023	FDS	-	From Lit.
15	348	43.5	6.5	40.3	0.968	0.824 $\pm$ 0.021	FDS less carnivores	-	From Lit.

See Table 14.1. Lit. = literature.

The scaling of birth weight or mass in an individual of any given species may vary with litter size. It therefore makes sense to look first at birth weight as a function of adult body weight where the litter size is one. Ideally these findings would be expressed as a function of maternal body weight. However, in most cases we are provided only with general values of body weight; these may often be larger than mean maternal body weight, owing to sexual dimorphism (Chapter 7).

The inferences from LSQ analyses of neonatal size (for a litter size of one) as a function of body size are depicted in Table 14.4. The outcomes of a prior study by Leitch *et al.* [17] together with those of the present analysis appear in rows one and two, respectively and in Figure 14.1, panels b and d. Fifty percent of the eutherian data employed in the analysis exhibited in row two are from Eisenberg [8].

The *marsupial* data shown in Figure 14.1, lower portion of panel b, are from Tyndale-Biscoe and Renfree [18]. Note that data shown in Figure 14.1d are restricted to eutherian mammals. The three very deviant species shown in panel d are the sun bear (*Helarctos malayanus*), the common duiker (*Sylvicapra grimmia*), and the bushy-tailed olingo (*Bassaricyon gabbi*). As is clear from Figure 14.1b, marsupials (in the given sample) have quite different and much lower birth weights than do eutherian mammals (i.e., the data are disjoint). For this reason, a single best-fit line is unlikely to provide an adequate account of the scaling of birth weight in mammals generally. More exactly, 170 species in the FDS (Table 14.5, row three) had a litter size of 1; 111 species had mean litter sizes of 1.1 to less than 3; 71 species had 3 to 4; and 64 species had mean litter sizes of 4.2 to 16.5.

The findings of prior (rows one and two) and present (rows three, four) LSQ analyses of neonatal mass as a function of adult body mass for varying litter sizes are given in Table 14.5. Data for the present study are from the literature.

Note that in Table 14.5 the slopes for the first two rows and for the last row do not differ significantly from the slope for the FDS (row three). In Table 14.5 (row three) the MPD for carnivores is 164%. When carnivores are omitted from the FDS (Table 14.5, row four) the MPD drops from 61 to 43.5%. The range in *neonatal* mass in eutherians in the present sample (Table 14.5, row three) is from 0.215 g (pygmy shrew, *Suncus etruscus*) to 120 kg (African elephant, *Loxodonta africana*). Possibly the smallest mammalian neonatal size is that for a marsupial, the Julia Creek dunnart (*Sminthopsis douglasi*), which has a birth weight of 17 mg [19]. On the other hand, bats have unusually large neonates when expressed in terms of neonatal weight (see above quote by Tuttle and Stevenson).

Testes size

Spermatogenesis appears to have been present in its essentials at the very beginning of metazoan evolution.

                                                                    Roosen-Runge 1977

Production of sperm in animals is of great antiquity (see above quote from Roosen-Runge). The testes in mammals are both reproductive (producing sperm), and endocrine, producing hormones such as testosterone. These two disparate functions appear to be mutually independent [20]. The conclusions of a prior LSQ study by Kenagy and Trombulak [21] of the scaling of testes weight, along with the present results, are shown in Table 14.6 and Figure 14.2.

The data shown in Figure 14.2, panel a, correspond to Table 14.6, row two. The rat and human ontogenetic data (Figure 14.2b) are from Hatai [22] and Krogman [23], respectively; both rat and human ontogenetic datasets start out below the adult interspecific line. The largest reported [24] testes are from a Right whale (*Eubalaena glacialis*); they weighed about 1,000 kg (no body size data given). This value of testes weight is about 65 times larger than predicted (i.e., 15.3 kg) for an animal of roughly this size (1,000/15.3 = 65). This estimate of testes mass was not included in the LSQ computations. The very divergent species in Figure 14.2c are the dolphin (*Pontoporia blainvillei*), the gorilla (*Gorilla gorilla*), and three species of hopping mice (*Notomys alexis, N. fuscus, N. mitchelli*).

Table 14.6. Scaling of testes mass as a function of adult body mass

N_{ord}	N_{sp}	MPD	pWR	DI (%)	r^2	Slope $\pm$ 99% CL	Sample	Source
12	130	103	7	29	0.859	0.718 $\pm$ 0.067	-	[21]
13	133	96	7.3	33	0.900	0.750 $\pm$ 0.057	FDS	From Lit.

See Table 14.1. Lit. = literature; of the 58 reference sources used in this study, 29 were cited by Kenagy and Trombulak [21].

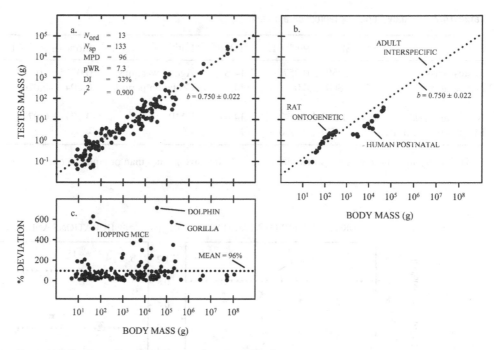

Figure 14.2 Scaling of testes mass as a function of body mass.

Egg size

The limited evidence found suggests that *egg diameter* in eutherian mammals is independent of body size ($b = 0.0$ or absolute invariance) and averages about 100 μm [25]. Marsupial eggs (in a few species) have somewhat larger diameters [18].

Summary

It was observed that reproduction plays a crucial role in evolution and natural selection. The main findings reported above are restated in Table 14.7 and in Figure 14.3. Note that panel a in Figure 14.3 refers to *structural* variables and panel b to *functional* variables. None of the *structural* slopes (panel a), with the possible exception of that for mammary gland mass, are consistent with a reference slope of 1 (relative invariance). In Figure 14.3b, age at weaning (item e) may (or may not) be consistent with absolute invariance.

Note that the dispersion (0.03 to 0.777) in *functional* slopes (Figure 14.3b) is much wider than that for *structural* slopes (panel a) (0.75 to 0.87). The tentative slope (0.0315 ± 0.071) (99% CL) for age at weaning (pWR = 3.9, panel b) is roughly consistent with absolute invariance ($b = 0$). Perhaps in no other mammalian organ system are phylogenetic differences as pronounced as in the reproductive system. Among the various systems discussed in Part II, the reproductive system offers the least scope for simple

Table 14.7 Summary slopes for reproductive system

Entity	Slope ± 99% CL	Table	Entity	Slope ± 99% CL	Table
Testes mass	0.750 ± 0.057	14.6	Age at weaning	0.0315 ± 0.071	14.3
Neonatal mass [a]	0.800 ± 0.023	14.5	Gestation period	0.204 ± 0.026	14.3
Neonatal mass [b]	0.857 ± 0.032	14.4	DMP	0.777 ± 0.077	14.3
Mammary gland mass	0.870 ± 0.090	14.3	-	-	-

DMP = daily milk production. [a] Litter size of one; [b] litter size greater than or equal to one.

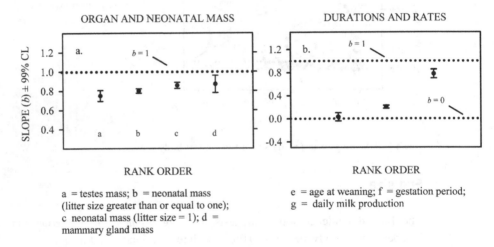

a = testes mass; b = neonatal mass
(litter size greater than or equal to one);
c neonatal mass (litter size = 1); d =
mammary gland mass

e = age at weaning; f = gestation period;
g = daily milk production

Figure 14.3 Summary slopes for reproductive system.

generalizations applicable to all mammals (see opening quote by Nalbandov). Thus, studies restricted to *eutherians* may provide the best results for a unified treatment of scaling in the reproductive system of selected mammals.

References

1. Reiss, M.J. (1989). *The Allometry of Growth and Reproduction*. Cambridge: Cambridge University Press.
2. Asdell, S.A. (1966). Evolutionary trends in physiology of reproduction. *Symposium of the Zoological Society London*, **15**:1–13.
3. Needham, J. (1931). *Chemical Embryology*. New York: MacMillan.
4. Sacher, G.A. and Staffeldt, E.F. (1974). Relation of gestation time to brain weight for placental mammals: implications for the theory of vertebrate growth. *American Naturalist*, **108**:593–615.
5. Western, D. (1979). Size, life history and ecology in mammals. *African Journal of Ecology*, **17**:185–204.

6. Zeveloff, S.I. and Boyce, M.S. (1980). Parental investment and mating systems in mammals. *Evolution*, **34**:973–982.

7. Millar, J.S. (1981). Pre-partum reproductive characteristics of eutherian mammals. *Evolution*, **35**:1149–1163.

8. Eisenberg, J.F. (1981). *The Mammalian Radiations*. Chicago, IL: University of Chicago Press.

9. Doyle, G.A. (1979). Development of behavior in Prosimians with special reference to the Lesser Bushbaby, Galago senegalensis moholi. In: Doyle, G.A. and Martin, R.D. (eds.) *The Study of Prosimian Behavior*. New York: Academic Press, pp. 157–206.

10. Gittleman, J.L. (1986). Carnivore brain size, behavioral ecology and phylogeny. *Journal of Mammalogy*, **67**:23–36.

11. Vaughan, T.A. (1978). *Mammalogy*. Philadelphia, PA: W.B. Saunders.

12. Martin, R.D. and MacLarnon, A.M. (1988). Comparative quantitative studies of growth and reproduction. *Symposium of the Zoological Society London*, **60**:39–80.

13. Millar, J.S. (1977). Adaptive features of mammalian reproduction. *Evolution*, **31**:370–386.

14. Linzell, J.L. (1972). Milk yield, energy loss in milk, and mammary gland weight in different species. *Dairy Science Abstract*, **34**:351–360.

15. Hanwell, A. and Peaker, M. (1977). Physiological effects of lactation on the mother. *Symposium of the Zoological Society London*, **41**:297–312.

16. Blackburn, D.G. (1991). Evolutionary origins of the mammary gland. *Mammal Review*, **21**:81–96.

17. Leitch, I., Hytten, F.E. and Billewicz, W.Z. (1959). The maternal and neonatal weights of some mammalia. *Proceedings of the Zoological Society London*, **133**:11–28.

18. Tyndale-Biscoe, H. and Renfree, M. (1987). *Reproductive Physiology of Marsupials*. Cambridge: Cambridge University Press.

19. Mortola, J.P., Frappell, P.B. and Wolley, P.A. (1999). Breathing through skin in a newborn mammal. *Nature*, **397**:660.

20. Goss, R.J. (1964). *Adaptive Growth*. New York: Academic Press.

21. Kenagy, G.J. and Trombulak, S.C. (1986). Size and function of mammalian testes in relation to body size. *Journal of Mammalogy*, **67**:1–22.

22. Hatai, S. (1913). On the weights of the abdominal and the thoracic viscera, the sex glands, ductless glands and the eyeballs of the Albino rat (*Mus Norvegicus albinus*) according to body weight. *American Journal of Anatomy*, **15**:87–119.

23. Krogman, W.M. (1941). Growth of Man. *Tabulae Biologicae*, **20**:1–963.

24. Payne, R. (1994). Among whales. *Natural History*, **103**:40–47.

25. Brody, S. (1945). *Bioenergetics and Growth*. New York: Reinhold.

15 Respiratory system

Comparative anatomy is largely the story of the struggle to increase surface in proportion to volume.

Haldane 1940

A lung is a within-the-body chamber connected to the body surface by a single channel... Schematically the lungs of fish, amphibians, reptiles and mammals may be viewed as hollow sacs whose walls are richly vascularized.

Dejours 1981

Introduction

Galileo (1564–1642) was impressed with the possible implications of physical scaling arguments for understanding how the skeletal system may scale with body size in adult mammals. (Chapter 12, Box 12.1). Although his remarks are still widely cited, it has turned out that he was mostly mistaken in the predictions he made. If he could have focused his attention on the respiratory system, he might have fared better. As it was, the microscopic structure of the lungs was first described by Malpighi (1628–1694) in 1661, almost 20 years after Galileo's death. The discovery of oxygen in 1774 is attributed to Priestley (1733–1804). The physical concept of energy did not arise until the nineteenth century. It follows that the necessary empirical facts and theoretical understanding that could have led to a similar argument to that of skeletal scaling for the respiratory system did not exist in Galileo's time. Only in the twentieth century did the needed facts and understanding emerge.

Much of the present analysis of scaling in the mammalian respiratory system is based on the work of Charles Taylor and Ewald Weibel and their colleagues, who carried out the most detailed and extensive scaling study yet reported in any mammalian organ system. The initial findings of this study occupied all of volume 44(#1) of the journal *Respiration Physiology* in 1981. This sweeping work, linking structure and function to variation in adult body size, is likely to remain an exemplary model of scaling studies in mammals for many years to come. The present study (see below) involves 12 respiratory variables, 14 instances, and 355 records in tabular form.

Alveolar diameter

In a prior study it was concluded that alveolar diameter in eight species of mammals (mouse to man) scales as the 0.148 ± 0.038 (s.e.) power of lung weight [1]. Thus *small* mammals are able to accommodate a proportionately larger number of alveoli per unit lung volume than are large mammals.

Alveolar number

The number of alveoli in the adult mouse is estimated at 2.3 million [2]. Hislop *et al.* [3] give a figure of 27 million for the number of alveoli in an adult macaque (*Macaca fascicularis*). According to a recent estimate [4] there are some 480 million alveoli in the two lungs of the adult human. The human infant is born with about 150 million alveoli [5]. Data for many more species are needed.

Alveolar surface area

The primary structural adjustment to enlarged body size in the lungs is an increase in the total alveolar surface area (see above quote from Haldane). This increase is accomplished by expanding lung volume and hence the number of alveoli. The inferences from a LSQ analysis of several pulmonary variables are given in Table 15.1, based on an analysis of a subset of the data of Gehr *et al.* [6]. Note that the slopes and 95% CL given in Table 15.1, column h, are taken directly from Gehr *et al.* [6]; they are based on a larger dataset than the one analyzed here (i.e., columns a to g) (see also Appendix B).

Table 15.1 Several pulmonary morphometric variables as a function of adult body mass

N_{ord}	N_{sp}	MPD	pWR	DI (%)	r^2	Slope $\pm$ 99% CL	Slope $\pm$ 95% CL	Entity
a	b	c	d	e	f	g	h [a]	i
6	20	27	5.4	6.9	0.991	0.936 ± 0.061	0.949 ± 0.0315	Alveolar SA (cm^2) [b]
6	20	12	5.4	6.9	0.638	0.056 ± 0.028	0.050 ± 0.0135	Barrier thickness (μm)
6	20	23	5.4	6.9	0.994	0.941 ± 0.050	0.952 ± 0.028	Capillary SA (cm^2) [b]
6	20	19	5.4	6.9	0.996	1.023 ± 0.045	1.000 ± 0.044	Capillary volume (cm^3)

N_{ord} = number of orders; N_{sp} = number of species; MPD = mean percent deviation; pWR = log(BW_{max}/BW_{min}); DI = diversity index; r^2 = coefficient of determination; CL = confidence limit(s).
[a] Results reported in column h are taken directly from Gehr *et al.* [6], who drew on a larger and more heterogeneous dataset than analyzed here (i.e., columns a to g); observe that CL in column g are 99% and in column h are 95%; [b] SA = surface area.

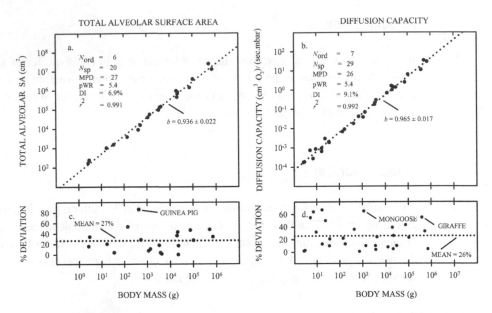

Figure 15.1 Total alveolar surface area (SA) and diffusion capacity as a function of adult body size.

The outcomes of a LSQ analysis of total alveolar surface area (SA) are shown in row one of Table 15.1, and in Figure 15.1 (panels a and c). Notice that total alveolar surface area may scale in accord with a repeating units model (Chapters 3, 24). The very deviant point in Figure 15.1c is for the guinea pig (*Cavia porcellus*).

Both rat and human *postnatal* data (not shown) for alveolar surface area are available from Burri *et al.* [7] and Dunnill [8], respectively. Most of the points for the postnatal rat are nearly collinear with the adult interspecific line. The human *ontogenetic* data approach the adult interspecific line from below. In the past it was estimated that total alveolar surface area in the adult human is about 70 m^2. The more recent estimate of Gehr *et al.* [6] is about twice as large (143 m^2), implying that alveolar surface area is some 84 times larger than the external body surface area (143/1.7 = 84). No doubt this substantial increase in surface area is required to maintain (especially maximal) energy metabolism. At rest in the adult human it is estimated that oxygen uptake is only 5–10% of the maximal level.

Barrier thickness

The second major structural determinant of pulmonary diffusion capacity, after alveolar surface area, is the mean thickness of the barrier between air in the alveoli and the hemoglobin in the red blood cells. This barrier to diffusion (e.g., of oxygen) consists of several closely apposed layers. The interior of an alveolus is lined by a thin (a few nm; 1 nm = 10^{-9} m) layer of surfactant beneath which lie flattened epithelial cells. For a study of the scaling of *surfactant* in mammals (mouse to cow) see Clements *et al.* [9]. This epithelial layer consists of little more than two apposed layers of plasma membrane. Beneath the epithelial cells lie two fused basement membranes, one secreted by the epithelial cells and one by the endothelial cells, which form the walls of the pulmonary capillaries. Like the epithelial cells, the endothelial cells consist of not much more than two closely apposed plasma membranes.

Finally, a respiratory gas (e.g., oxygen) must cross a thin layer of plasma before traversing the erythrocyte plasma membrane. From shrews to cows the barrier thickness ranges from 0.25 to 0.60 µm [10] (differing by a factor of 2.4). It is useful in thinking about barrier thickness to keep in mind that the diffusion coefficient for oxygen is about ten thousand times greater in air than in water.

The fruits of the present analysis of the scaling of barrier thickness based on a *subset* of the data of Gehr *et al.* [6] are given in Table 15.1, row two, columns a to g. It was found that barrier thickness scales as the 0.056 ± 0.028 (99% CL). Gehr *et al.* [6] reported a slope of 0.050 ± 0.0135 (95% CL) for a larger and more heterogeneous dataset (see Table 15.1, row two, column h). Scaling of barrier thickness approximates absolute invariance ($b = 0.0$).

Breath holding

For a brief discussion of breath holding in adult mammals, see Chapter 18.

Capillary surface area

In a histological tissue section of the alveolar wall, examined at the light microscopic level of resolution, one has the subjective impression that the entire alveolar wall is occupied by capillaries. Furthermore, most capillaries are seen to lie between two alveoli. Thus, the pulmonary capillary bed has sometimes been described (incorrectly) as forming a sheet. At the electron microscopic level it is evident that the alveolar wall is not everywhere completely lined by capillaries. Nevertheless, one can reasonably expect that capillary surface area will scale nearly in the same way as does alveolar surface area. The results of the present analysis of a subset of the total capillary surface area data of Gehr *et al.* [6] are brought together in Table 15.1, row three. The best-fit LSQ slope for capillary surface area (0.941 ± 0.050) (99% CL) is not significantly different from that for total alveolar surface area (see Table 15.1, row one).

Capillary volume

If capillaries may be treated as cylinders of constant diameter and constant length (Chapter 9), then one anticipates that the ratio of capillary volume to capillary surface area is a constant. The outcomes of the present analysis of a subset of the data of Gehr *et al.* [6] suggest that this is possible, albeit not certain (compare slopes and CLs in Table 15.1, columns g and h, rows three, four) (see Chapter 23 for a discussion of

Table 15.2 Scaling of diaphragm mass as a function of adult body mass

N_{ord}	N_{sp}	MPD	pWR	DI (%)	r^2	Slope $\pm$ 99% CL
7	11	32	6.4	7.6	0.995	1.015 ± 0.079

See Table 15.1.

special pleading). The slope for capillary volume (1.000 ± 0.044) (Table 15.1, row four) is consistent with relative invariance $(b = 1.0)$. Be reminded that the slopes for capillary surface area (0.888 ± 0.056) (99% CL) and volume (0.970 ± 0.050) (99% CL) presented in Chapter 9, Table 9.12, were expressed as a function of *lung volume*, not body mass as here.

Diaphragm

The respiratory process of inspiration is driven primarily by contraction of a domed-shaped sheet of skeletal muscle termed the diaphragm, a structure peculiar to mammals (Chapter 2, Table 2.1). In the adult human the diaphragm weighs about 200 g. Slopes for the scaling of diaphragm weight on adult body weight of 0.915 and 0.865 have been reported by Stewart [11] and Mathieu *et al.* [12], respectively (no confidence limits given). The findings of the present LSQ analysis of diaphragm mass on body mass based on data taken from the literature are given in Table 15.2. These results ($b = 1.015 \pm 0.079$) (99% CL) suggest (but are not persuasive, owing to the rather wide 99% CL) scaling in accord with relative invariance ($b = 1.0$).

For a LSQ analysis of the scaling of diaphragm myofiber diameter on body weight see Chapter 12, "Myofiber diameter". A discussion of selective factors possibly influencing diaphragmatic evolution is given by Ruben *et al.* [13].

Diffusion capacity

[The diffusion capacity for oxygen]... *estimates the global conductance of the lung for* O_2 *transfer to the blood by diffusion.*

Weibel 1984

Pulmonary diffusion capacity refers to the volume of gas (usually oxygen or carbon dioxide) transported per unit time between the alveoli and the red blood cells for a given partial pressure gradient. It would perhaps better be termed pulmonary conductance. Physiological conductance may be measured by the single-breath test, wherein a test gas containing a small amount of carbon monoxide (0.3%) diluted in a non-absorbing gas (such as 10% helium and air) is inhaled and then held in the lungs for ten seconds. Following an analysis of the exhaled gas it is possible to calculate the conductance. See above quote from Weibel.

Table 15.3 Scaling of (maximum) diffusion capacity (cm^3 O$_2$)/(sec.mbar) as a function of adult body mass

N_{ord}	N_{sp}	MPD	pWR	DI (%)	r^2	Slope $\pm$ 99% CL	Source
6	22	23	5.4	7.2	0.995	0.991 $\pm$ 0.044	[6]
7	29	26	5.4	9.1	0.992	0.965 $\pm$ 0.046 ^{a}	From Lit.

See Table 15.1. ^{a} 22 of the records in this dataset are from [6] (see row one); Lit. = literature.

From a structural standpoint, pulmonary conductance depends chiefly on total alveolar surface area and the mean barrier thickness (see above). Morphometric determination of these parameters allows one to make an estimate of the maximum pulmonary conduction [10]. The present analysis of (maximum) pulmonary conductance as determined by morphometry is displayed in Table 15.3 and Figure 15.1 (panels b, d). The data given in row one are based on a *subset* of the data of Gehr *et al.* [6].

One sees from Table 15.3, row two, that the best-fit LSQ 99% confidence interval for diffusion capacity (cm^3 O$_2$/(sec.mbar)) lies between 0.919 and 1.011. Because of the wide confidence limits, diffusion capacity may or may not scale in accord with relative invariance ($b = 1.0$). Two deviant species identified in Figure 15.1d are the mongoose (*Mungos mungo*) and the giraffe (*Giraffa camelopardalis*). The four deviant species in the upper left corner (Figure 15.1d), ordered by body mass, are the common pipistrelle *Pipistrellus pipistrellus*, the shrews *Crocidura crossei* and *Crocidura poensis*, and the Malayan free-tailed bat *Tadarida mops*; the first and last species fall above the best-fit line, the other two below.

Lung volume or mass

For a simplified view of the lungs see the above quote from Dejours. Lung mass or volume is the simplest measure of the aggregate size of the respiratory system (exclusive of the trachea; see below). Ogiu *et al.* [14] put lung weight at 809 g in adult Japanese men weighing on average 62.3 kg. Brody [15] gave an exponent of 0.986 $\pm$ 0.012 (s.e.) for the scaling of lung weight on adult interspecies body weight. A LSQ analysis of lung volume derived from a *subset* of the data of Gehr *et al.* [6] and an analysis of lung mass based on data taken from the literature generally are brought together in Table 15.4 and Figure 15.2. The slope for lung mass (1.000 $\pm$ 0.026, (99% CL) is consistent with relative invariance.

The intercept derived from this LSQ analysis (Table 15.4, row two) implies that on average the lungs comprise 1.01% of body mass in adult mammals in the given sample (see Chapter 19, Table 19.4, row 23). The evidence presented in Table 15.4, bottom row, accords with relative invariance ($b = 1.0$) of lung mass scaling on body mass.

The human *fetal* and rat *postnatal* data shown in Figure 15.2b are from Shepard (personal communication) and Hatai [16], respectively. One finds that the human fetal data lie mostly above the adult interspecific line, and the rat data are almost collinear.

Table 15.4 Lung volume or mass as a function of adult body mass

N_{ord}	N_{sp}	MPD	pWR	DI (%)	r^2	Slope ± 99% CL	y-variable	Source
6	20	21.6	5.4	24.3	0.995	1.051 ± 0.052	Lung volume	[6]
12	115	32.3	7.7	31.3	0.989	1.000 ± 0.026	Lung mass	From Lit.

See Table 15.1. Lit. = literature.

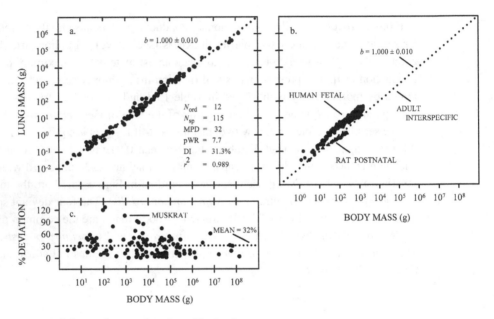

Figure 15.2 Lung size as a function of body size.

The two most deviant points in Figure 15.2c (not labeled), at body weights of about 100 g, represent the common hamster (*Cricetus cricetus*) and the golden hamster (*Mesocricetus auratus*), both of which lie below the best-fit LSQ line. The taxonomic name for the muskrat identified in panel c is *Ondatra zibethicus*.

Assuming that adult human lungs have an alveolar surface area of 143 m² and a weight of 0.809 kg, we infer a surface density of 177 m²/kg (143/0.809 = 177). This is about 4,500 times less than the surface density of a typical clay [17]. Both clays (soils) and mammalian lungs depend for their normal functioning on relatively high surface densities.

Minute volume

The volume of air inspired per minute at rest is known as the *minute volume*. In a normal young human adult, it is about 6 liters/min. The inferences from the present LSQ analysis of minute volume as a function of adult body weight (mouse to elephant) are given in Table 15.5. Stahl [18] gives an exponent of 0.80 ± 0.01 (s.e.) for minute

Table 15.5 Scaling of minute volume as a function of adult body mass

N_{ord}	N_{sp}	MPD	pWR	DI (%)	r^2	Slope ± 99% CL	Units	Source
10	28	48	5.6	13.3	0.957	0.739 ± 0.085	cm³/min	From Lit.

See Table 15.1. Lit. = literature.

Table 15.6 Resting (rows one, two) and maximal respiratory rates (row three) as a function of adult body mass

N_{ord}	N_{sp}	MPD	pWR	DI (%)	r^2	Slope ± 99% CL	Entity	Sample [a]
10	76	29	6.15	18.9	0.866	−0.265 ± 0.032	Resting	Eutherians
11	103	37	6.15	22.3	0.773	−0.265 ± 0.037	Resting	FDS
5	8	26	5.25	3.9	0.758	−0.117 ± 0.100	Maximal [b]	-

See Table 15.1. [a] Data for resting respiratory rate in cetaceans (three species), monotremes (two species), and an edentate were omitted from calculations in rows one and two; [b] human data, which differ from the initial best-fit LSQ line by 279%, were omitted from the calculations. The slopes in rows one and two are, coincidentally, the same.

volume on adult body mass or weight. Because of the wide 99% CL (±0.085) in the present analysis, these two slopes (0.80 and 0.739) are not significantly different.

Partial pressure of oxygen in lungs

For a brief discussion of P_{50} in pulmonary blood, see Chapter 9.

Respiratory rate

The *respiratory rate* is usually reported as the number of breaths per minute. In the human adult at rest it is normally in the range of 8–20. Small mammals (e.g., bats and shrews) have maximal respiratory rates in the range of 500 to 1,000 per minute. Jürgens *et al.* [19] reported a maximal respiratory rate of 894 breaths per minute in the Etruscan shrew (*Suncus etruscus*). In addition, Thomas and Suthers [20] observed respiratory rates of 576 breaths per min in the spear-nosed bat (*Phyllostomus hastatus*) in flight.

The outcomes of a LSQ analysis of resting and maximal respiration rates (based on data taken from the literature) are given in Table 15.6 and Figure 15.3. The point for the fin whale (*Balaenoptera physalus*) in Figure 15.3 was not included in the LSQ computations. The respiratory data for this 45 foot (13.7 m) fin whale are from Kanwisher and Senft [21] and the body weight, as derived from body length, is from Lockyer [22]. This whale was beached, seemed to be breathing primarily from the right lung, and, as based on the SBT (Chapter 4), was probably subadult. More physiological data on (adult) whales are much to be desired. The three deviant species in Figure 15.3c are Balston's

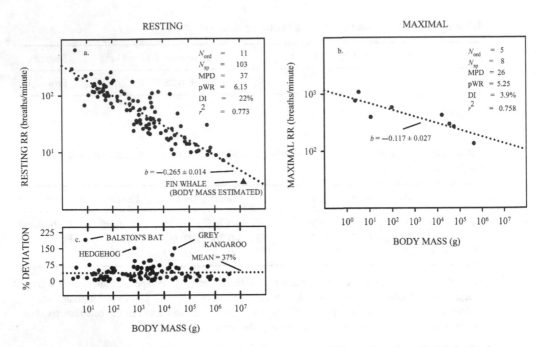

Figure 15.3 Resting and maximal respiratory rates (RR) as a function of adult body size.

bat (*Scotorepens balstoni*), the gray kangaroo (*Macropus giganteus*), and the hedgehog (*Erinaceus europeaus*). The data represented in Table 15.6, row one, are the same as for row two, except that 27 species of marsupials are not included in row one (the 99% CL for slope of the LSQ best-fit marsupial line are excessively wide (>0.1)). The slopes shown in Table 15.6 (rows one, two) agree reasonably well with a hypothetical reference slope for ubiquity ($b = -1/4$; Chapter 3, Table 3.5).

Note in Figure 15.3b and Table 15.6, row three, that the slope (-0.117 ± 0.100) (99% CL) for *maximal respiration rate* is much shallower than the slope (-0.265 ± 0.037) (99% CL) for resting respiration rate. That is, maximal respiratory rate (in a small sample) deviates much less from absolute invariance ($b = 0.0$) than does resting respiratory rate. The 99% CL (± 0.100) on maximal respiratory rate are, however, very wide (see Table 15.6, bottom row).

Recall, from Chapter 9, that resting and maximal heart rates scale as the -0.220 ± 0.033 (99% CL) and -0.146 ± 0.025 (99% CL) powers of body size, respectively. The slopes are not significantly different for either *resting* heart or respiratory rates, nor for *maximal* heart or respiratory rates. Note the wide 95% CL (± 0.10) on the slope for maximal respiratory rate, so that the actual slope is very uncertain (see Table 15.6, row three).

Trachea

...dead space volume is very nearly a constant fraction of lung volume.

Tenney and Bartlett 1967

Table 15.7 Scaling of tracheal diameter and length as a function of adult body mass

N_{ord}	N_{sp}	N_{pnts}	MPD	pWR	DI (%)	r^2	Slope $\pm$ 99% CL	Tracheal	Source
10	24	24	18	5.7	12.9	0.9785	0.411 $\pm$ 0.0365	Diameter	From Lit. [a]
-	-	35	23	4.2	-	0.932	0.397 $\pm$ 0.051	Length	[23] [b]

See Table 15.1. [a] Note: about 80% of the records are from [23]; [b] data obtained by graphical interpolation.

All of the air that enters and leaves the adult mammalian lung does so through the trachea. Recall that after an expiration of air a proportion of waste gas (chiefly carbon dioxide) remains in the trachea and hence is inspired during the next breath. This volume of air, termed the "dead space," which is predominantly (but not exclusively) associated with the trachea, and which does not participate in gas diffusion, represents a partial inefficiency in the "design" of the mammalian respiratory system with respect to oxygen transport (see above quote from Tenney and Bartlett). The fruits of a LSQ analysis of the scaling of tracheal diameter (species identified by common name) and length (species not identified) with adult body mass are summarized in Table 15.7. If one takes the mean exponent for tracheal diameter as 0.411 (see Table 15.7, row one), and 0.397 for tracheal length (row two), then it is found that tracheal volume scales as the 1.22 power ($2 \cdot 0.411 + 0.397 = 1.22$) of body mass. This exponent for tracheal volume is almost certainly too high to apply to mammals generally (see Chapter 3). Still, it is of interest that Daniels and Pratt [24] estimated the respiratory dead space in the Late Jurassic 30-ton dinosaur (*Mamenchisaurus hochuanensis*) (thought to have had a neck at least 11 m long) at 150–500 liters. They conclude that the respiratory system for this dinosaur must have more nearly resembled those of contemporary birds rather than extant mammals.

Mortola and Fisher [25] found that the *diameter*, *length*, and *volume* of the trachea in newborn mammals (cat, dog, dolphin, guinea pig, human, mouse, pig, rabbit, rat) scale with body mass with exponents of 0.357 $\pm$ 0.0139 (s.e.), 0.336 $\pm$ 0.0134 (s.e.), and 1.066 $\pm$ 0.0304 (s.e.), respectively. These findings are reasonably self-consistent ($2 \cdot 0.357 + 0.336 = 1.05$ rather than 1.066).

Water loss

On average, in a process known as *insensible water loss*, a sedentary adult human loses an estimated 0.25–0.35 liters of water per day through expiration from the respiratory system.

Summary

Maximal flow of a gas across the alveolar membrane, for a given gradient in partial pressure, is proportional to total alveolar surface area and inversely proportional

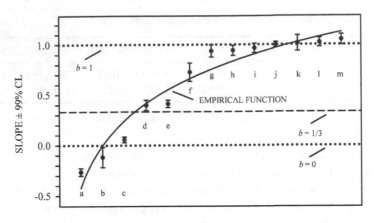

RANK ORDER

a = RRrest; b = RRmax; c = barrier thickness; d = tracheal length;
e = tracheal diameter; f = minute volume; g = total alveolar SA;
h = total capillary SA; i = diffusing capacity; j = lung weight;
k = diaphragm weight; l = total capillary volume; m = lung volume

RRrest = resting respiratory rate; RRmax = maximal respiratory rate

Figure 15.4 Summary slopes for respiratory system.

to barrier thickness. It was found empirically that barrier thickness varies only slowly (or possibly not at all) with body size ($b = 0.056 \pm 0.028$) (99% CL). Thus, the main variable governing maximal gas flow is alveolar surface area. Of the diverse organ systems in the body, perhaps none illustrate better than the respiratory system Haldane's epigram (see above quote) to the effect that comparative anatomy is largely the story of nature's need to increase surface area in proportion to volume. Whether a "struggle" was involved in this adaptation to increased body size is unclear (since repeating units appear to provide a simple and presumably easily implemented method to achieve large increases in surface area relative to body surface area [1]).

The main slopes discussed above are shown in Figure 15.4. The regular distribution of 13 slopes seen in this plot is conspicuous. Most of the slopes for the respiratory system fall near the reference slopes of: $-1/4$ (for ubiquity), 0 (absolute invariance), 1/3 (dimension of length), and 1 (for relative invariance) (see Chapters 3 and 24). More specifically, seven slopes cluster around relative invariance ($b = 1.0$), two accord with a slope of 1/3, one with absolute invariance ($b = 0.0$) and one with ubiquity ($b = -1/4$). That is to say, 11 out of 13 (or 85%) of the slopes for the respiratory system accord, to a first approximation, with one or another of the reference slopes. This finding may reflect the fact that mammalian respiration is strongly dependent on physical factors such as convection and diffusion. The distribution of slopes for the respiratory system resembles that for the digestive system (see Figure 10.5).

The smooth curve shown in Figure 15.4 is merely an empirical function. It differs from the observed data points by a MPD of 45%. If the third point (with a slope for barrier thickness near 0) is omitted, the MPD drops to 21%.

References

1. Prothero, J. (1996). Scaling of organ subunits in adult mammals and birds: a model. *Comparative Biochemistry and Physiology*, **113A**:97–106.
2. Knust, J., Ochs, M., Jørgen, H., Gundersen, G. and Nyengaard, J.R. (2009). Stereological estimates of alveolar number and size and capillary length and surface area in mice lungs. *Anatomical Record*, **292**:113–122.
3. Hislop, A., Howard, S. and Fairweather, D.V.I. (1984). Morphometric studies on the structural development of the lung in *Macaca fascicularis* during fetal and postnatal life. *Journal of Anatomy*, **138**:95–112.
4. Ochs, M., Nyengaard, J.R., Jung, A. *et al.* (2004). The number of alveoli in the human lung. *American Journal of Respiratory Critical Care Medicine*, **169**:120–124.
5. Hislop, A.A., Wigglesworth, J.S. and Desai, R. (1986). Alveolar development in the human fetus and infant. *Early Human Development*, **13**:1–11.
6. Gehr, P., Mwangi, D.K., Ammann, A. *et al.* (1981). Design of the mammalian respiratory system. V. Scaling morphometric pulmonary diffusing capacity to body mass: wild and domestic mammals. *Respiratory Physiology*, **44**:1–86.
7. Burri, P.H., Daly, J. and Weibel, E.R. (1974). The postnatal growth of the rat lung. I. Morphometry. *Anatomical Record*, **178**:711–730.
8. Dunnill, M.S. (1962). Postnatal growth of the lung. *Thorax*, **17**:329–333.
9. Clements, J.A., Nellenbogen, J. and Trahan, H.J. (1970). Pulmonary surfactant and evolution of the lungs. *Science*, **169**:603–604.
10. Weibel, E.R. (1970). Morphometric estimation of pulmonary diffusion capacity. I. Model and method. *Respiratory Physiology*, **11**:54–75.
11. Stewart, D.M. (1972). The role of tension in muscle growth. In: Goss, R.J. (ed.) *Regulation of Organ and Tissue Growth*. New York: Academic Press, pp. 77–100.
12. Mathieu, O., Krauer, R., Hoppeler, H. *et al.* (1981). Design of the mammalian respiratory system: VII. Scaling mitochondrial volume in skeletal muscle to body mass. *Respiratory Physiology*, **44**:113–128.
13. Ruben, J.A., Bennett, A.F. and Hisaw, F.L. (1987). Selective factors in the origin of the mammalian diaphragm. *Paleobiology*, **13**:54–59.
14. Ogiu, N., Nakamura, Y., Ijiri, I. *et al.* (1997). A statistical analysis of the internal organ weights of normal Japanese people. *Health Physics*, **72**:368–383.
15. Brody, S. (1945). *Bioenergetics and Growth*. New York: Reinhold.
16. Hatai, S. (1913). On the weights of the abdominal and the thoracic viscera, the sex glands, ductless glands and the eyeballs of the Albino rat (*Mus norvegicus albinus*) according to body weight. *American Journal of Anatomy*, **15**:87–119.
17. Hillel, D.J. (1991). *Out of the Earth. Civilization and the Life of the Soil*. New York: Macmillan.
18. Stahl, W.R. (1967). Scaling of respiratory variables in mammals. *Journal of Applied Physiology*, **22**:453–460.
19. Jürgens, K.D., Fons, R., Peters, T. and Sender, S. (1996). Heart and respiratory rates and their significance for convective oxygen transport rates in the smallest mammal, the Etruscan shrew *Suncus etruscus*. *Journal of Experimental Biology*, **199**:2579–2584.
20. Thomas, S.P. and Suthers, R.A. (1972). The physiology and energetics of bat flight. *Journal of Experimental Biology*, **57**:317–335.

21. Kanwisher, J. and Senft, A. (1960). Physiological measurements on a live whale. *Science*, **131**:1379–1380.

22. Lockyer, C. (1976). Body weights of some species of large whales. *Journal Conseil International pour l'Exploration de la Mer*, **36**:259–273.

23. Tenney, S.M. and Bartlett, D. Jr. (1967). Comparative quantitative morphology of the mammalian lung: trachea. *Respiratory Physiology*, **3**:130–135.

24. Daniels, C.B. and Pratt, J. (1992). Breathing in long necked dinosaurs: did sauropods have bird lungs? *Comparative Biochemistry and Physiology*, **101A**:43–46.

25. Mortola, J.P. and Fisher, J.T. (1980). Comparative morphology of the trachea in newborn mammals. *Respiratory Physiology*, **39**:297–302.

16 Urinary system

The complex function of the kidney in man and other vertebrates would suggest that this organ has an extraordinarily complex structure. On close examination, however, it is found to be made up of a very large number of structurally similar and anatomically simple functional units.

Smith 1951

The urinary system bears primary responsibility for the *relative constancy* of the internal fluid environment in mammals. It operates on a metabolically extravagant but inherently fail-safe principle. All free-living organisms face the possibility of ingesting potentially *poisonous* or *toxic* materials. The presence of these materials in the environment cannot always be known in advance. It follows that a fail-safe strategy is to discard everything in the blood plasma (below the size of small proteins) and then to reabsorb only those substances (e.g., glucose, amino acids) that the body requires for its normal metabolism. This approach may be likened to a strategy of "throwing out the baby with the bath water." It is energetically expensive (see below); nevertheless, one may reasonably assume that if it were not an effective and affordable strategy, natural selection would long since have discarded it.

In the *adult human* about one-fifth of the total blood volume passes through the two kidneys every minute. The blood plasma, excepting the plasma proteins – which are generated internally – is extruded through a thin membrane bounding each renal or glomerular corpuscle. This extruded fluid passes down the nephron, where those materials specifically needed by the body are reabsorbed. Poisons, toxins, and waste products such as urea (see below) are sent on to the bladder via the ureter for temporary storage prior to voluntary evacuation through the urethra. In this way, in a human adult, a volume some ten times that of the body's total extracellular fluid (of about 15.5 liters) is filtered and reabsorbed each day. The metabolic cost of the adult human kidneys at rest amounts to 20–25% of cardiac output and about 6% of resting energy metabolism. This is remarkable, given that the two kidneys comprise less than 0.5% of adult body weight. Per gram of tissue, the human kidneys consume roughly as much oxygen as the brain, but appreciably less than the heart. The *efficiency* of the human kidneys, as a function of the ratio of work done to energy consumed, has been put at 1 to 5%.

Ingested proteins are broken down into amino acids in the gastrointestinal tract. The absorbed amino acids undergo deamination in the liver, giving rise to an amino group (NH_2) which, by attaching to a hydrogen ion, produces ammonia (NH_3). Ammonia (highly toxic) is converted to the much less toxic urea ($CO(NH_2)_2$) via the ornithine biochemical cycle in the liver. Urea makes its way through the circulatory system to the kidneys. On a normal diet a human adult will produce some 13–33 g of urea per day, requiring the *co-excretion* of about 800 cm^3 of water.

It has been stated [1] that about three-quarters (78%) of nitrogenous waste on average is excreted as urea, based on a study of eight species of mammals: these are the bat (apparently an African fruit bat of uncertain species), camel (*Camelus bactrianus*), cat (*Felis silvestris*), dog (*Canis familiaris*), human (*Homo sapiens*), llama (*Lama guanicoe*), spiny anteater (*Tachyglossus aculeatus*), and whale (*Balaena mysticetus*). In effect, the removal of nitrogenous waste from the body requires a plentiful supply of exogenous fresh water (for an exception see "Water intake" in Chapter 10). By excreting variable amounts of water and salts in the urine, the kidneys play a major role in maintaining the body's water, electrolyte, and acid–base balance.

The mammalian kidney is divided into two zones, an *outer cortex* and an *inner medulla*. The primary structural and functional unit of the kidney is the nephron, which consists of a microscopic glomerulus or renal corpuscle connected to a serial arrangement of tubules (proximal convoluted tubule, straight descending portion of the U-shaped loop of Henle, the straight ascending portion, and finally the distal convoluted tubule). The roughly spherical glomeruli are restricted to the cortex. Each glomerulus contains a tuft of capillaries uniquely interposed between afferent and efferent arterioles. Consequently the mean driving blood pressure (of about 50 mm Hg) in the glomerular capillaries is nearly twice that at the arteriolar end of capillaries generally. This accounts for the high rate of hydrostatic filtration across the walls of glomerular capillaries and into the capsular or urinary space.

The protein-free filtrate then flows sequentially into the proximal convoluted tubule, the loop of Henle, and the distal tubule, before entering a collecting duct and thence the ureter. The loop of Henle and the collecting ducts are found in the medulla. The loop of Henle participates in a *countercurrent multiplier system*, without which mammals would be unable to excrete urine hypertonic to the plasma. In the urine's course through the renal tubules, especially in the *proximal tubule*, the bulk of the amino acids, bicarbonate, and glucose are reabsorbed; the amounts of chloride, potassium, and sodium reabsorbed vary with diet. At the microscopic and submicroscopic levels of resolution, the structure of the nephron is a matter of appreciable complexity (see above quote from Smith).

This chapter reviews data for five urinary *y*-variables, in eight instances, for a total of 373 records, all expressed as a function of adult *body mass* in tabular form. In addition, two other urinary *y*-variables (along with data for two of the *y*-variables just cited) in five instances and 60 records are expressed in terms of *kidney mass* (Table 16.3). Pooling the data expressed in terms of either *organ* or *body* mass, we have data for seven distinct *y*-variables, 13 instances, and a total of 433 records, all in tabular form.

Renal morphometric variables in the adult human

Quantitative morphometric data on the adult human kidneys are provided in a booklet published in 1966 by Abbott Laboratories [2]. The source of these data is not stated;

Table 16.1 Lengths and surface areas of various renal structures in the adult human [2, 3]

Entity	Parameter	Single nephron	Both kidneys
Glomerulus	Total capillary length	2–4 mm	4–8 km
	Capillary filtering area	1 mm^2	2 m^2
PCT ^{a}	Length	14 mm	28 km
	Absorptive area	84 mm^2	1.7 m^2 ^{b}
Descending limb of Henle	Length	9 mm	18 km
	Internal area	27 mm^2	0.5 m^2
Ascending limb	Length	5–10 mm	10–20 km
	Internal area	45 mm^2	0.9 m^2
Distal convoluted tubule	Length	5 mm	10 km
	Internal area	25 mm^2	0.5 m^2
Collecting duct	Length	20 mm	2 km
	Internal area	180 mm^2	0.18 m^2
All segments	Length	58 mm	60–80 km
	Internal area	181 mm^2	4–5 m^2

^{a} PCT = proximal convoluted tubule; ^{b} area of brush border not included.

however, the same data are given by Abbrecht [3]. These findings are brought together in Table 16.1. Note that the dimension of length is modest for a single nephron, but large (compared with whole body length) when aggregated for both kidneys.

In a footnote [2], it is stated that if the dimension of surface area of the brush border were included then the *total surface area* of the proximal convoluted tubules would be larger by a factor of at least 100. If true, that would make the total renal surface area roughly comparable to the total alveolar area of the adult human lungs (see Chapter 15). At the same time, the total length of all nephrons in the adult human is estimated at about 70 km (Table 16.1, second row from bottom). For a précis of exponents (and their standard errors) relating to renal scaling in mammals see Calder [4].

GLOMERULAR VARIABLES

Glomerular diameter

The results of a LSQ analysis of data bearing on *glomerular diameter* as a function of body mass are brought together in Table 16.2, row one. These analytical findings are derived from a *subset* of the data of Rytand [5]; it was found that mean glomerular diameter scales as the 0.109 ± 0.035 (99% CL) power of body mass.

Glomerular number

The *number of glomeruli* in one kidney of a human adult is estimated at 0.8 to 1.2 million. Glomerular number is reported by Holt and Rhode [6] to scale in a

Table 16.2 Scaling of three glomerular variables as a function of adult body mass

N_{ord}	N_{sp}	MPD	pWR	DI (%)	r^2	Slope ± CL	CL	Entity	Source
5	10	8	5.4	4.4	0.932	0.109 ± 0.035	99%	Glom. diameter [a]	[5]
9	16	-	6.2	11	0.96	0.62 ± 0.06	95%	Glom. number [a]	[6]
5	10	-	3.5	2.9	0.98	0.79 ± 0.05	95%	GFR [b]	[6]
9	23	-	4.1	8.2	0.98	0.72 ± 0.04	95%	GFR [b]	[7]

N_{ord} = number of orders; N_{sp} = number of species; MPD = mean percent deviation; DI = diversity index; r^2 = coefficient of determination; CL = confidence limit(s); [a] Glom. = glomerular; [b] GFR = glomerular filtration rate; sources [6, 7] do not provide "raw" data.

Table 16.3 Scaling of four renal variables as a function of adult kidney mass

N_{ord}	N_{sp}	MPD	pWR	DI (%)	r^2	Slope ± 99% CL	Entity	Source
5	11	10	4.5	3.8	0.908	0.124 ± 0.043	Glom. diameter [a]	[5]
6	19	16	4.7	5.9	0.906	0.163 ± 0.037	Glom. diameter [a]	From Lit.
7	16	28.5	4.5	6.2	0.973	0.641 ± 0.086	Glom. number [a]	From Lit.
7	7	5	3.7	3.6	0.550	0.025 ± 0.041	PT diameter [b]	[8]
7	7	11	3.7	3.6	0.830	0.113 ± 0.092	PT length [b]	[8]

See Table 16.2. [a] Glom. = glomerular; [b] PT = proximal tubule; Lit. = literature.

small sample of mammals as the 0.62 ± 0.06 (95% CL) power of body mass; see Table 16.2, row two.

Glomerular filtration rate

The average *glomerular filtration rate* (GFR) for young adult human kidneys is given as 180 liters/day (125 cm^3/min). The outcomes of LSQ analyses of GFR (cm^3/min) expressed as a function of adult body mass are reviewed in Table 16.2, rows three and four. The best-fit mean slopes vary between 0.72 ± 0.04 and 0.79 ± 0.05 (both 95% CL).

For most organ systems one is compelled (by the absence of needed data) to express scaling relations as a function of body mass or weight. Since body mass is a highly heterogeneous quantity, it would often be preferable to express scaling relations for a given system in terms of *organ mass*. Some data of this type are available for the kidneys, as partly presented in Table 16.3.

Observe that the slopes for *glomerular diameter* and *glomerular number* (Table 16.3, rows two and three) may be judged roughly consistent with the slopes predicted by a repeating units model (Chapter 3, "A repeating units model"). In addition, the slope (0.025 ± 0.041) (99% CL) for the scaling of *proximal tubule diameter* may (or may not) prove to be consistent with absolute invariance ($b = 0.0$) (Table 16.3, bottom row).

Kidney size

In the human adult, the paired kidneys range in size on average from 240 to 280 g. Brody [9] (see Box 16.1) reported a slope for kidney weight on body weight of 0.846 ± 0.010 (s.e.) (Appendix C). The conclusions drawn from two LSQ analyses of kidney

---

Box 16.1 Samuel Brody

Brody (1890–1956) was born in Lithuania [B1]. In 1920 he joined the staff of the Department of Dairy Husbandry at the University of Missouri in Columbia, where he remained, for the most part, the rest of his life. While on sabbatical he was awarded a Ph.D. from the University of Chicago in 1928. Throughout his long career he was primarily interested in growth and energetic efficiency, chiefly in farm animals. He is best known for his monumental work [B2] entitled *Bioenergetics and Growth*. This work runs to 1,023 pages, and contains more than 540 figures (including chemical formulae and photos, but mainly data plots) and more than 100 tables. Most of the plots are expressed as a function of time (age) rather than body size. But log–log plots expressed in terms of body size, particularly those in Chapters 13, 17, and 24 of his book, are still of interest in size-based mammalian scaling studies. His book contains what is probably the widest survey of scaling in mammals up to that time. Scaling data on birds are also reported.

Brody was surely aware of the importance of size range in determining slopes representative of mammals generally (see Appendix C). Whenever possible he includes data for the smallest and largest mammals available. Many of the datasets he employs are based on measurements made by himself and his colleagues. Thus Brody's classic work was the first to present a broad array of scaling variables in a uniform manner based largely on calculations made mainly by himself and his colleagues. Brody routinely gives slopes and their standard errors, along with standard errors of estimate and correlation coefficients. He often identifies the species involved (by common name). Subsequent general works on scaling have tended to simply collate the analytical inferences of various different workers; but these different workers have published varying amounts of detail. Some give confidence limits or standard errors on slopes, but many do not. Some present measures of scatter in the data, but most do not. Others provide correlation coefficients. Only a few provide information about the species represented. For these and other reasons Brody's book set a standard; it is one of the few works from that era that continues to be cited in contemporary studies [B3].

B1. Morgan, A.F. (1960). Samuel Brody: A biographical sketch. *Journal of Nutrition*, **70**:1–9.
B2. Brody, S. (1945). *Bioenergetics and Growth. With Special Reference to the Efficiency Complex in Domestic Animals*. New York: Reinhold.
B3. Blaxter, K. (1986). Bioenergetics and growth: the whole and the parts. *Journal of Animal Science*, **63**:1–10.

Table 16.4 Scaling of kidney mass as a function of adult body mass

N_{ord}	N_{sp}	MPD	pWR	DI (%)	r^2	Slope $\pm$ 99% CL	Source
a	b	c	d	e	f	g	h
15	119	42	6.9	35	0.980	0.880 $\pm$ 0.030	[10]
15	159	33	7.7	42	0.987	0.880 $\pm$ 0.021 ^{a}	From Lit.

See Table 16.2. ^{a} 93 of the records in row two are cited by Beuchat [10]; Lit. = literature.

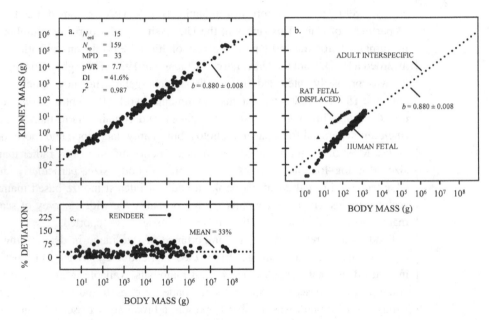

Figure 16.1 Scaling of kidney size as a function of body size.

mass scaling as a function of body mass in adult mammals are shown in Table 16.4 and Figure 16.1. The LSQ analysis of kidney mass shown in Table 16.4, row one, is based on a subset (see Appendix B) of the data of Beuchat [10] (taken from the literature). See also Prothero [11]. A problem with some kidney data in the literature is that it is not always clear whether one or both kidneys were weighed.

As it happens, the slopes in the two rows (Table 16.4) are identical. The exponent (0.880 $\pm$ 0.021) (99% CL) for *kidney mass* on adult body mass in Table 16.4, row two, is close to the exponent (0.895 $\pm$ 0.018) (99% CL) for the scaling of *liver mass* on adult body mass (see Chapter 10). A slope of 0.88 for kidney scaling implies that the kidneys in a blue whale are smaller by a factor (f) of 9 than would be the case if the slope were 1 ($\Delta b = 1 - 0.880 = 0.12$; $N_d = 26.6 \cdot 0.12 = 3.2$; $f = 2^{3.2} = 9$) (Chapter 3).

Figure 16.1 is based on the data summed up in Table 16.4, row two. In panel b, the human and rat ontogenetic data are from T.H. Shepard (personal communication) and Hatai [12], respectively. Observe that the rat data have been *displaced upwards* by a factor of 10 for clarity. When plotted correctly the points for the rat are virtually collinear with the adult interspecific line; however, they are then obscured by the human data. The human fetal data start out well below the adult interspecific line and then become virtually collinear with it, finally ending above it. In panel c, the single very deviant point corresponds to the reindeer (*Rangifer tarandus*).

Urine production

The daily *production of urine* in the normal human adult amounts to 1–2 liters per day. By contrast, an Indian elephant (*Elephas maximus*), weighing 3,700 kg, produces about 50 liters of urine per day [13]. The results of an earlier LSQ analysis of daily urine production [7] are displayed in Table 16.5, row one. The findings of the present LSQ analysis are synopsized in Table 16.5 (row two) and in Figure 16.2. Note that the slopes for daily urine production are not significantly different from the slopes for glomerular filtration rate (see Table 16.2, rows three and four).

Holt and Rhode [6] reported a slope of 0.79 ± 0.05 (95% CL) for daily urine production based on data drawn from five orders and ten species (see Table 16.2, row

Table 16.5 Scaling of urine production (cm³/day) as a function of adult body mass

N_{ord}	N_{sp}	N_{pnts}	MPD	pWR	DI (%)	r^2	Slope ± CL	CL	Source
9	23	35	-	4.2	8.4	0.90	0.75 ± 0.10	(s.e.)	[7]
8	13	13	46	5.3	7.7	0.977	0.764 ± 0.109	(99%)	From Lit.

See Table 16.2. N_{pnts} = number of points; Lit. = literature.

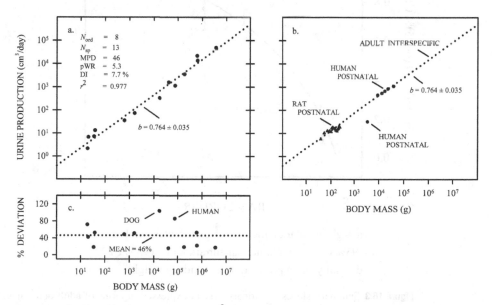

Figure 16.2 Scaling of urine production (cm³/day) as a function of body size.

three). The human and rat ontogenetic data shown in Figure 16.2b are from Watson and Lowrey [14] and Donaldson [15], respectively. Except for one point for the human, both datasets are essentially collinear with the adult interspecific line. In Figure 16.2c, we find that dog (*Canis familiaris*) and human (*Homo sapiens*) are highly deviant from the best-fit LSQ line.

Recall that in Chapter 10 we found that water intake scales as the 0.862 ± 0.047 (99% CL) power of adult body mass. Because of the very wide CL this figure is not statistically different from the figure (0.764 ± 0.109) for urine production as reported in Table 16.5, bottom row. Finally, as might be anticipated, it is known that cetaceans are able to secrete strongly hypertonic urine [16]. No doubt this is an adaptation to life in a salty environment.

Summary

The urinary system is concerned with maintaining the constancy of the internal fluid environment. This homeostasis is achieved mainly by the selective export of liquid waste products to the external environment. The main five slopes (based on *adult body mass* rather than *organ mass*) derived in this chapter are displayed in Figure 16.3. The distribution of slopes seen in Figure 16.3 is broadly similar to those found in Chapters 10

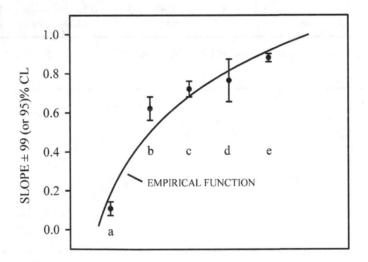

RANK ORDER

a = glomerular diameter; b = glomerular number
(95% CL); c = glomerular filtration rate (95% CL);
d = daily urine production; e = kidney weight

Figure 16.3 Summary slopes for urinary system expressed in terms of adult body mass.

(digestive system) and 15 (respiratory system). The scaling of the urinary system, as with most other systems, would benefit greatly from the coherent and systematic approach taken by Taylor and Weibel and colleagues to the scaling of the respiratory system (see Chapter 15).

The smooth curve shown in Figure 16.3 is an empirical function. This function differs from the data (at five points) by a mean of 21%.

References

1. Baldwin, E. (1952). *Dynamic Aspects of Biochemistry.* Cambridge: Cambridge University Press.
2. Abbott Laboratories (1966). *Microstructure of the Nephron.* North Chicago, IL: Abbott Laboratories.
3. Abbrecht, P.H. (1968). An outline of renal structure and function. In: Dedrick, R.L., Bischoff, K.B. and Leonar, E.F. (eds.) *The Artificial Kidney.* New York: American Institute of Chemical Engineers, pp. 1–14.
4. Calder, W.A. (1984). *Size, Function, and Life History.* Cambridge, MA: Harvard University Press.
5. Rytand, D.A. (1937). The number and size of mammalian glomeruli as related to kidney and to body weight, with methods for their enumeration and measurement. *American Journal of Anatomy,* **62**:507–520.
6. Holt, J.P. and Rhode, E.A. (1976). Similarity of renal glomerular hemodynamics in mammals. *American Heart Journal,* **92**:465–472. (These authors provide analytical results but no "raw" data.)
7. Edwards, N.A. (1975). Scaling of renal functions in mammals. *Comparative Biochemistry and Physiology,* **52A**:63–66. (No "raw" data provided.)
8. Oliver, J. (1968). *Nephrons and Kidneys: A Quantitative Study of Developmental and Evolutionary Mammalian Renal Architectonics.* New York: Harper & Row.
9. Brody, S. (1945). *Bioenergetics and Growth.* New York: Reinhold. (The kidney weight (1,200 g) given for the elephant (p. 642) should be 18,100 g.)
10. Beuchat, C.A. (1996). Structure and concentrating ability of the mammalian kidney: correlations with habitat. *American Journal of Physiology,* **271**:R157–R179.
11. Prothero, J. (1984). Organ scaling in mammals: the kidneys. *Comparative Biochemistry and Physiology,* **77A**:133–138.
12. Hatai, S. (1913). On the weights of the abdominal and the thoracic viscera, the sex glands, ductless glands and the eyeballs of the Albino rat (*Mus norvegicus albinus*) according to body weight. *American Journal of Anatomy,* **15**:87–119.
13. Benedict, F.G. (1936). *The Physiology of the Elephant.* Washington, DC: Carnegie Institution of Washington.
14. Watson, E.H. and Lowrey, G.H. (1967). *Growth and Development of Children,* 5th edn. Chicago, IL: Year Book Medical Publishers.
15. Donaldson, H.H. (1924). *The Rat: Reference Tables and Data.* Philadelphia, PA: Wistar Institute.
16. Birukawa, N., Ando, H., Goto, M. *et al.* (2005). Plasma and urine levels of electrolytes, urea and steroid hormones involved in osmoregulation of cetaceans. *Zoological Science,* **22**:1245–1257.

17 Function

The principle that function is an expression of structure is, however, unassailable.

Brody 1945

...detailed structure alone is never a unique basis for any function or description.

Pattee 1973

In prior chapters of Part II we briefly discussed *functional* scaling in the context of the structure and function of individual organ systems. However, there are important functions associated with the *whole body* rather than primarily with any one organ system. These organism-wide functions include body temperature regulation, energy metabolism (in various forms), maximal lifespan, and thermal conduction. However, a discussion of some functions, including locomotion (which in its particulars varies substantially among bats, cetaceans, and land mammals) and most life-history traits, falls outside the scope of this work. It seems likely that structure in mammals does in fact set determinable limits to maximal function, but may not determine lower levels of function in any direct or simple fashion (see above quotes from Brody and Pattee). The present chapter involves six y-variables, 25 instances, and a total of 2,850 records, all in tabular form.

BODY TEMPERATURE

...without homeothermy, the development and functioning of intricate biological mechanisms such as the human brain would be inconceivable.

Hardy 1972

Mammalian endothermy must be seen as a highly successful, if somewhat extravagant, means simply of maintaining a constant body temperature within fine limits and under a wide range of environmental conditions.

Kemp 1982

In searching for habitable planets elsewhere in our galaxy and beyond, astronomers focus on distant objects having surface temperatures resembling those on earth. Comparable temperatures are highly atypical of the universe at large, as may be seen from Table 17.1. Temperatures suitable for life as we know it are sandwiched roughly between the boiling and freezing points of water – a very narrow range from a cosmic standpoint. Seemingly only a large object relative to our planet, like our sun, and one at a much higher surface

Table 17.1 Selected temperatures on earth and elsewhere, just after the Big Bang and now

Temperature (°C)	Place, time or condition
10^{32}	10^{-43} seconds after Big Bang
15×10^6	Sun's core temperature
5,500	Sun's mean surface temperature
100	Boiling point of water
15	Earth's mean surface temperature
0	Freezing point of water
−270	Cosmic background radiation

temperature and appropriately remote, could continuously maintain the life-promoting surface temperatures that the earth has enjoyed for billions of years.

Adult mammals and birds are remarkable among animals generally in their ability to maintain a sensibly constant *core temperature* in the face of relatively wide variations in ambient temperature (see above quotes by Hardy and Kemp). Perhaps the most out-standing example is the arctic fox (*Alopex lagopus*), which is able to preserve its core body temperature at ambient temperatures of −50 °C with no increase in metabolic rate and no electric blanket. A time-averaged constant body temperature requires that heat production be equal to heat loss. Heat production reflects metabolically expensive exothermic biochemical reactions, responsive to hormones including adrenaline (short-term), thyroxine and triiodothyronine (both long-term), muscle contraction (including shivering), and in some instances activation of brown fat. Heat loss occurs by convection and also by conduction, radiation, and evaporation (sweating or panting) of water (from mouth, respiratory system, skin). A relatively small amount of heat is also lost in many species via excretion of urine and feces. In non-sweating species, an increase in ambient temperature may be accompanied by an increase in respiration rate (panting).

The hypothalamus is the primary controller of body temperature in vertebrates. Core body temperature may be regarded as one of the more precisely regulated (±1 °C) of the various homeostatic variables in humans and many other mammalian species (see below). In any case, it is not surprising that the measurement of body temperature is one of the simplest and most direct steps to assessing the state of health in man and other mammals. As we will see in a moment, core body temperature in most mammals studied thus far averages around 36 °C.

It should be said that not all mammals maintain a constant body temperature. Some mammals (heterotherms), including many species of bats, allow their body temperature to approach ambient temperature when they are inactive or during daily torpor. In flight, the body temperature of bats may rise to 41.5 °C. Other mammals, including bears, permit a modest drop in body temperature during hibernation [1]. The naked mole rat (*Heterocephalus glaber*) shows little or no regulation of body temperature [2]. Some relatively large land mammals, for example the camel and the elephant (when water is in short supply), allow body temperature to rise during the heat of day (no doubt to

Table 17.2 LSQ analyses of body temperature (°C) as a function of adult body mass [4] [a]

Col./Row	N_{ord}	N_{sp}	MPD	pWR	DI (%)	r^2	Slope ± 99% CL	Intercept (°C) LSQ	Intercept (°C) Mean [b]	Entity
	a	b	c	d	e	f	g	h	i	j
1	1	25	2	2.2	0.5	0.007	−0.002 ± 0.012	38.2	37.7	Carnivores
2	1	42	4.6	2.1	0.5	0.040	0.009 ± 0.020	34.0	35.2	Chiroptera
3	13	263	4	5.1	26	0.017	0.003 ± 0.004	36.0	36.6	Eutherians
4	2	27	4.4	2.5	0.6	0.193	−0.015 ± 0.017	37.8	36.1	Insectivores
5	2	54	2.6	3.6	16	0.013	0.002 ± 0.007	34.7	35.2	Marsupials Monotremes
6	15	317	3.9	5.1	32	0.0045	0.001 ± 0.003	36.0	36.3	FDS

N_{ord} = number of orders; N_{sp} = number of species; MPD = mean percent deviation; pWR = log(BW_{max}/BW_{min}); DI = diversity index; r^2 = coefficient of determination; CL = confidence limits; LSQ = least squares; Col. = column; FDS = full dataset.
[a] Just those body size data were employed which could be checked against the standardized body weight table (SBT) (Chapter 4 and Appendix B) and found to be satisfactory; [b] arithmetic mean of body temperatures for each sample.

conserve water and energy) and then to drop during the cool of night, for a daily temperature swing of up to 5–6 °C. In addition, mammals are able to conserve heat by introducing insulation. For example, fat has about three times the insulation value of water. On the whole, despite blood circulation, the tissues of the body provide effective heat insulation. For a discussion of the possible role of body temperature in determining SMR in mammals, see Clarke *et al.* [3] and below.

The results from LSQ analyses of logarithmically transformed coordinates of *subsets* of body temperature data afforded by White and Seymour [4] are presented in Table 17.2 (columns c to h, and rows one to six) and Figure 17.1, panel a. Remember that Figure 17.1a is a semi-log plot, with body temperature on a linear (ordinate) scale plotted against body mass on a logarithmic scale (abscissa). However, in panel b, the *ontogenetic* data (recovered by graphical interpolation) for the white-footed mouse (*Peromyscus leucopus*) are plotted on a linear–linear scale [5]. Observe, as is believed to be normative in developing mammals, that body temperature starts out well below the mean adult value. Figure 17.1c is also a semi-log plot. The two most deviant species in panel c which fall well below the mean (see panel a) are monotremes: the short-nosed echidna (*Tachyglossus aculeatus*) and the long-nosed echidna (*Zaglossus bruijni*).

The LSQ slopes given in Table 17.2, column g (rows one to six), are consistent with the assumption that body temperature in these various samples is essentially independent of body size. Notice in Table 17.2, bottom row, that the *LSQ estimate* of mean body temperature (i.e., the LSQ intercept, column h) and the *arithmetic mean* body temperatures (column i) differ by only 0.3 °C (36.3 − 36.0) for 317 different species. See also Figure 17.1, panels a and b, where the *dotted lines* represent the arithmetic mean of the various body temperatures.

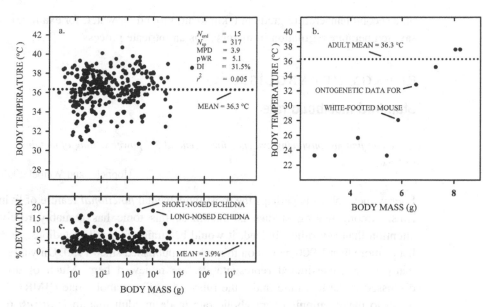

Figure 17.1 Body temperature as a function of body size in mammals [4].

The greatest spread (1.7 °C) between the LSQ mean temperature (computed from log–log coordinates) and the arithmetic mean temperature is shown by insectivores (Table 17.2, row four). Considering different *taxonomic* orders of mammals, the lowest mean body temperature reported here is that for bats (column h, row two), the highest those of carnivores (columns h, i, row one). The mean percent deviations (MPD) of 2 to 4.6% shown in Table 17.2 (column c) are among the lowest found in this work. In larger samples than those employed here, White and Seymour [4] did find correlations between body temperature and body size in marsupials, eutherians, and mammals generally.

It is of interest to inquire into the factors which may enter into the "selection" of a species-specific value for body core temperature. From the viewpoint of endotherms unloading heat, it is advantageous to have a body temperature somewhat above mean ambient temperature. If core temperature is significantly below the mean ambient temperature, a likely consequence is a substantial water loss due to evaporation, a loss which in arid environments may be lethal (see Chapter 18). So there is a plausible argument from the standpoint of water and energy economy for body temperature to be slightly higher than mean ambient temperature. However, small nocturnal mammals, which are active in cooler environments than may be the case for diurnal mammals, tend to have lower body temperatures than other mammals [6].

It seems likely that the original "choice" of core temperature involved various trade-offs, which during the early evolution of any given mammalian species became, over time, "frozen in," so to speak. If true, the values of core temperature one observes today need not be adaptive to contemporary ambient temperatures [7]. One should also emphasize that heat flow from the core of the body to the surface depends on a gradient in temperature. Likewise, heat flow from the body surface to the environment depends on a second temperature gradient (skin to air, ground, or water, for example). Suffice to say, temperature regulation in mammals is an intricate process.

ENERGY METABOLISM

Standard metabolic rate

...we can find no convincing evidence for a general 3/4 power scaling of metabolic rate...

Hoppeler and Weibel 2005

Energy metabolism is perhaps the most fundamental *functional* feature of living organisms. Among scaling studies in mammals, this topic has arguably received more attention than any other. Indeed, it would be straightforward, if tedious, to draw up a list of more than 1,000 papers on this subject published since 1930. Only a few of these many papers, considered representative, are reviewed here. Much of this work is discussed in the literature under the rubric "basal" metabolic rate (BMR). This usage refers to measurements of metabolic rate made in adult mammals *at rest* (during the inactive phase of the circadian rhythm), in the *post-absorptive* state (after 10 to 12 hours of fasting in the adult human), and in the ambient *thermoneutral* temperature range.

Although the acronym BMR is in wide use, it is somewhat misleading. It is known (certainly in humans) that energy metabolism is below "basal" during sleep [8]. The same is true of those heterothermic mammals which hibernate or enter daily torpor [9]. For this reason the phrase "standard metabolic rate" (SMR) is used here in preference to BMR. For a brief history of the study of energetics in mammals, see Hulbert and Else [10]. For a different terminology than used here see Withers [11]. A synopsis of three earlier LSQ analyses of SMR (from two laboratories) in adult mammals is provided in Table 17.3.

The first row of Table 17.3 reports the findings from a re-analysis of a dataset published by Kleiber [12] in 1932 (see Box 17.1). Using his full dataset, including birds and multiple points for some species and not others, one obtains a slope of 0.738 ± 0.022 (99% CL) (not shown in Table 17.3). Next, one omits his data for birds and uses only one point per mammalian species. From this reduced dataset, the best-fit LSQ slope is found to be 0.725 ± 0.047 (99% CI is 0.678 to 0.772) (Table 17.3, row one). Observe that the 99% confidence interval (CI) includes 3/4 and nearly 2/3 ($0.678 - 0.667 = 0.011$).

Table 17.3 LSQ analyses of SMR in selected earlier studies as a function of adult body mass

Row	N_{ord}	N_{sp}	N_{pnts}	MPD	pWR	DI (%)	r^2	Slope $\pm$ 99% CL	Year	Source
1	4	5	5	3	3.5	1.6	1.000	0.725 ± 0.047	1932	[12] [a]
2	6	11	11	9	5.3	5.4	0.998	0.723 ± 0.037	1945	[13] [a]
3	5	12	26	9	4.5	4	0.997	0.756 ± 0.025	1947	[14]

See Table 17.2; N_{pnts} = number of points. [a] One point per species, birds omitted.

Row two of Table 17.3 brings together the outcomes of a LSQ analysis of a *subset* of the data Brody [13] published in 1945. The LSQ slope derived from these data is 0.723 $\pm$ 0.037 (99% CL), not significantly different from the slope derived from the data of Kleiber [12] (row one). In his book Brody [13] often rounded his best-fit LSQ slope down to 0.7, in accord with his pragmatic stance on this issue.

The modern period of interspecific bioenergetic studies in adult *mammals* began with a well-known paper by Kleiber [14], published in 1947 (see Table 17.3, bottom row). In this work he collated measurements of standard energy metabolism from 14 different

---

Box 17.1 Max Kleiber

Max Kleiber (1893–1976) was born in Zurich, Switzerland. He earned a Doctor of Science degree from the Federal Institute of Technology in 1924 and then joined the Department of Agricultural Husbandry at the University of California Davis in 1929. Two papers on basal energy metabolism in mammals that he published in 1932 [B1] and 1947 [B2] are still often cited. In these papers he argued that basal metabolism as determined in adult mammals scales not according to external body surface area, as then widely believed, implying an exponent of 2/3 for geometrically similar animals of constant density, but rather according to the three-quarters power of body weight. Part of his rationale for choosing an exponent of 3/4 was that it could be conveniently calculated on a slide-rule (a then common mechanical device for doing numerical calculations, now displaced by the electronic pocket calculator) and in any case was not statistically different from the empirical exponents of about 0.73 derived at that time by himself and others, such as Brody [B3]. Thus he saw the choice of a specific exponent, in part, as merely a matter of practical convenience.

Conceivably Kleiber would have been bemused had he lived to see the vast attention given to what he may privately have regarded as little more than a rule-of-thumb. Whatever may be the fate of his three-quarters power rule (see text), there is no question that it has exerted an enormous influence on bioenergetic studies in mammals and birds. In addition to his well-known contributions to mammalian energetics, Kleiber was also an early proponent of the use of isotopes for the study of metabolic processes in intact (alive and undissected) animals.

In 1961 Kleiber published a book [B4] that remains one of a small number of books on scaling and energetics from that era still meriting study with respect to scaling in mammals.

B1. Kleiber, M. (1932). Body size and metabolism. *Hilgardia*, **6**:315–353.
B2. Kleiber, M. (1947). Body size and metabolic rate. *Physiological Review*, **27**:511–541.
B3. Brody, S. (1945). *Bioenergetics and Growth: With Special Reference to the Efficiency Complex in Domestic Animals*. New York: Reinhold. (e.g., see Figure 13.7, p. 370 and Table 13.1, p. 388.)
B4. Kleiber, M. (1961). *The Fire of Life: An Introduction to Animal Energetics*. New York: John Wiley & Sons.

papers plus his own unpublished data for the rat (presumably *Rattus norvegicus*) (Kleiber identified species by common names only). All of his data pertained to laboratory or domesticated species, drawn from five taxonomic orders. In his data sample of 26 records (i.e., N_{pnts} = 26) (see Table 17.3) there were multiple records for five species: six rabbits, four cows, four dogs, three humans, and two sheep. In effect these five species were given a greater statistical weight (19 points) than were data for seven other species, each assigned only one point. The values of pWR are 2.6 for the five multiple species and 3.3 for the seven single species; pWR for the whole sample is 4.5 (Table 17.3, bottom row). Daily activity cycles were apparently ignored [15].

In seven cases (four rabbits, two sheep, one cow) in Table 17.3, bottom row, the body sizes fall outside the "normal" range as defined by the SBT (Chapter 4, Appendix B). The diversity index for Kleiber's sample is 4%. As McNab [16] noted, in retrospect Kleiber's dataset represents a biased sample from the perspective of mammals generally. The inferences from a LSQ analysis of the full dataset (FDS) (N_{sp} = 12, number of points N_{pnts} = 26) provided by Kleiber [14] are reviewed in Table 17.3, row three. The best-fit LSQ slope for these data is 0.756 ± 0.025 (99% CL).

If, instead, one uses a single point per species, choosing the body mass closest to the logarithmic mean in each case of multiple points per species, one obtains a LSQ slope of 0.758 ± 0.037 (99% CL). Thus the effect of using multiple values for single species is not, as one might suppose, to alter the slope but rather to narrow the confidence limits (from 0.037 for 12 records to 0.025 for 26 records). Kleiber [14] went on to state that "the 3/4 power of body weight is therefore recommended as representative of metabolic body size." Kleiber's affirmation of the three-quarters power rule for the scaling of SMR in adult mammals was generally accepted for the next 35 years. The following synoptic analysis of SMR in adult mammals takes a broadly similar approach to that of Dodds *et al.* [17]. Data derived from six different laboratories are reviewed. See also above quote from Hoppeler and Weibel.

Data of Bartels (1982)

There is an exponent [for SMR] *of 0.66 for mammals in a body weight range from 2.5 g up to 3.8 tons, whereas mammals within the range of 2.5–100 g show an exponent of only 0.43...*

Bartels 1982

LSQ analyses of SMR over varied adult body size ranges by Bartels [18] in 1982 led, over time, and with further work, to growing doubts as to the general validity of the three-quarters power rule. His inferences are restated in Table 17.4 (unfortunately neither "raw" data nor confidence limits on the slopes were provided).

Possibly his most significant finding (Table 17.4, row one) is that a slope of only 0.42 obtains over the narrow size range from 2.4 to 260 g (pWR = 2.0). This inference is however weakened by the low coefficient of determination (0.76). For body sizes of 260 g to 3.80 tonnes he reported a slope of 0.76 (row two). Finally, for his full size range (2.4 g to 3.8 tonnes) (pWR = 6.2), the slope dropped back to 0.66 (Table 17.4, row three). (Observe that Kleiber's 1947 dataset – mouse to cow – extended from 21 g

Table 17.4 Summary of SMR findings of Bartels [18]

Row	Mass range (g)	pWR	r^2	Slope
1	2.4–260	2.0	0.76	0.42
2	260–3,800,000	4.2	0.98	0.76
3	2.4–3,800,000	6.2	0.96	0.66

See Table 17.2.

to 0.6 tonnes, i.e., pWR = 4.5; Table 17.3, row three.) Remember that the difference in slope between 3/4 and 2/3 (i.e., 0.083), if applicable over the whole mammalian size range (pWR = 8.0), implies a difference in SMR of 2.2 doublings or a factor (f) of 4.6 (0.083·26.6 = 2.2 doublings; $f = 2^{2.2} = 4.6$) (Chapter 3). Stated otherwise, in large cetaceans, SMR would be about 4.6 times larger if it scaled as the three-quarters power of body size rather than the two-thirds power.

Bartels [18] was perhaps the first to argue persuasively – and, it is argued here, correctly – that the slope for SMR varies significantly with the body size range considered (see above quote from Bartels). This finding implies that SMR is a *non-linear* function of body mass on a log–log plot. Slopes may also vary with activity cycles, a fact often overlooked [15].

Data of McNab (1988)

McNab [19] published an especially useful paper on SMR in 1988, wherein he presented "raw" SMR data for 320 species (see also Dawson and Hulbert [20]). McNab [19] systematically excluded data for laboratory and domesticated species. Five records from his dataset, where species names were not given or appeared to be duplicates were omitted, giving a dataset of 315 records. McNab [19] was himself involved in about one-quarter of the measurements reported. Few biologists were then better qualified to select suitable data on SMR from the literature (see also McNab [21] and Blaxter [22]).

McNab [19] concluded from his 1988 study that SMR in adult mammals with body weights less than 300 g scales with an exponent of about 0.60 and an exponent of about 0.75 in mammals weighing more than 300 g (see also Table 17.4, rows one and two). I carried out an extensive LSQ re-analysis of McNab's data, validating the (mainly) robust character of his findings; a small portion of this (unpublished) work is précised in Table 17.5. In carrying out these re-analyses McNab's data were changed from the (relative) mass-specific form of $cm^3 O_2/g \cdot h$ to the (absolute) mass-free form of $cm^3 O_2/h$. (Plotting a y-variable as a ratio involving mass in the denominator against mass (x-variable) may introduce spurious correlations.)

If we restrict our attention to mammals with resting *body temperatures* in the range of 36–37 °C, we obtain a slope of 0.671 ± 0.041 (99% CL) (Table 17.5, row one) (see below). The best-fit slope for *eutherians* is 0.716 ± 0.025 (99% CL) (Table 17.5, row two); that for *mammals generally* ($N_{sp} = 315$) is virtually the same (Table 17.5, row

Table 17.5 LSQ analysis of McNab's 1988 SMR dataset [19]

Row	N_{ord}	N_{sp}	MPD	MPD	DI (%)	r^2	Slope ± 99% CL	99% CI	Entity
1	7	47	16	3.3	6.3	0.977	0.671 ± 0.041	0.630–0.712	BT = 36.5 ± 0.5 °C
2	16	267	31	5.3	33	0.955	0.716 ± 0.025	0.691–0.741	Eutherians
3	18	315	30	5.3	39	0.955	0.713 ± 0.023	0.690–0.736	FDS

See Table 17.2. BT = body temperature; CI = confidence interval; FDS = full dataset.

Table 17.6 LSQ analysis of subset of the SMR data of White and Seymour [4]

N_{ord}	N_{sp}	MPD	MPD	DI (%)	r^2	Slope ± 99% CL	99% CI	Sample
15	586	39	5.1	35	0.912	0.677 ± 0.022	0.654–0.699	FDS [a]

See Table 17.2; [a] 45% of the species in this sample are rodents and 78% of the species have body masses less than 1 kg. Size range is from a 2.4 g shrew to a 325 kg moose. No data for cetaceans, elephants or pinnipeds are included.

three). This slope (0.713 ± 0.023 (99% CL) does not convincingly support a hypothetical slope of either 2/3 or 3/4.

Data of White and Seymour (2003)

In 2003 White and Seymour [4] assembled a much larger dataset (619 species) than that of McNab [19]. They corrected all their metabolic rates to a common core body temperature of 36.2 °C (see also Figure 17.1), using throughout a fixed Q_{10} of 3.0. (Q_{10} is the measure of the rate of change in a biological or chemical system when the temperature increases by 10 °C.) The results of a LSQ analysis of their data (after eliminating data for apparently duplicate species and for one indeterminate species, and correcting an apparent misprint) are displayed in Table 17.6 and Figure 17.2. The best-fit slope for 586 species is 0.677 ± 0.022 (99% CL) (Table 17.6). This slope is consistent with a hypothetical slope of 2/3 but not of 3/4. Note that the diversity index (DI) is 35% in this sample (Table 17.6). The added data for the killer whale (*Orcinus orca*), weighing 3.6 tonnes, shown in Figure 17.2a, are from Kasting *et al.* [23]. This datum was not included in the LSQ computations. The human ontogenetic data shown in Figure 17.2b are from Holliday [24].

In Figure 17.2a, the rightmost upper seven points at the highest body sizes all lie above the best-fit LSQ line. If we omit these seven outliers the resultant slope of 0.658 ± 0.025 (99% CL) is not significantly different from the value of 0.677 ± 0.022 (99% CL) (Table 17.6). The two highly deviant species seen in Figure 17.2c are the Indian flying fox (*Pteropus giganteus*) and the ghost bat (*Macroderma gigas*).

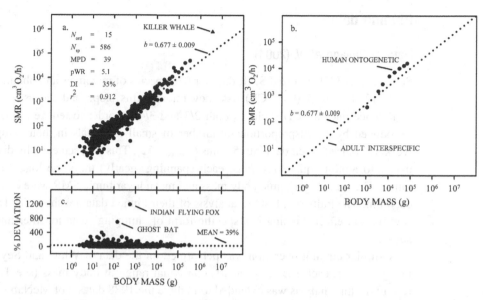

Figure 17.2 Scaling of SMR in mammals as a function of body size based on a subset of the data of White and Seymour [4].

There is a simple test to see whether core body temperature *may* play the role in SMR that White and Seymour [4] assume. One simply takes their published body temperature data and combines them with the metabolic rate data of McNab [19], which were not corrected for body temperature. One finds 47 species in the McNab dataset with body temperatures in the range of $36.5 \pm 0.5\,°C$. This range includes the body temperature ($36.2\,°C$) which White and Seymour [4] employed as a common standard (see above). Thus, on average, no correction to SMR for body temperature is required.

For example, consider the marsupial species Mulgara (*Dasycercus cristicauda*). White and Seymour [4] give a body temperature of $36.9\,°C$ for this species. Thus this species falls into the body temperature range of $36.5 \pm 0.5\,°C$. We find from McNab [19] that this species has a body mass of 89 g and a BMR (or SMR) of $0.52 \text{ cm}^3 \text{ O}_2/\text{g}\cdot\text{h}$. Converting BMR to absolute units, we obtain $46.3 \text{ cm}^3 \text{ O}_2/\text{h}$ ($89\cdot0.52 = 46.3$). Recall that the only data coming from White and Seymour [4] are for species identification and body temperature. No metabolic data taken from White and Seymour [4] are employed in this exercise. Proceeding in the same way, body masses and BMR (or SMR) values were obtained for another 46 species in the McNab [19] dataset.

The body masses in this *subset* of McNab's data [19] ranged from 15 g to 29.3 kg (pWR = 3.3). A LSQ analysis of these data gave a slope of 0.671 ± 0.041 (99% CL) (see Table 17.5, row one), roughly consistent with the conclusion of White and Seymour [4] that SMR scales with body size to the two-thirds power in the given sample. This result does *not* depend on correcting SMR to a given core body temperature using a constant Q_{10}. This finding supports the contention of White and Seymour that body temperature may be useful in assessing SMR. For a contrary view, see below.

Binning data

Data of Savage *et al.* (2004)

Savage *et al.* [25] analyzed SMR data for 626 species of mammals and obtained a slope of 0.712 ± 0.0125 (95% CL). As they remark, this slope and the associated 95% confidence interval (CI) excludes both 2/3 and 3/4. In order to reduce a possible bias introduced by the disproportionate number of small mammals in their sample, these workers "binned" their data into 52 bins (N_{bins} = 52). This was achieved by dividing the data into almost equal-sized *bins*, each spanning (nearly) the same logarithmic mass range. The mean logarithmic body mass and mean logarithmic SMR were computed for each bin. The fruits of a LSQ re-analysis of their binned data are shown in Table 17.7, row one. In effect, binning is a specific form of "lumping" data together into disjoint subsets.

A similar binning operation was performed on the data of White and Seymour [4], using 21 bins, each spanning (nearly) an equal range in body mass (see Table 17.7, row two); this analysis was extended to include the 1988 dataset of McNab [19], using 10 bins (Table 17.7, row three) and the more recent 2008 dataset (N_{sp} = 638) of McNab [26] using 15 bins (Table 17.7, row four). (An alternate analysis of his 2008 data is provided below.) The consequences of the LSQ analyses of the binned data of White and Seymour [4] and Savage *et al.* [25] are displayed in Figure 17.3, panels a and b. Note that none of the slopes shown in Table 17.7 is significantly different from the *arithmetic mean* slope of 0.729 ((0.737 + 0.719 + 0.732 + 0.727)/4 = 0.729).

Keep in mind that panel a of Figure 17.3 shows the binned data of White and Seymour [4] and panel b that of Savage *et al.* [25]. In both panels the best-fit LSQ line has been displaced downwards for clarity. Careful examination of Figure 17.3, panels a and b, shows that in both panels the binned data are *curvilinear*. The slopes tend to be lower at small body masses, a little steeper at higher body weights, and steeper yet at the highest body weights. This curvature is less obvious when, as is normally the case, a best-fit line is centered on the data. Panels c and d in Figure 17.3 show that the slope from point-to-point increases linearly with the logarithm of body size (see simulation below). The two slopes in panels c and d are numerically the same (see below).

Table 17.7 Scaling of SMR in adult mammals for "binned" data

Row	N_{sp}	N_{bins}	MPD	MPD	r^2	Slope ± 99% CL	99% CI	Source
1	626	52	40	6.2	0.986	0.737 ± 0.034	0.703–0.770	[25]
2	586	21	19	5.1	0.990	0.719 ± 0.047	0.673–0.766	[4]
3	315	10	14	4.6	0.996	0.732 ± 0.054	0.678–0.786	[19]
4	638	15	17	4.9	0.992	0.727 ± 0.054	0.673–0.781	[26]

See Table 17.2. N_{bins} = number of bins.

Table 17.8 Simulation of binned SMR data for four datasets

Number of bins	Numerical constants		MPD	Year	Source
	α	β			
Column	a	b	c	d	e
10	0.943	0.2235	15	1988	[19]
21	0.949	0.223	8	2003	[4]
52	0.929	0.227	25	2004	[25]
15	0.950	0.223	17	2008	[26]
Means	0.943	0.224	16	-	-

MPD = mean percent deviation.

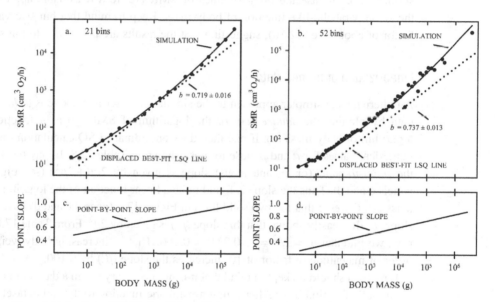

Figure 17.3 Scaling of SMR (for "binned" data) as a function of adult body size [4, 25].

Simulation of binned data by a parabolic function

The four binned datasets of McNab [19, 26], White and Seymour [4], and Savage *et al.* [25] are reasonably well simulated (represented) by a generalized parabolic (*non-linear*) function of the form $y = (\alpha + \beta x)^2$. The findings of these simulations (carried out with commercial software) using this function are brought together in Table 17.8 (see also solid curves – labeled "simulation" – in Figure 17.3, panels a, b). Notice that in these simulations (x, y) are in *logarithmic* form.

The values of the numerical constants (α, β) for each of the four simulations are presented in Table 17.8, columns a and b, respectively. Thus, in the first row, the 10 empirical binned (logarithmic) values of SMR from McNab [19] are simulated by the function $(0.943 + 0.2235x)^2$, where x is a binned value of body mass, expressed in logarithmic coordinates. Compare the values of α and β in rows one to four of Table 17.8. The *mean* values of α and β in Table 17.8 are 0.943 ± 0.008 (STD) and 0.224 ± 0.002 (STD), respectively (bottom row). The given function is *non-linear*, implying that the assumptions underlying ordinary least squares need not obtain.

The mean percent deviations (MPD) between the simulated and binned values of SMR, expressed in native (not logarithmic) coordinates, are shown in Table 17.8, column c. For example, the 10 *simulated* values of binned SMR, in native coordinates, deviate from the 10 *empirical* binned values (also in native coordinates) by a mean of 15% (range is 1.3 to 25.1%) for the data in Table 17.8, row one. The mean value (16%) of the four MPD is given in Table 17.8, bottom row.

The main implication of the findings presented in Table 17.8 is that the four separate empirical datasets of binned values of body mass and SMR are numerically very similar. In each case the binned values of SMR are, to a reasonable approximation, the same parabolic-like function of body mass. Keep in mind that bin size varies by a factor of about five (52/10), suggesting that the results are *insensitive* to bin size.

Linearization of binned data

An alternate and simpler approach to the simulation described above is to *linearize* the data by plotting the *square root* of the logarithm of SMR (y) as a function of the logarithm of body mass (x). If one then does an ordinary LSQ calculation one obtains numerical values for α and β close to those shown in Table 17.8. It may be shown that the *point-to-point* (or instantaneous) slope is given by $2\alpha\beta + 2\beta^2 x$ (see Figure 17.3, panels c and d). Thus the slope is found to be a linear function of the logarithm of body mass (x). Observe that the two solid lines in Figure 17.3 (panels c and d) have the same slope, β'. It is easily shown that this slope (β') is given by $2\beta^2$. From Table 17.8, bottom row, we find that $\beta' = 2\beta^2 = 2 \times 0.224^2 = 0.100$. That is, increasing body weight by an order of magnitude (a factor of 10) increases the slope (β') by 0.100.

Finally, it should be kept in mind that it is not necessary to bin a dataset in order to fit a parabolic function to it. After fitting a parabolic function to the full dataset (FDS) of McNab [26] one obtains by simulation values of $\alpha = 0.943$ and $\beta = 0.227$. These are close to the values given in Table 17.8, bottom row. This exercise suggests that the several outcomes shown in Table 17.7 are not due to binning.

End-samples

We have FDS for SMR for two datasets derived from McNab (see Tables 17.5, row three and Table 17.10, row four) and another FDS derived from the data of White and Seymour [4] (see Table 17.6), each having more than 100 records. I constructed end-samples (Chapter 5) for these three FDS consisting of the ten smallest and ten largest species for each dataset. I linearized the above parabolic function by expressing the square root of the logarithm of SMR as a function of the logarithm of body mass. Then using ordinary LSQ I computed the values of the parameters α and β. The inferences

Table 17.9 Numerical constants (α, β) as calculated from end-samples (N_{sp} = 20) for three datasets

Numerical constants			
α	β	Year	Source
0.848	0.240	1988	[19]
0.956	0.221	2003	[4]
0.885	0.231	2008	[26]
0.896	0.231	Means	-

Table 17.10 LSQ partition analysis of McNab's 2008 SMR dataset [26]

Row	N_{sp}	MPD	pWR	r^2	Slope $\pm$ 99% CL	99% CI	Entity
1	306	31	1.6	0.601	0.647 $\pm$ 0.078	0.568–0.725	BM < 100 g
2	332	32	4.5	0.919	0.772 $\pm$ 0.033	0.739–0.805	BM > 100 g
3	170	36	3.5	0.911	0.834 $\pm$ 0.053	0.782–0.887	BM > 1,000 g
4	638	32	6.2	0.959	0.721 $\pm$ 0.016	0.705–0.737	FDS

See Table 17.2. BM = body mass; FDS = full dataset.

from these LSQ calculations for the three given datasets are shown in Table 17.9, where rows one and three are derived from the data of McNab [19, 26] and row two from White and Seymour [4].

The constants (α, β) in Table 17.9 are in fair agreement with each other and with the parameter values given in Table 17.8. Compare the means of α and β given in the bottom rows of Tables 17.8 and 17.9. Observe that in the three examples given in Table 17.9, the end-samples constitute only 3–6% of the respective full datasets. This finding is consistent with the hypothesis that the *non-linearity* in the SMR data is at most weakly influenced by ecological and phylogenetic effects. That is, the values of the constants α and β are not strongly affected when data for most species are omitted.

Partitioning data

Data of McNab (2008)

An alternative to binning data to reveal potential *non-linearity* is to partition the data into subsets spanning different body size ranges (see also discussion of Bartels' data above, and Chapters 21, 23). As mentioned previously, McNab [26] (2008) provided SMR data for 638 species of adult mammals. I used his data to examine the effects of partitioning the data. In carrying out this analysis we assume that if the sample size is large enough (here $N_{sp} \geq 170$) and the size range sufficiently large (here pWR ≥ 1.6), then the slopes for the end-sample subsets will not differ significantly from the slope for the full dataset (FDS) if the data are truly distributed *linearly or nearly linearly* on a log–log plot.

The fruits of LSQ analyses of three different *subsets*, as well as the *full* dataset (FDS) of McNab [26], are outlined in Table 17.10 (see last column). Observe that the slopes (0.647 and 0.834) for the first and third rows in Table 17.9 are statistically different at the 99% CL. The slope ($b = 0.721$) for the FDS (Table 17.10, bottom row) is inconsistent with a slope of either 2/3 or 3/4. These findings are consistent with the results obtained by "binning," in pointing to inherent *non-linearity* in the SMR data.

Inferences from above calculations

1. When the logarithm of SMR is plotted against the logarithm of body mass, the resultant plot is *non-linear*. However, the plot is well linearized by plotting the *square root* of the logarithm of SMR against the logarithm of body mass. It is then found that the point-to-point slope is a *linear* function of the logarithm of body mass. The higher the body mass, the greater the slope (Figure 17.3, c and d).
2. The fact that the four sets of numerical constants (α, β) in Table 17.8 are in good agreement with one another implies that these four datasets provide us with essentially the *same* inferences regarding the scaling of SMR. Probably additional data from species lying within the same body size range (as represented in the above datasets) will provide information of only limited usefulness insofar as scaling of SMR in mammals generally is concerned.
3. The numerical constants in Table 17.8, row two, were derived from values of SMR corrected to the same body temperature by White and Seymour [4]. Since the constants (α, β) are much the same as those for three other datasets not corrected for body temperature, it is likely that body temperature is not of primary relevance to these calculations [27].
4. The fact that the end-samples (Table 17.9) ($N_{sp} = 20$) for three datasets give constants (α, β) fairly similar to one another and to the constants in Table 17.8 suggests that *ecology* and *phylogeny* play at most a modest role in determining the slope of the LSQ line governing the scaling of SMR as a function of adult body mass in mammals generally.

Further problems associated with the scaling of SMR

Although BMR has been widely used in comparisons among species, its biological significance is obscure.

Ricklefs *et al.* 1996

There is no reason to suspect that differences in metabolic levels [in particular groups of homeotherms] *are due to differences in body temperature.*

Poczopko 1980

The three-quarters power rule of Kleiber [14] as applied to SMR has provoked extensive discussion and criticism (e.g., [27–30]). Bartels *et al.* [31] noted that the standard

Table 17.11 Species with similar body masses and body temperatures but different values of SMR [4]

Species	Body mass (g)	BT (°C)	SMR (cm³ O₂/h)	Species	Body mass (g)	BT (°C)	SMR (cm³ O₂/h)
Rhinonycteris aurantius	8.27	36.1	9.4	*Mus minutoides*	8.06	36.3	24
Scotinomys xerampelinus	15.2	36.2	31.9	*Carollia perspicillata*	14.9	36.4	108
Burramys parvus	44.3	36.1	36.8	*Artibeus jamaicensis*	45.2	36.4	78
Ptilocerus lowii	58	36.5	43.5	*Tamias amoenus*	57.1	37	96.3
Tamias merriami	75	37	78.8	*Dobsonia minor*	73.7	36.5	174

BT = body temperature; SMR = standard metabolic rate.

conditions prescribed for SMR in large mammals (adult, resting, post-absorptive, thermoneutral state, and respect for activity cycles) may not be pertinent to small active mammals, including some species of insectivores. For example, Newman and Rudd [32] found that members of the species *Sorex sinuosus* died after 5 to 23 hours of fasting. As it turns out, when one deletes insectivores from the dataset of White and Seymour [4] the slope is not significantly changed (i.e., the slope without insectivores is 0.691 ± 0.023 (99% CL), rather than a slope of 0.677 ± 0.022 (99% CL) (Table 17.6)). It is also the case that the biological significance of SMR (or BMR) is unclear (see above quote from Ricklefs *et al.*).

The constant Q_{10} correction factor (of 3.0) for body temperature that White and Seymour [4] applied to all their metabolic data might be taken to imply that there is a simple *physical* relation between metabolic rate and body temperature, with higher body temperatures necessarily implying higher metabolic rates. Examples showing that an obligatory relationship between SMR and body temperature need not obtain in all mammals are provided in Table 17.11 [4]. We see that mammals with similar body weights and body temperatures may have values of SMR differing by a factor of 2 to 3. For instance, in row two, different species with body weights of 15.1 and 14.9 g, and body temperatures of 36.3 and 36.4 °C nevertheless have values of SMR that vary by a factor of 3.4 (108/31.9 = 3.4). For a critique of the method White and Seymour [4] employed to correct metabolic rate for variations in body temperature see Savage *et al.* [25]; see also Heldmaier and Ruf [33], and Blaxter [22].

It is straightforward to construct a table (not shown) similar to Table 17.11, wherein two species with similar body sizes have different body temperatures, yet the species with lower body temperatures have higher values of SMR. Also, it is known that some species of desert rodents can increase their body temperatures significantly without increasing their oxygen consumption [34]. Although hibernating bears show a significant decrease in metabolic rate, this decrease is not primarily attributable to a drop in body temperature [35]. It is evident that body temperature is not necessarily a direct determinant of SMR (see above quote from Poczopko). As implied above, it is difficult to know how uniform the physiological conditions actually are when SMR is measured in large and small species [22].

In fact uniform physiological conditions for determination of SMR across many species of very different body sizes may not be possible. Moreover, we cannot know how SMR scales in mammals generally until we have comparable information on energy metabolism in large pinnipeds and cetaceans weighing from 1 to 150 tonnes; see also Agutter and Wheatley [36]. In diving mammals, blood flow to the visceral organs appears to be severely restricted during a dive, so that SMR might be lower during a dive (with no external oxygen available) than when "resting" at the surface.

Relative organ metabolic rate in the adult human and organ size in selected mammals

The expensive-tissue hypothesis… emphasizes the essential interrelationship between the brain, BMR and other metabolically expensive tissues.

Aiello and Wheeler 1995

It is well known that four organs (brain, heart, kidney, and liver), constituting less than 6.5% of adult human body size (see Table 17.12, column e), account for some 54

Table 17.12 Organ mass in selected mammals and organ oxygen uptake in the resting adult human

Organ mass (as % of adult body mass)						Organ VO$_2$ (% total)
Species	Shrews			Mouse	Human	Blue whale	
	Sorex araneus	*Sorex araneus*	*Suncus etruscus*	*Mus musculus*	*Homo sapiens*	*Balaenoptera musculus*	*Homo sapiens*
Column/ Organ	a	b	c	d	e	f	g
Brain	2.5	3.77	2	1.8	2.1	0.007 (0.008) [a]	18 (16)
Fat/ blubber	4.2	-	-	10.3	13.4	14.6 (21)	-
Heart	1.3	1.1	1.2	0.6	0.7	0.3 (0.5)	12 (11)
Kidney	2.0	2.1	1.9	1.8	0.6	0.3 (0.4)	7.2 (7.7)
Liver	6.3	7.8	6.8 [b]	6.2	2.9	0.7 (0.8)	20 [c] (19)
Lungs	-	-	0.94	0.72	2.6 [d]	0.8 (1.2)	4.4
Skeletal muscle	38.2	-	-	34.2	41.4	45 (46)	20 (15)
Skin	12.4	-	-	14.3	8	-	4.8 (1.7)
Body mass (g)	7.7	6.8	2.5	23	67,750	136×10^6 (122×10^6)	63,000
Source	[37]	[38]	[39]	[40]	[41]	[42] ([43])	[44] ([45])

[a] Brain weight assumed to be less than 10 kg (see Chapter 13); [b] includes gall bladder; [c] hepatic portal VO$_2$; [d] estimates of normal human adult lung mass vary by a factor of 2 (or more). References in parentheses refer to results in parentheses.

(16 + 11 + 7.7 + 19 = 54) to 57% of total oxygen consumption (based on *in vivo* measurements) in the resting adult human (see Table 17.12, column g). These organs may be categorized as metabolically "expensive" (see above quote from Aiello and Wheeler). Masses for these organs scale with exponents of body size of 0.759 ± 0.013 (brain, Chapter 13), 0.953 ± 0.022 (heart, Chapter 9), 0.880 ± 0.021 (kidney, Chapter 16), and 0.895 ± 0.018 (liver, Chapter 10) (99% CL in each case).

One sees from column f in Table 17.12 that these four organs account for less than 1.4% of body mass in an adult blue whale (0.007 + 0.3 + 0.3 + 0.7 = 1.3). That is, mass for these four organs as a percent of total body mass drops by a factor of 4.6 between human and blue whale (6.5/1.4 = 4.6). It seems likely that the energy metabolism of these four organs will be a smaller fraction of the total in a blue whale than in a human. To wit, the contribution of each of these four organs is likely to be a decreasing fraction of total SMR with increasing body size.

In the adult mouse (Table 17.12, column d) these four tissues account for 10.4% of body mass (e.g., 1.8 + 0.6 + 1.8 + 6.2 = 10.4) whereas these same four organs comprise 12.1–14.8% of body mass in an adult shrew (see Table 17.12, columns a, b, c), about twice as high as the figure (<6.5%) for the adult human. If anything, one expects these four organs to utilize an even larger fraction of total oxygen consumption in the shrew [18]. Note that for the data in Table 17.12, the most metabolically "expensive" single organ (per g tissue) in the adult human is the heart; skin is the least expensive.

Wang *et al.* [46] reported in 2001 an analysis of organ metabolic rate for brain, heart, kidney, and liver in the rat and human (plus additional data for the cat and dog). They infer that these four metabolically active organs may account for about two-thirds of SMR in a 100 g mammal and about one-third in a mammal weighing 1,000 kg. These important distinctions tend to be ignored when a lumped-parameter concept of "metabolic body size" is invoked.

Scaling of maximal oxygen consumption (VO$_2$max)

VO$_2$max is the (apparent) upper limit to oxygen consumption that can be achieved without incurring an oxygen debt. In humans, one can be reasonably certain that VO2max is achieved in a given individual when requested, but in other species this may be less certain. VO$_2$max is of particular interest from a scaling standpoint because it may prove possible to uncover necessary relations between the empirical slope for VO$_2$max and structural features of mammals (see below). That is, biological structure implies complementary functional limits (see opening quotes from Brody and Pattee).

A number of different methods are available to assess VO$_2$max. These include exposure to low ambient temperatures (cold-induced), breathing He-O$_2$ (which is highly conductive), and various forms of exercise, including flying (in bats), rowing (in humans), swimming (in many species), or running on a treadmill for bipeds and quadrupeds.

Table 17.13 LSQ analysis of VO$_2$max as a function of adult body mass

N_{ord}	N_{sp}	N_{pnts}	MPD	pWR	DI (%)	r^2	Slope $\pm$ 99% CL	Method	Source
6	53	53	29	3	5	0.936	0.649 $\pm$ 0.063	Cold-induced	From Lit.
2	9	9	14	2.6	1	0.989	0.797 $\pm$ 0.109	Treadmill	[49]
5	30	34	44	4.8	5.8	0.962	0.870 $\pm$ 0.083	Treadmill	[50]
5	25	25	48	4.8	5.5	0.968	0.916 $\pm$ 0.097	Treadmill [a]	[50]

See Table 17.2. N_{pnts} = number of points; Lit. = literature.
[a] Removed duplicate species and those records where body weights were found to be small as judged by the SBT (see Chapter 4 and Appendix B).

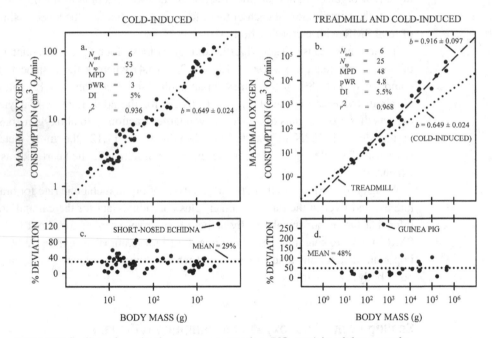

Figure 17.4 Scaling of maximal oxygen consumption (VO$_2$max) in adult mammals.

It was decided to first analyze the findings based on exposure to cold because more data are available than for the other methods. It is known that treadmill exercise in chipmunks, mice, and rats gives values of VO$_2$max that are 16–23% higher than those for cold-induced measurements [47, 48]. For this reason, cold- and exercise-induced VO$_2$max are here treated as separate y-variables. The conclusions of a LSQ analysis of VO$_2$max as determined by exposure to cold are displayed in Table 17.13, row one, and Figure 17.4 (panel a) (data from the literature). The best-fit LSQ slope for cold-induced VO$_2$max is 0.649 $\pm$ 0.063 (99% CL). The results in Figure 17.4a for VO$_2$max were obtained by exposing animals to low ambient temperatures. The percent deviations from the best-fit LSQ line are shown in panel c.

From the data of Taylor *et al.* [49] based on a study of nine species of wild African mammals (size range from mongoose to eland), using one point per species, one finds a best-fit LSQ slope of 0.797 ± 0.109 (99% CL) (see Table 17.13, row two). Bear in mind the wide 99% CL (±0.109). The results shown in Table 17.3, rows three and four, are derived from the data of Weibel *et al.* [50].

In Figure 17.4b, the dashed line represents the best-fit LSQ inferences for VO_2max obtained by exercise on a treadmill. The dotted line in panel b represents the cold-induced trajectories seen in panel a (where MPD = 29). The very deviant point in Figure 17.4c corresponds to the short-nosed echidna (*Tachyglossus aculeatus*) and that in panel d to the guinea pig (*Cavia porcellus*). The best-fit slope (dashed line) has a slope of 0.916 ± 0.097 (observe the rather wide 99% CL).

One might construe the findings in Table 17.13, row one, as being consistent with a slope of 2/3 and those in row two as consistent with a slope of 3/4. Given the rather wide 99% CL (±0.06 and ±0.109, respectively) neither of these inferences is secure. In a more recent publication, based on data taken from the literature, Weibel *et al.* [50] reported data for VO_2max as determined in 30 different species (see Table 17.13, row three). After removing data for duplicate species and some cases where body weights were judged to be low (with reference to the SBT, Chapter 4), one obtains the results depicted in Table 17.13, bottom row. The slopes appearing in rows three and four are not significantly different; but both differ significantly from 3/4. Be aware that VO_2max, like SMR, may also prove to be *non-linear* on a log–log plot when data over a wider size range (say pWR > 7) become available.

Weibel *et al.* [50] correlate the exponent for VO_2max with the empirical scaling exponents for muscle mass, muscle mitochondrial volume, and muscle capillary volume. It seems likely that partly different physiological mechanisms are at work between cold-induced and exercise-induced VO_2max. Exercise tends to increase body temperature, whereas cold induces a peripheral vasoconstriction.

The reader may have noticed that the slopes for maximal oxygen consumption ($b = 0.916$), total alveolar surface area ($b = 0.936$) (Chapter 15, Table 15.1), and for the model-based total area of repeating units ($b = 8/9$) (Chapter 3, Table 3.4) are roughly comparable to one another (recall that lung mass scales empirically as the first power of body mass (Table 15.4)). As a final point it is worth noting that, perhaps contrary to expectations, some species of plants of the Arum family, undergoing inflorescence, have VO_2max levels comparable to some small mammals of similar size [51].

MAXIMAL LIFESPAN IN MAMMALS

In 1987 we reported on the conclusions of an extensive study of maximal lifespan in mammals [52] (see also Jürgens and Prothero [53]). The overall results of this study are displayed in Table 17.14. It was found that maximal lifespan scales as the 0.187 ± 0.023 (99% CL) power of body size. The MPD (63%) is high and the coefficient of determination ($r^2 = 0.41$) is very low. Accordingly this best-fit LSQ line has limited usefulness for predicting maximal lifespan in individual species. However, the exponent of 0.187 ± 0.023 (99% CL) for maximal lifespan does *not* support a one-quarter power rule.

Table 17.14 LSQ analysis of maximal lifespan (years) in mammals

N_{ord}	N_{sp}	MPD	pWR	DI (%)	r^2	Slope $\pm$ 99% CL	Source
20	578	63	6.2	56	0.41	0.187 ± 0.023	[52]

See Table 17.2.

THERMAL CONDUCTANCE

Thermal conductance is an important property of mammals in relation to understanding the scaling of thermoregulation [54, 55]. It is convenient to start with Ohm's law, relating the current (I) that flows through a resistance (R) subject to a given voltage drop (ΔV) across it. Ohm's law states that:

$$R = \Delta V / I \tag{17.1}$$

Here one thinks of the difference in electrical potential (ΔV) as the cause of current flow. The reciprocal of resistance is conductance (C). A high resistance (or insulation) implies a low conductance. Thus, one has:

$$C = I / \Delta V \tag{17.2}$$

When, by analogy, one applies this rule to physical heat flow one has:

$$C = \text{heat flow} / \Delta T \tag{17.3}$$

where the driving force ΔT is a temperature drop (or gradient) and C is termed the *thermal conductance*. Thermal conductance is a measure of the rate at which an animal loses heat to the environment. The term is somewhat of a misnomer, because heat flow from the core to the body surface in a mammal is a direct function of both conduction (atoms jostling their neighbors) and convection (bulk fluid flow), as well as being influenced at the body surface by convection (of air), evaporation (of sweat), and radiation. These different modes of heat dissipation are unlikely to scale with the same exponents of body size [56]. The specific conductance governing the flow of heat from the skin to the environment may be termed the "coefficient of heat transfer" [57].

In a steady state, at rest, one may assume that heat flow is directly proportional to oxygen consumption (VO_2). Thus, assuming core temperatures are above ambient temperatures, one writes:

Thermal conductance (C) $= VO_2/($core temperature $-$ ambient temperature$)$
$= VO_2/\Delta T$

$$\tag{17.4}$$

where C has units of the form $cm^3\ O_2/h \cdot {}^\circ C$. It is convenient to think of this as a *lumped-parameter* model, which aggregates the different modalities of heat transfer into a single measure. Thermal conductance may be measured in two different ways. When the ambient temperature to which a mammal is subjected varies over a temperature range

Table 17.15 Scaling of thermal conductance as a function of body mass in adult mammals

N_{ord}	N_{sp}	MPD	pWR	DI (%)	r^2	Slope $\pm$ 99% CL	Source
12	136	22	4.1	17	0.93	0.568 $\pm$ 0.036	[60] [a]
8	35	27	3.9	7.8	0.96	0.570 $\pm$ 0.056	[61, 62, 63, 64]
13	100	24	4.8	20.3	0.95	0.544 $\pm$ 0.032	From Lit.

See Table 17.2. [a] 74 of the 136 species in this sample are rodents.

below the thermoneutral zone, the metabolic rate tends to increase linearly with decrease in ambient temperature. The slope of the line relating metabolic rate to ambient temperature is a measure of thermal conductance (see Equation (17.3)).

Alternatively, thermal conductance may be measured as metabolic rate (VO_2) at the critical temperature defined by the lower end of the thermoneutral zone (Equation (17.4)). The mass-specific thermal conductance is obtained by dividing thermal conductance values by body mass. Since it is a poor practice to compute best-fit LSQ equations to data of the form y/x vs x (as noted above), in the following analyses mass-specific measures of thermal conductance were multiplied by body mass, thereby giving an absolute measure of thermal conductance. For a critical review of thermal conductance, and its measurement, see McNab [58].

In 1967 Herreid and Kessel [59] reported a slope for thermal conductance on body size of 0.495 $\pm$ 0.141 (99% CL) for N_{ord} = 3, N_{sp} = 24 (including 18 species of rodents), pWR = 2.3, DI = 1.6%. However, the 99% CL ($\pm$0.141) are much too wide to support useful scaling inferences. The yields of a LSQ analysis of a *subset* of the thermal conductance data of Bradley and Deavers [60] are brought together in Table 17.15, row one. LSQ analyses on subsets of data on thermal conductivity collated from four papers by McNab [61–64] were performed. Since there stands to be a greater consistency among measurements made in one laboratory, those measurements in which McNab was directly involved were analyzed separately. The outcomes of this LSQ analysis are presented in Table 17.15, row two. The LSQ analyses of thermal conductance from the literature generally are synopsized in Table 17.15, row three. The mean and standard deviation of the slopes specified in Table 17.15 in rows one, two, and three are 0.561 $\pm$ 0.012, respectively. The three slopes in Table 17.15 are not significantly different.

Recall that Aschoff [65] showed that mass-specific thermal conductance in mammals and birds depends on *activity* cycles. He reported that *weight-specific* thermal conductance during rest and activity scale with body mass raised to the power –0.52; however, active thermal conductance was 50% higher than resting conductance.

Summary

In this chapter, empirical data bearing on the scaling of body temperature, SMR, maximal oxygen consumption, organ size in several species and organ metabolic rate

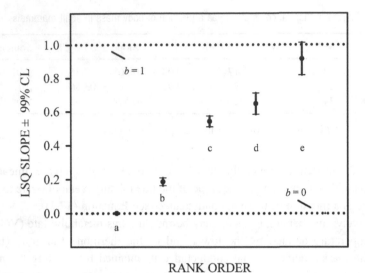

RANK ORDER

a = body temperature; b = maximal lifespan;
c = thermal conductance; d = cold-induced
VO₂max; e = exercise-induced VO₂max

Figure 17.5 Summary slopes for five functions.

in the adult human, maximal lifespan, and thermal conductance were analyzed using LSQ. The results suggest that body temperature is, to a first approximation, independent of body size in mammals in the given samples (see however White and Seymour [30]). On balance, after reviewing the evidence cited above, it is concluded that SMR in mammals *generally* does not scale in accord with Kleiber's three-quarters power rule. More importantly, the relation between SMR and body size is *non-linear*, as shown by both binning and partitioning.

The slope for the square root of the logarithm of SMR on body size is found to be a *linear* function of the logarithm of body mass. In small- and medium-sized mammals (BW < 10 kg) SMR appears to scale at about the two-thirds power of body weight. In larger mammals, SMR may scale roughly as the three-quarters power. However, the more important point is that there is a *linear* variation in slope from small to large mammals (see Figure 17.3, panels c and d) when expressed as a function of the *logarithm* of body mass. No single slope applies to mammals generally, although a slope of about 2/3 may apply nearly to some small mammals and a slope of about 3/4 to at least some large mammals.

Maximal oxygen consumption, of greater intrinsic scaling interest than SMR, scales at about the two-thirds power of body mass when cold-induced and at a significantly higher exponent of 0.916 ± 0.097 (99% CL) (see Table 17.13, row four) when exercise-induced. None of the outcomes for VO₂max are consistent with a three-quarters power rule. Maximal lifespan in a large sample was found to scale as the

0.187 $\pm$ 0.023 power of body weight, *inconsistent* with a one-quarter power rule. Finally thermal conductance was determined to scale as the 0.544 $\pm$ 0.032 (99% CL) power of body weight.

The major slopes found in this chapter are brought together in Figure 17.5. The range in slopes is from near zero (for body core temperature) to 0.916 (for maximal oxygen consumption). Only one slope (that for body temperature, where $b \approx 0.0$) corresponds nearly to an invariance (here, absolute). No slope for SMR is included, as the present evidence (see above) suggests that no single slope characterizes SMR in mammals generally.

References

1. Hock, R.J. (1958). Hibernation. In: Ferrer, M.I. (ed.) *Cold Injury*. New York: Transactions Fifth Conference Josiah Macy Jr. Foundation, pp. 61–133.
2. Johansen, K., Lykkeboe, G., Weber, R.E. *et al.* (1976). Blood respiratory properties in the naked mole rat *Heterocephalus glaber*, a mammal of low body temperature. *Respiratory Physiology*, **28**:303–314.
3. Clarke, A., Rothery, P. and Isaac, N.J. (2010). Scaling of basal metabolic rate with body mass and temperature in mammals. *Journal Animal Ecology*, **79**:610–619.
4. White, C.R. and Seymour, R.S. (2003). Mammalian basal metabolic rate is proportional to body mass$^{2/3}$. *Proceedings of the National Academy of Sciences*, **100**:4046–4049.
5. Hill, R.W. (1976). The ontogeny of homeothermy in neonatal *Peromyscus leucopus*. *Physiological Zoology*, **49**:292–306.
6. Kemp, T.S. (1982). *Mammal-like Reptiles and the Origin of Mammals*. London: Academic Press.
7. Irving, L. (1972). *Arctic Life of Birds and Mammals including Man*. New York: Springer-Verlag.
8. Fraser, G., Trinder, J., Colrain, I.M. *et al.* (1989). Effect of sleep and circadian cycle on sleep period energy expenditure. *Journal of Applied Physiology*, **66**:830–836.
9. Geiser, F. (1988). Reduction of metabolism during hibernation and daily torpor in mammals and birds: temperature effect or physiological inhibition? *Journal of Comparative Physiology*, **158**:25–37.
10. Hulbert, A.J. and Else, P.L. (2004). Basal metabolic rate: history, composition, regulation, and usefulness. *Physiological Biochemical Zoology*, **77**:869–876.
11. Withers, P.C. (1992). *Comparative Animal Physiology*. Fort Worth, TX: Saunders College Publishing.
12. Kleiber, M. (1932). Body size and metabolism. *Hilgardia*, **6**:315–353.
13. Brody, S. (1945). *Bioenergetics and Growth*. New York: Reinhold.
14. Kleiber, M. (1947). Body size and metabolic rate. *Physiological Reviews*, **27**:511–541.
15. Prothero, J. (1984). Scaling of standard energy metabolism in mammals: neglect of circadian rhythms. *Journal of Theoretical Biology*, **106**:1–8.
16. McNab, B.K. (1997). On the utility of uniformity in the definition of basal rate of metabolism. *Physiological Zoology*, **70**:718–720.
17. Dodds, P.S., Rothman, D.H. and Weitz, J.S. (2001). Re-examination of the "3/4-law" of metabolism. *Journal of Theoretical Biology*, **209**:9–27.

18. Bartels, H. (1982). Metabolic rate of mammals equals the 0.75 power of their body weight? *Experimental Biology Medicine*, **7**:1–11. (The question mark in the title was inadvertently omitted during printing, thus undermining its critical intent. [Personal communication].)

19. McNab, B.K. (1988). Complications inherent in scaling the basal rate of metabolism in mammals. *Quarterly Review Biology*, **63**:25–54. (McNab indicates there are data for 321 species, but data for only 320 species are given. Data omitted appear to be for the red kangaroo (*Macropus rufus*). See Reference 20.)

20. Dawson, T.J. and Hulbert, A.J. (1970). Standard metabolism, body temperature, and surface area of Australian marsupials. *American Journal of Physiology*, **218**:1233–1238.

21. McNab, B.K. (2002). *The Physiological Ecology of Vertebrates: A View from Energetics*. Ithaca, NY: Cornell University Press.

22. Blaxter, K. (1989). *Energy Metabolism in Animals and Man*. Cambridge: Cambridge University Press.

23. Kasting, N.W., Adderley, S.A.L., Safford, T. and Hewlett, K.G. (1989). Thermoregulation in Beluga (*Delphinapterus leucas*) and Killer (*Orcinus orca*) whales. *Physiological Zoology*, **62**:687–701.

24. Holliday, M.A. (1986). Body composition and energy needs during growth. In: Falkner, F. and Tanner, J.M. (eds.) *Human Growth: A Comprehensive Treatise*. New York: Plenum Press, pp. 101–117.

25. Savage, V.M., Gillooly, J.F., Woodruff, W.H. *et al.* (2004). The predominance of quarter-power scaling in biology. *Functional Ecology*, **18**:257–282.

26. McNab, B.K. (2008). An analysis of the factors that influence the level and scaling of mammalian BMR. *Comparative Biochemistry and Physiology*, **A151**:5–28.

27. Capellini, I., Venditti, C. and Barton, R.A. (2010). Phylogeny and metabolic scaling in mammals. *Ecology*, **91**:2783–2793.

28. Packard, G.C. and Birchard, G.F. (2008). Traditional allometric analysis fails to provide a valid predictive model for mammalian metabolic rates. *Journal of Experimental Biology*, **211**:3581–3587.

29. da Silva, J.K.L., Garcia, G.J.M. and Barbosa, L.A. (2006). Allometric scaling laws of metabolism. *Physics of Life Reviews*, **3**:229–261.

30. White, C.R. and Seymour, R.S. (2005). Allometric scaling of mammalian metabolism. *Journal of Experimental Biology*, **208**:1611–1619.

31. Bartels, H., Bartels, R., Baumann, R. *et al.* (1979). Blood oxygen transport and organ weights of two shrew species (*S. etruscus* and *C. russula*). *American Journal of Physiology*, **236**: R221–R224.

32. Newman, J.R. and Rudd, R.L. (1978). Minimum and maximum metabolic rates of *Sorex sinuosus*. *Acta Theriologica*, **23**:371–380.

33. Heldmaier, G. and Ruf, T. (1992). Body temperature and metabolic rate during natural hypothermia in endotherms. *Journal of Comparative Physiology*, **B162**:696–706.

34. Hart, J.S. (1971). Rodents. In: Whittow, G.C. (ed.) *Comparative Physiology of Thermoregulation*. New York: Academic Press, pp. 1–149.

35. Toien, Ø., Blake, J., Edgar, D.M. *et al.* (2012). Hibernation in black bears: independence of metabolic suppression from body temperature. *Science*, **331**:906–909.

36. Agutter, P.S. and Wheatley, D.N. (2004). Metabolic scaling: consensus or controversy. *Theoretical Biology Medical Modelling*, **1**:13–38.

37. Welcker, H. and Brandt, A. (1903). Gewichtswerthe der Körperorgane bei dem Menschen und den Thieren. *Archiv für Anthropologie*, **28**:1–89.

38. Pucek, Z. (1965). Seasonal and age changes in the weight of internal organs of shrews. *Acta Theriologica*, **10**:369–438.

39. Jürgens, K.D., Bartels, H. and Bartels, R. (1981). Blood oxygen transport and organ weights of small bats and small non-flying mammals. *Respiratory Physiology*, **45**:243–260. (Brain size data by personal communication.)

40. Martin, A.W. and Fuhrman, F.A. (1955). The relationship between summated tissue respiration and metabolic rate in the mouse and dog. *Physiological Zoology*, **28**:18–34.

41. Forbes, R.M., Cooper, A.R. and Mitchell, H.H. (1956). Further studies of the gross composition and mineral elements of the adult human body. *Journal of Biological Chemistry*, **223**:969–975.

42. Winston, W.C. (1950). The largest whale ever weighed. *Natural History*, **59**:392–398.

43. Laurie, A.H. (1933). Some aspects of respiration in blue and fin whales. *Discovery Reports*, **7**:365–406.

44. Bard, P. (1961). Blood supply of special regions. In: Bard, P. (ed.) *Medical Physiology*. Saint Louis, MO: C.V. Mosby, pp. 239–265.

45. Aiello, L.C. and Wheeler, P. (1995). The expensive-tissue hypothesis: the brain and the digestive system in human and primate evolution. *Current Anthropology*, **36**:199–221.

46. Wang, Z.M., O'Connor, T.P.O., Heshka, S. and Heymsfield, S.B. (2001). The reconstruction of Kleiber's law at the organ-tissue level. *Journal of Nutrition*, **131**:2967–2970.

47. Seeherman, H.J., Taylor, C.R., Maloiy, G.M.O. and Armstrong, R.B. (1981). Design of the mammalian respiratory system. II. Measuring maximum aerobic capacity. *Respiratory Physiology*, **44**:11–23.

48. Hayes, J.P. and Chappell, M.A. (1986). Effects of cold acclimation on maximum oxygen consumption during cold exposure and treadmill exercise in deer mice, *Peromyscus maniculatus*. *Physiological Zoology*, **59**:473–481.

49. Taylor, C.R., Maloiy, G.M., Weibel, E.R. *et al.* (1981). Design of the mammalian respiratory system: III. Scaling maximum aerobic capacity to body mass: wild and domestic mammals. *Respiratory Physiology*, **44**:25–37.

50. Weibel, E.R., Bacigalupeb, L.D., Schmitt, B. and Hoppeler, H. (2004). Allometric scaling of maximal metabolic rate in mammals: muscle aerobic capacity as determinant factor. *Respiratory Physiology Neurobiology*, **140**:115–132.

51. Prothero, J.W. (1979). Maximal oxygen consumption in various animals and plants. *Comparative Biochemistry and Physiology*, **63A**:463–466.

52. Prothero, J. and Jürgens, K.D. (1987). Scaling of maximal lifespan in mammals: a review. In: Woodhead, A.D. and Thompson, K.H. (eds.) *Evolution of Longevity in Mammals: A Comparative Approach*. New York: Plenum Press, pp. 49–74.

53. Jürgens, K.D. and Prothero, J. (1991). Lifetime energy budgets in mammals and birds. *Comparative Biochemistry and Physiology*, **100A**:703–709.

54. Cossins, A.R. and Bowler, K. (1987). *Temperature Biology of Animals*. London: Chapman and Hall.

55. Calder, W.A. (1984). *Size, Function, and Life History*. Cambridge, MA: Harvard University Press.

56. Turner, J.S. (1988). Body size and thermal energetics: how should thermal conductance scale? *Journal of Thermal Biology*, **13**:103–117.

57. Burton, A.C. (1934). The application of the theory of heat flow to the study of energy metabolism. *Journal of Nutrition*, **7**:497–533.

58. McNab, B.K. (1980). On estimating thermal conductance in endotherms. *Physiological Zoology*, **53**:145–156.

59. Herreid, C.F. and Kessel, B. (1967). Thermal conductance in birds and mammals. *Comparative Biochemistry and Physiology*, **21**:405–414.

60. Bradley, S.R. and Deavers, D.R. (1980). A re-examination of the relationship between thermal conductance and body weight in mammals. *Comparative Biochemistry and Physiology*, **65A**:465–476.

61. McNab, B.K. (1984). Physiological convergence amongst ant-eating and termite-eating mammals. *Journal of Zoology*, **203**:485–510.

62. McNab, B.K. (1992). The comparative energetics of rigid endothermy: the Arvicolidae. *Journal of Zoology*, **227**:585–606.

63. McNab, B.K. (1995). Energy expenditure and conservation in frugivorous and mixed-diet carnivorans. *Journal of Mammalogy*, **76**:206–222.

64. McNab, B.K. (2000). The standard energetics of mammalian carnivores: Felidae and Hyaenidae. *Canadian Journal of Zoology*, **78**:2227–2239.

65. Aschoff, J. (1981). Thermal conductance in mammals and birds: its dependence on body size and circadian phase. *Comparative Biochemistry and Physiology*, **69A**:611–619.

18 Lethal limits

...what we observe is not nature in itself but nature exposed to our method of questioning.

Heisenberg 1962

Unbelievably, bacteria can even survive the vacuum of space.

Ashcroft 2000

This chapter focuses on the scaling of lethal limits in adult mammals. Lethal limits, as defined here, arise from externally induced perturbations of one or more homeostatic variables which, if sufficiently prolonged, result in death. These limits for a given homeostatic variable may, or may not, vary with body size. The study of lethal bounds stands to tell us how tolerant mammals are to perturbations in homeostatic variables. Fatal limits may vary from individual to individual for any given species as well as from species to species. For a general discussion of "extremes" in biology see Ashcroft [1], the source of the above quote. A schematic conception of lethal limits is shown in Figure 18.1 (see Wright [2]). Lethal limits constitute a somewhat unusual but nevertheless potentially useful way of observing mammals (see above quote from Heisenberg).

Three *zones*, denoted by N (normal), R (reversible), and L (potentially lethal) are identified in Figure 18.1. *Zone N* is the range of normal endogenous variation in any given homeostatic variable. For example, core body temperature (e.g., as measured orally or rectally) in the adult human typically exhibits a circadian rhythm between 36.5 °C in sleep and 37.5 °C around midday. Thus, the N zone for human core body temperature is about 1 °C wide. The width of the lower (LR) and upper (UR) *reversible* zones provides a quantitative measure of the safety margins associated with the N zone. For example, in a fever, say following an infection, body temperature may rise to 40 °C, resulting in an increased rate of protein degradation and other changes. After a fever wanes, any tissue injury incurred can be repaired, usually leaving no long-term detectable effects. Thus, the UR zone, or safety margin, for human body temperature is at least 2.0 °C wide (40 − 37.5 = 2.5). Mortal limits have been much more widely studied in plants than in animals, and in ectotherms rather than in homeotherms such as mammals.

If body temperature in the adult human rises above 43.5 °C or so and stays there for some period of time, the outcome is likely to be fatal. This premise points to an UR zone in the adult human that is actually about 6 °C wide (43.5 − 37.5 = 6) rather than 2 °C. The upper and lower reversible zones are shown in Figure 18.1 as symmetric; in most cases

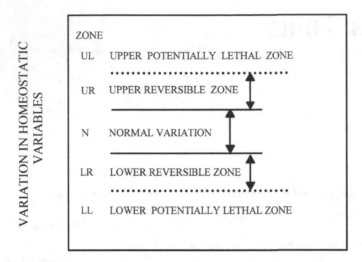

Figure 18.1 A schematic conception of lethal limits in mammals.

this is unlikely (see below). Note that whether a perturbation of a homeostatic variable is actually *fatal* usually depends on the exposure time. Often lethal bounds are based on 50% survival rates from exposure to a given perturbation for a defined time. As a rule, the longer the exposure to a taxing perturbation the more certain it is to be fatal. Even if the perturbation is removed before death, an animal may die after some hours have passed. Despite pervasive cultural denial, it is our common fate, along with all other living creatures, animal or plant, to finally take leave via a one-way excursion into a lethal zone.

It is important to keep in mind that the magnitude of potentially fatal perturbations of homeostatic variables may vary with acclimation, age of the organism, ambient temperature, diet, exposure time, methods of perturbation, water availability, and numerous other factors. Generally, the actual cause of death when a fatal limit is exceeded is not known with certainty. In some instances, cardiac arrest or brain damage (for example due to anoxia) may be suspected. In any event, the study of lethal limits in mammals stands to give us some insight into potentially important aspects of homeostasis and its safety margins. Next, we cite some known performance limits that fall short of being fatal, as well as some empirically determined lethal limits.

Breath-holding

The need for oxygen is the most insistent requirement for human life… The brevity of duration of life without oxygen leaves only a narrow margin of oxygen reserve to separate life from death.

Irving 1934

According to Irving [3], based on data taken from the literature (reported in 1882), the world-record (two hours) for breath-holding in mammals is held by the northern bottlenose whale (*Hyperoodon ampullatus*). Northern elephant seals (*Mirounga angustirostris*) may be submerged for up to 48 minutes [4]. The current remarkable human record for breath-holding is about 22 minutes. Putting the total reserves of available oxygen stores in the adult human body at about $1,500 \, cm^3$, and assuming a resting oxygen consumption of $250 \, cm^3/min$, it follows that the estimated average survival time without breathing is about 6 minutes ($1,500/250 = 6$) [5]. More acutely, it is known that blockage of the cerebral circulation in the human produces unconsciousness in 8–10 seconds [6] and death in about 4 minutes [7]. See above quote from Irving.

The mechanisms permitting long periods of breath-holding in diving mammals represent the implementation, not of new strategies so much as extensions of existing ones. These include slowed heart rate (bradycardia), curtailed energy metabolism, enlarged blood volume, hyperventilation, increased myoglobin content of skeletal muscle, reduced cardiac output, and selective redistribution of blood flow, all the while protecting the brain and heart [8]. For a recent discussion of the important role of myoglobin in diving endurance in marine mammals see Mirceta *et al.* [9].

Fasting and starvation

If fasting continues, a lethal weight [in the vampire bat *Desmodus rotundus*] *is reached at about 72 per cent of the original weight...*

McNab 1973

Fasting and starvation studies are usually carried out by withholding food while providing access to water, although this is not always stated explicitly. The results of studies of this kind in seven mammalian species are summarized in Table 18.1. The respective species names, by row, are *Sorex cinereus, Sorex sinuosus, Desmodus rotundus, Rattus norvegicus, Felis silvestris, Canis familiaris,* and *Homo sapiens.* For a discussion of fasting endurance as a function of body size in mammals see, for example, Millar and Hickling [10].

One sees a tendency in Table 18.1 for larger mammals to survive fasting for longer periods than smaller ones; however, the data for the dog (117 days; Table 18.1, row six) are nearly double the longest reported time for a human (63 days). We see in Table 18.1 (first row) that a masked shrew died after 11 hours of fasting. It is also notable from Table 18.1 that larger mammals are viable after greater percentage losses of body mass than are small mammals. Part of this difference no doubt reflects differences in standard metabolic rate (SMR). If one mammal is 10 times larger than another in body size, its SMR will only be about 5 times larger ($10^{0.7} = 5$).

Another factor is that larger mammals are likely to contain more fat (see Chapter 8, Table 8.4, where it is stated that $b = 1.05$ (for pWR = 7.5) for adipose tissue). The mean mass loss for the seven cases reported in Table 18.1 is 34% (STD = 20%). From the human data of Benedict [17] we find that body mass dropped steadily (by 21%) over the 31 days of a fast, whereas SMR dropped 31% (from about 1,600 kcal/day to about 1,100 kcal/day at 24 days) and then rose slightly during the last week of the fast to 1,130 kcal/day. The average SMR over the fast was about 1,300 kcal/day

Table 18.1 Survival period or time-to-death following fasting or starvation [a]

Row	Common name	Estimated initial body mass (g)	Survival period or time-to-death		Mean mass loss (%)	Source
			Survival (days/hours)	Death (days/hours)		
1	Masked shrew	3.5	-	0/11	12	[11]
2	Suisun (ornate) shrew	3.7–7.4	-	0/(5–23)	9	[12]
3	Common vampire bat	42	-	2/2	28	[13]
4	Rat	211–227	-	25/0	54	[14]
5	Cat	3,310–3,590	(52–57)/0	-	53	[15]
6	Dog	26,330	117/0	-	63	[16]
7	Human	59,900	31/0	-	20.7	[17, 18]
8	Human	-	63/0	-	-	[19]

[a] Free access to water is assumed for rows one and two; in row three animals were given no food or water; in rows four, five, and six water was freely available; in row seven subject was given about 900 cm^3 of water daily; for row eight subject lived on water only (quantity unknown).

Table 18.2 Utilizable energy reserves in a normal adult human (70 kg male)

Substrate	Tissue	Substrate mass (g)	Conversion factor (kcal/g)	Reserve substrate energy (kcal)
Fat (lipid)	Adipose	15,000	9.4	141,000
Protein	Chiefly muscle	10,000/2 [a]	4.3	21,500
Glycogen	Muscle	560	4.1	2,300
Glycogen	Liver	100	4.1	410
Total	-	-	-	165,000

[a] Only about 50% of total body protein can be utilized as an energy source.

(1 kcal = 4,184 joules). A weight loss of 28% due to fasting is likely to be lethal in a vampire bat (see above quote from McNab).

An upper limit to the duration of fasting can, in principle, be calculated from an inventory of energy reserves. The data required to make this type of calculation in adult humans are summarized in Table 18.2, as adapted from Cahill *et al.* [20].

Several interesting deductions may be drawn from Table 18.2. First, the energy from carbohydrate (mainly glycogen) present in the adult human body adds up to only about 2,700 kcal. If one assumes that SMR for an adult male at rest is about 2,000 kcal/day, then carbohydrates will support energy demands for only about 1.4 days (2,700/2,000 = 1.35). Because the brain is "addicted" to glucose, the body resists serious depletion of its carbohydrate stores; during a fast, the body produces glucose from proteins, lactate, etc. By far the largest contributor to energy reserves is *fat*. Taking the total energy reserves at 165,000 kcal, and a mean daily SMR of 2,000 kcal, then the calculated upper limit to the duration of fasting in the *adult human* is about 83 days (165,000/2,000 = 82.5). If one uses only fat in the calculations, and assumes a mean fasting SMR of 1,300 kcal/day (see above), then one projects a hypothetical survival time of 108 days (141,000/1,300 = 108). Either estimate suggests an upper limit to fasting endurance in an adult human of about three months. For a careful physiological study of a Buddhist monk, who went for eight days with no food and little if any water, see Yoshimura *et al.* [21] (Chapter 10). The record for human survival without food or water is put at 18 days [22].

Hyperacceleration

Hyperacceleration (due to forces greater than that of gravity at earth's surface) is an abnormal form of stress, produced in a centrifuge, which organisms would not experience in the wild. Nonetheless, the results of this special stress, as summarized in Table 18.3, are of interest [23]. The species used in these experiments (only common names are given) are mouse (*Mus musculus*), rat (*Rattus norvegicus*), guinea pig (*Cavia porcellus*), rabbit (*Oryctolagus cuniculus*), cat (*Felis silvestris*), monkey (*Macaca mulatta*), and dog (*Canis familiaris*).

The mean survival time in these experiments is 6.6 minutes. There is, apparently, no significant correlation between body size and survival time in this small sample. These workers found no post-mortem evidence of pathological symptoms in most of their animals. A plausible cause of death might be anoxia due to reduced cerebral blood flow. See also Economos [24].

Lethal body temperatures

Cells can never survive if their internal fluid freezes...

Ball 1999

Denaturation of vital cellular proteins occurs so rapidly at 44–46°C that... tissue death occurs after a few hours.

Hardy 1971

Table 18.3 Survival times in a centrifuge at an acceleration of about 6 g in selected mammals

Common name	Lethal time limit (min.)	Common name	Lethal time limit (min.)
Mouse	5–6	Cat	4–5
Rat	8–12.5	Monkey	5–10
Guinea pig	8–12.5	Dog	5–6
Rabbit	2.5–3	-	-

min. = minute.

Every living organism has a definite range of body temperatures within which it must remain in order to maintain its functional integrity.

Wright 1976

Body temperature appears to be the most tightly regulated of the various homeostatic variables (Chapter 17). It may be for this reason that fatal perturbations in body temperature have come in for more scrutiny than other perturbations, especially in plants and fish, and also in mammals. A fundamental limit is reached on the cooling side when cell contents freeze (see above quote by Ball). A similar limit, on the upper side, is reached when the temperature is sufficiently high to denature proteins (see quote by Hardy). In order to remain viable, mammals must maintain their body within a certain range (see quote by Wright). Schmidt-Nielsen [25] noted that monotremes, marsupials, and eutherians have normal mean body temperatures of about 30.5, 35.5, and 37 °C, respectively. The approximate upper fatal temperatures for these same groups are 37, 40.5, and 43 °C [25]. Thus, he inferred that what is here termed the upper reversible zone for each group is, on average, about 6 °C wide. The lower and upper lethal body temperature limits for selected mammals found in the literature are summarized in Table 18.4. For a chart summarizing many aspects of cold-induced stress on mammals, see Folk [26].

In Table 18.4 the species names, by row, are *Myotis yumanensis, Perognathus longimembris, Peromyscus leucopus, Blarina brevicauda, Dipodomys panamintinus, Rattus norvegicus, Cavia porcellus, Oryctolagus cuniculus, Macaca mulatta, Canis familiaris,* and *Homo sapiens.*

Observe in Table 18.4 that the *lower* fatal body temperatures range from 3.5 to 24 °C across species, whereas the upper lethal temperatures extend only from 41.7 to 45.2 °C. We *assume* that the mean core temperature in mammals generally is near 36 °C

Table 18.4 Lower and upper lethal body temperatures in selected mammals

Row	Common name	Estimated BM (g) [a]	Lethal body temperature °C		Exposure time (Hours/min)	Source
			Lower	Upper		
1	Yuma myotis	5–7	-	43.5	1/0	[27]
2	Little pocket mouse	7–9	-	41.9–42.5	0/15	[28]
3	White-footed mouse	10–43	3.5–5	-	0/45 – 8/30	[29]
4	Northern short-tailed shrew	16–28.6	-	43	≤0/60	[30]
5	Panamint kangaroo-rat	65	-	44.3–45.2	3/0	[31]
6	Rat	200–260	15	-	2/0	[32]
7	Guinea pig	450–700	17.5–21	42.8 [b]	-	[33]
8	Rabbit	960–2,420	4–14	43.4 [b]	-	[33, 34]
9	Rhesus monkey	5,400–7,700	14	-	-	[33]
10	Dog	10,000	15–18	41.7 [b]	-	[33]
11	Human	70,000	24–26	43.0	-	[33]
12	Mean	-	14.7	43.4	-	-

[a] BM = body mass; [b] 50% survival.

(see Chapter 17), and that the N zone (see Figure 18.1) in mammals generally is $\pm 1\,°C$ wide [35]. See from the bottom row of Table 18.4 that the mean *upper* lethal temperature is $43.4\,°C$. It follows that the upper reversible zone in the present sample is about $6\,°C$ wide $(43.4 - (36 + 1) = 6.4)$, in agreement with the figure of $6\,°C$ given by Schmidt-Nielsen [25]. Apparently the lower reversible zone is about $20\,°C$ wide $((36 - 1) - 14.7 = 20.3)$, nearly three times wider that the upper reversible zone (Figure 18.1).

There are numerous exceptions to the above picture. The lowest reported survivable core temperature in a human, as far as known, is $13.7\,°C$ [1], well below the figures (24–$26\,°C$) given in Table 18.4, row 11. (This recovery was made possible by expert treatment in a hospital.) During hibernation, body temperature may fall to 2 or $3\,°C$ [36]. With respect to the upper fatal temperature range, Taylor [37] reported that Grant's gazelle (estimated body size of 38–82 kg) and the oryx (65–75 kg) survived for six hours having rectal temperatures of $46.5\,°C$ with "no observable ill effects." For additional data on lower fatal body temperatures in (infant) mammals see, for example, Adolph [38].

Water deprivation and dehydration

The only means of dissipating heat in an environment warmer than the body surface is by the evaporation of water.

Schmidt-Nielsen 1954

In a hot climate a man who is working may lose 10–14 liters of water a day by the skin...

Widdowson and Dickerson 1964

An adequate supply of fresh water is a much more critical requirement for most (but not all) mammals than is food. As a rule, access to fresh water is the most critical requirement for a mammal, after oxygen. This is especially true in hot environments, such as deserts (see quotes from Schmidt-Nielsen, and from Widdowson and Dickerson). Table 18.5 summarizes the lethal bounds for time without water in selected mammals. The data are not strictly comparable, because the data for the mouse and ground squirrel were acquired at significantly lower ambient temperatures. Even so, it is worth observing (Table 18.5, bottom row) that over this substantial size range (pWR 5.0), the mean survival time-without-water is 11 days, with a standard deviation of 12 days. The mean weight loss is 29 ± 9 (STD) percent of initial body weight (Table 18.5, bottom row).

The implication is that for this sample of mammals, the lethal limit to dehydration is in the (rather wide) range of 20 to 45% of initial body mass. The respective species names (extrapolated mostly from common names), by row, are *Peromyscus polionotus*, *Ammospermophilus leucurus*, *Ovis orientalis*, *Homo sapiens*, *Camelus ferus* (?), *Bos primigenius*.

Recall that some species of small desert rodents, for example Merriam's kangaroo-rat (*Dipodomys merriami*), are able to survive simply on water derived from their normal diet, without drinking free water [40]. Small mammals, by virtue of their relatively large

Table 18.5 Survival time without water in selected mammals

Species common name	Estimated body mass (g)	Ambient temperature (°C)	Survival time (days)	Mass loss (% initial mass)	Source
Oldfield mouse	11.7	24.5–28 (range)	2.7	35.5	[39]
Ground squirrel	92.5	19–25 (range)	36	45	[40]
Sheep	30,000	40 (daily max.)	6–8	24–31	[41]
Human	70,000	36 (daily mean)	2 (predicted)	20	[42]
Camel	450,000–650,000	40 (daily max.)	12–15	22–26	[41]
Cow	450,000–900,000	40 (daily max.)	3–4	20–25	[41]
Mean	-	-	11	29	-

Table 18.6 Miscellaneous lethal limits

Row	Variable	Species	Normal range	Lethal range	Source
1	Anoxia	Cat, dog	-	2–4 minutes	[5]
2	Total plasma calcium	Human	9–11 mg% [a]	>20 mg%	[44]
3	Plasma glucose	Human	65–105 mg% [a]	<45 mg%	[44]
4	Plasma pH	Mammals	7.35–7.45	<7 or >8	[45]
5	Blood pressure	Dog	95 mm Hg	<40 mm Hg	[46]
6	Hyperbaric oxygen	Mammals	0.2 atmospheres	4–5 atmospheres [b]	[47]

[a] 1 mg% calcium = 0.25 mmol/l; 1 mg% glucose = 0.056 mmol/l; [b] produces convulsions.

surface-to-volume ratio, cannot afford to control their body temperature by evaporating water [43]. They must avoid relatively high ambient temperatures and the consequent dehydration by suitable behavior (such as by being nocturnal and/or by burrowing). Presumably, marine mammals also get their fresh water from their diet. It follows that the findings summarized in Table 18.5 cannot be generalized to all mammals.

Miscellaneous lethal limits

Some data on other lethal bounds are summarized in Table 18.6.

It is reasonably certain that among the major organs the brain is the one most sensitive to anoxia (Table 18.6, row one). Plasma calcium levels (row two) are potentially lethal when the normal level is doubled, and likewise halving plasma glucose levels (row three) may be fatal. If one expresses plasma pH (Table 18.6, row four) as hydrogen ion concentration, rather than as the logarithm of concentration, then one finds that decreasing the concentration by a factor of 2 or increasing it by a factor of 3.5 is likely to be fatal. As a rough-and-ready rule, it appears from these data that doubling or halving homeostatic levels of various variables may be fatal over some time period. Hyperbaric oxygen (row six) may be an exception; still, this perturbation would seldom, if ever, be encountered in nature.

Summary

The zone of normal physiological homeostatic control appears to be bounded above and below by reversible zones, wherein, depending upon the duration of a perturbation, some non-lethal damage may be incurred, being repairable when the disturbing condition is removed (Figure 18.1). As noted above, the two reversible zones (LR, UR) are usually asymmetric. For example, the upper reversible (UR) zone for core body temperature is perhaps 6 °C wide, whereas the lower reversible zone (LR) is about 21 °C wide (in several species of small mammals). On average, in a small sample of mammals, a roughly one-third loss in body mass due to fasting (with access to water) is likely to be lethal.

At the same time, it should be kept in mind that some perturbations that might be thought to be fatal may not be in the short-to-medium run, in certain species. For example, it has been reported that four marmots underwent bilateral nephrectomy in winter and survived 28 to 35 days [48]. Likewise black bears during winter sleep are able (possibly owing, in part, to reduced blood flow to kidneys and gut) to survive without food or water and without urinating or defecating for three to five months [49].

The changes that occur in mammalian physiology following perturbations of the homeostatic variables are not well understood. It is a curious and unexplained fact that many mammalian subsystems seem to operate near lethal limits. The safety margins are often surprisingly narrow. It is as if nature were unaware of the mortal dangers lurking so near at hand (see above quote from Irving). Perhaps this is the price to be paid for narrow homeostatic zones, since so-called "lower" forms, such as fish, appear to have much wider (and behaviorally rather than homeostatically mediated) zones (e.g., for body temperature) than do mammals.

It would be useful if more studies, under standardized conditions, could be carried out in the reversible zones, in a wider variety of mammalian species. With careful design, it should be possible to carry out these studies with little or ideally no harm or pain to the animals involved.

References

1. Ashcroft, F. (2000). *Life at the Extremes.* Berkeley, CA: University of California Press.
2. Wright, G.L. (1976). Possible mechanisms involved in death in laboratory animals due to thermal stresses (heat and cold). In: Johnson, H.D. (ed.) *Progress in Biometeorology.* Amsterdam: Swets & Zeitlinger, pp. 167–173.
3. Irving, L. (1939). Respiration in diving mammals. *Physiological Reviews,* **19**:112–134.
4. Le Boeuf, B.J., Costa, D.P. and Huntley, A.C. (1988). Continuous deep diving in female northern elephant seals, *Mirounga angustirostris. Canadian Journal of Zoology,* **66**:446–458.
5. Irving, L. (1934). On the ability of warm-blooded animals to survive without breathing. *Scientific Monthly,* **38**:422–428.
6. Blinkov, S.M. and Glezer, I.I. (1968). *The Human Brain in Figures and Tables: A Quantitative Handbook.* New York: Plenum Press.

7. Craig, A.B.J. (1976). Summary of 58 cases of loss of consciousness during underwater swimming and diving. *Medicine Science Sports*, **8**:171–175.

8. Elsner, R. and Gooden, B. (1983). *Diving and Asphyxia: A Comparative Study of Animals and Man.* Cambridge: Cambridge University Press.

9. Mirceta, S., Signore, A.V., Burns, J.M. *et al.* (2013). Evolution of mammalian diving capacity traced by myoglobin net surface charge. *Science*, **340**:1303–1321.

10. Millar, J.S. and Hickling, G.J. (1990). Fasting endurance and the evolution of mammalian body size. *Functional Ecology*, **4**:5–12.

11. Morrison, P.R. (1946). The metabolism of a very small mammal. *Science*, **104**:287–289.

12. Newman, J.R. and Rudd, R.L. (1978). Minimum and maximum metabolic rates of Sorex sinuosus. *Acta Theriologica*, **23**:371–380.

13. McNab, B.K. (1973). Energetics and the distribution of vampires. *Journal of Mammalogy*, **54**:131–144.

14. Horst, K., Mendel, L.B. and Benedict, F.G. (1930). The metabolism of the albino rat during prolonged fasting at two different environmental temperatures. *Journal of Nutrition*, **3**:177–200.

15. Prentiss, P.G., Wolf, A.V. and Eddy, H.A. (1959). Hydropenia in cat and dog: ability of the cat to meet its water requirements solely from a diet of fish or meat. *American Journal of Physiology*, **196**:626–632.

16. Howe, P.E., Mattill, H.A. and Hawk, P.R. (1912). Fasting Studies: VI. Distribution of nitrogen during a fast of one hundred and seventeen days. *Journal of Biological Chemistry*, **11**:103–127.

17. Benedict, F.G. (1916). The relationship between body surface and heat production especially during prolonged fasting. *American Journal of Physiology*, **41**:292–308.

18. Young, V.R. and Scrimshaw, N.S. (1971). The physiology of starvation. *Scientific American*, **225**:14–21.

19. Kleiber, M. (1961). *The Fire of Life. An Introduction to Animal Energetics.* New York: John Wiley & Sons.

20. Cahill, G.F. Jr., Owen, O.E. and Morgan, A.P. (1968). The consumption of fuels during prolonged starvation. *Advances in Enzyme Regulation*, **6**:143–150.

21. Yoshimura, H., Inoue, G., Yamamoto, M. *et al.* (1953). A contribution to the knowledge of dehydration of human body. *Journal of Biochemistry*, **40**:361–375.

22. McFarlan, D. (ed.) (1989). *The Guinness Book of Records 1990.* Enfield: Guinness Publishing, p. 19.

23. Britton, S.W., Corey, E.L. and Stewart, G.A. (1946). Effects of high acceleratory forces and their alleviation. *American Journal of Physiology*, **146**:33–51.

24. Economos, A.C. (1979) Gravity, metabolic rate, and body size of mammals. *Physiologist,* **22**: S71–S72.

25. Schmidt-Nielsen, K. (1977). Problems of scaling: locomotion and physiological correlates. In: Pedley, T.J. (ed.) *Scale Effects in Animal Locomotion.* London: Academic Press, pp. 1–21.

26. Folk, G.E. (1966). *Introduction to Environmental Physiology.* Philadelphia, PA: Lea & Febiger.

27. Licht, P. and Leitner, P. (1967). Physiological responses to high environmental temperatures in three species of microchiropteran bats. *Comparative Biochemistry and Physiology*, **22**:371–387.

28. Chew, R.M., Lindberg, R.G. and Hayden, P. (1967). Temperature regulation in the little pocket mouse, Perognathus longimembris. *Comparative Biochemistry and Physiology*, **21**:487–505.

29. Sealander, J.A. Jr. (1953). Body temperatures of white-footed mice in relation to environmental temperature and heat and cold stress. *Biological Bulletin*, **104**:87–99.

30. Neal, C.M. and Lustick, S.I. (1973). Energetics and evaporative water loss in the short-tailed shrew *Blarina brevicauda*. *Physiological Zoology*, **46**:180–185.

31. Dawson, W.R. (1955). The relation of oxygen consumption to temperature in desert rodents. *Journal of Mammalogy*, **36**:543–552.

32. Adolph, E.F. (1948). Lethal limits of cold immersion in adult rats. *American Journal of Physiology,***155**:378–387.

33. Precht, H., Christophersen, J., Hensel, H. and Larcher, W. (1973). *Temperature Limits of Life*. New York: Springer-Verlag.

34. Fisher, K.C. (1958). An approach to the organ and cellular physiology of adaptations to temperature in fish and small mammals. In: Prosser, C.L. (ed.) *Physiological Adaptations*. Washington, DC: American Physiological Society, pp. 3–48.

35. Schmidt-Nielsen, K. (1984). *Scaling, Why is Animal Size so Important?* Cambridge: Cambridge University Press.

36. Dill, D.B. and Forbes, W.H. (1941). Respiratory and metabolic effects of hypothermia. *American Journal of Physiology*, **132**:685–697.

37. Taylor, C.R. (1970). Dehydration and heat: effects on temperature regulation of East African ungulates. *American Journal of Physiology*, **219**:1136–1139.

38. Adolph, E.F. (1963). How do infant mammals tolerate deep hypothermia? In: Herzfeld, C.M. and Hardy, J.D. (eds.) *Temperature, Its Measurement and Control in Science and Industry*. New York: Reinhold, pp. 511–515.

39. Glenn, M.E. (1970). Water relations in three species of deer mice (Peromyscus). *Comparative Biochemistry and Physiology*, **33**:231–248.

40. Hudson, J.W. (1962). The role of water in the biology of the antelope ground squirrel *Citellus leucurus*. *California University Publications Zoology*, **64**:1–51.

41. Ingram, D.L. and Mount, L.E. (1975). *Man and Animals in a Hot Environment*. New York: Springer-Verlag.

42. Adolph, E.F. (1969). *Physiology of Man in the Desert*. New York: Hafner.

43. Schmidt-Nielsen, K. (1964). *Desert Animals: Physiological Problems of Heat and Water*. New York: Dover Publications.

44. Cannon, W.B. (1929). Organization for physiological homeostasis. *Physiological Reviews*, **9**:399–431.

45. Lockwood, A.P.M. (1963). *Animal Body Fluids and their Regulation*. Cambridge, MA: Harvard University Press.

46. Kaihara, S., Rutherford, R.R., Schwentker, E.P. and Wagner, H.N. Jr. (1969). Distribution of cardiac output in experimental hemorrhagic shock in dogs. *Journal of Applied Physiology*, **27**:218–222.

47. Arieli, R. (1988). Oxygen toxicity is not related to mammalian body size. *Comparative Biochemistry and Physiology*, **91**:221–223.

48. Britton, S.W. and Silvette, H. (1937). Survival of marmots after nephrectomy and adrenalectomy. *Science*, **85**:262–263.

49. Nelson, R.A. (1973). Winter sleep in the black bear: a physiologic and metabolic marvel. *Mayo Clinic Proceedings*, **48**:733–737.

Part III

Survey of results

19 Structural summary

The olfactory bulb and its associated structures are the only [telencephalic] *ones to undergo an absolute size reduction in phylogeny.*

Stephan and Andy 1964

We want to see how, in some cases at least, the forms of living things, and of the parts of living things, can be explained by physical considerations . . .

D'Arcy Thompson 1943

This chapter brings together in tabular form empirical exponents for 52 *structural* y-variables presented separately in Part II. (Numerous y-variables in Part II were omitted because of small sample size, too wide confidence limits, desire to reduce double counting, etc.) Thus we compare and contrast the exponents for 52 structural y-variables, having an average *sample size* (N_{sp}) of 101 ± 196 (STD) and a range in sample sizes from 10 to 1,266. The mean *body size range* for these 52 y-variables is given by pWR = 6.8 ± 1.6 (STD). The total number of records analyzed for these 52 y-variables is 5,277; each of these 5,277 records is unique.

We have two primary aims. The first is simply to list the 52 structural y-variables in terms of the dimensions of *length* (Table 19.3) (16 instances), *surface area* (Table 19.4) (6 instances), and *volume* or *mass* (Table 19.5) (30 instances), where each table gives empirical exponents expressed in rank order. Second, and more broadly, our aim is to examine how structural exponents are influenced by two main constraints, *dimensionality* and *invariance* (both absolute – including scaling at constant shape – and relative). Before introducing the empirical exponents for 52 structural y-variables, it may be useful to state briefly a concept of *dimensionality*. Invariance will be discussed shortly.

An abstract conception of dimensionality

Given a straight line, it takes one coordinate to specify the position of a point located on it. Similarly, given a plane or a volume, it takes two or three coordinates, respectively, to specify the position of a point. Hence, we define *dimensionality* in terms of the minimum number of coordinates required to specify the position of a point in 1-, 2- or 3-dimensional space. This slightly abstract conception of dimensionality is independent of considerations of either size or shape. Hence it logically precedes our concepts of structural scaling. See also Table 24.6.

The concept of physical dimensions

On the other hand, the concepts of *physical* length, area, and volume relate directly to tangible objects. Structural scaling is concerned mainly with how physical lengths and areas vary with volume (or more often mass) in a series of similar objects (Chapter 3). Scaling at *constant shape* is of special interest, partly because length and area then become simple functions of volume.

x-Variables

In most mammalian scaling studies, the *x*-variable, sometimes referred to as the *independent* variable, is body mass (BM) or body volume (BV). The *x*-variable is often not independent, as when it includes the *y*-variable. For example, when we plot organ mass (*y*) against body mass (*x*) we should keep in mind that *x* includes y. In such cases *x* is not strictly independent of *y*. (See scaling of muscle and skeleton in Chapter 12.) Bear in mind that BV and BM are numerically the same when body density is 1.

y-Variables

In this chapter I selected out those structural *y*-variables in Part II with the physical dimensions of length, surface area, volume, or mass. This selection was made in order to study in detail the dual roles of *dimensionality and invariance* in shaping empirical scaling exponents in adult mammals (see Table 19.1). We want to know whether dimensionality (column d) strongly influences structural exponents, or whether dimensionality and/or other constraints (such as invariance or constancy) are jointly at work. We also consider briefly, for the sake of completeness, the partly hypothetical scaling of *component number*. The term "component" refers to any body part, including atoms, ions, molecules, macromolecules, cells, organs, and tissues. Number is a quantity with zero dimensions (see Table 19.1, column d, last row).

A classification of reference exponents

In Table 19.1 (columns b, c, and d), the selected *reference* exponents are expressed in terms of body volume (BV) or mass (BM) (see also Chapter 3, Table 3.5). In Table 19.1, column b (absolute invariance), all five reference exponents are zero. But in column c (relative invariance), the two exponents for component volume or component number are each 1. Finally, in column d (dimensionality), the reference exponents for the

Table 19.1 Simplified summary of reference exponents for the scaling of structural variables expressed as a function of body volume or mass

| Constraints | Reference exponents expressed in terms of body volume or mass for: | | Dimensionality |
| | Invariance | | |
	Absolute	Relative	
a	b	c	d
Physical dimension			
Diameter, girth, length	0	-	1/3 [a]
Surface area	0	-	2/3 [a]
Volume	0	1	1
Diameter or length ratio [b]	0	-	0
Component number	0	1	0

[a] Each exponent applies only under a special scaling constraint (see text).
[b] Either the ratio of the diameter of a given bone (say the humerus) in a small mammal to the diameter of the humerus in a large mammal; or the ratio of the length of a given bone (say the femur) in a small mammal to the length of the femur in a large mammal.

dimensions of length, area, and volume are 1/3, 2/3, and 1, respectively (Table 19.1, column d, rows one to three). Observe that the reference exponents for dimensionality given in column d, two bottom rows, as associated with diameter/length ratios and with the scaling of component number, are both zero. Notice that all the reference exponents in Table 19.1 lie between 0 and 1.

Be aware of certain special features of Table 19.1 relating to numerical equivalence. First, an empirical exponent for component *volume* (column a, row three) equal to or near 1 may be attributable either to relative invariance (column c) or to dimensionality (column d). Second, an empirical exponent equal (or close) to 0 (column a, row four) pertaining to a *ratio* of lengths may be attributable to either absolute invariance (column b) or to dimensionality (column d). Third, an empirical exponent for component *number* (column a, row five) at or near 0 may be chalked up either to absolute invariance (column b) or to dimensionality (column d). It is clear from Table 19.1 that in several instances given empirical exponents may be associated with reference exponents for either dimensionality or invariance.

Constraints on empirical scaling exponents

Throughout the following discussion we encounter two classes of exponents: *reference* and *empirical*. Reference exponents are exact whereas empirical exponents (for continuous variables) are always burdened with some degree of uncertainty. We ask to what degree empirical exponents are associated (correlated) with one or more reference exponents; no causality is implied.

Table 19.2 Examples showing how absolute and relative invariance are correlated with component size

Type of invariance	Component	
	Diameter, SA, volume, or mass	Number
a	b	c
Absolute ($b = 0$)	H_2O, Hb, RBC	Brain, kidney
Relative ($b = 1$)	Mass of lung or skeletal muscle, blood volume	H_2O, Hb, RBC

SA = surface area; Hb = hemoglobin; RBC = red blood cell.

Absolute invariance

Invariance implies some form of constancy. There are two primary forms of scaling invariance in adult mammals: *absolute* and *relative*. Absolute invariance means that a *y*-variable scales with BM or BV with an ideal exponent (*b*) of 0 (Table 19.1, column b). It applies primarily to molecular and microscopic structures (see row one, Tables 19.2 and 19.5) (see also Table 3.3). Recall that absolute invariance with respect to the size and shape of components such as ions, atoms, and small molecules such as water, is imposed by physics. In effect, absolute invariance of atoms and small molecules is an inherent feature of our universe. On the other hand, there may be as many as 10,000 different proteins (macromolecules) in the adult mammalian body (the exact number is not yet known); the size and shapes of the great majority of these proteins are probably independent of organ and body size. This internal constraint is most likely attributable to natural selection rather than to physics per se.

At the *microscopic* level, the same argument applies to at least some cell types. It is likely, for example, that the empirical scaling exponents for the diameter, surface area, and volume of red blood cells (and possibly other cell types such as sperm and ova) accord reasonably well with absolute invariance (see Table 19.2, column b, first row; see also Chapter 9). When we consider the scaling of many macromolecules individually, we may conclude that an exponent near 0 will prove eventually to be the most common reference exponent in mammalian scaling, even though we cannot yet cite empirical data in diverse mammals to support this contention.

On the other hand, at the *macroscopic* level, adult mammals have one heart, one liver, two lungs, and two kidneys (Table 19.2, column c, first row). Hence the *number* of organs of a given type in the adult mammal no doubt scales in accord with absolute invariance (Table 19.2, column c, row one). It is true, however, that some species of adult mammals have more than one stomach (see Chapter 10) and variable numbers of teeth (including none). Most, but not all organs scale *exactly* at constant number ($b = 0.0$).

Relative invariance

The second form of scaling constancy in adult mammals is termed *relative invariance* ($b = 1.0$) (see Table 19.1, column c, rows three and five, as well as Table 19.2, bottom row). That is, a y-variable expressed in terms of volume or mass and subject to relative invariance scales in direct proportion to BV or BM ($b = 1.0$). At the macroscopic level, relative invariance applies very nearly to the empirical exponents for the volume (or mass) of several organs and tissues, such as blood volume, lung, and total muscle mass (see Table 19.2, column b, bottom row and Table 19.5, column b, rows 22 to 26). It is probable that relative invariance accounts for at least 50% of the scaling of total volume (or mass) in adult mammals (visit Table 19.5; compute the antilogs of the intercepts for rows 22, 23, 24; then sum the antilogs ($0.08 + 0.01 + 0.42 = 0.51$)). Recall that scaling at constant shape need not, and usually does not, entail scaling in accord with relative invariance (Chapters 3, 24).

At the microscopic level the empirical exponent for the *number* of red blood cells appears to scale in accord with relative invariance (as inferred from the scaling of red blood cell concentration and blood volume (see Table 19.2, column b, bottom row, and Chapter 9)). In the same way, at the molecular level, the empirical exponents for the *number* of hemoglobin (Chapter 9) and water molecules (Chapter 8), among others, likely scale in adult mammals in agreement with relative invariance ($b = 1.0$) (Table 19.2, column c, bottom row).

Scaling at constant shape

Organ and body shape are important topics in the "design" of mammals. Scaling at constant shape may be viewed as a special case of scaling in accord with *absolute invariance* (see Chapters 3, 24). We may enforce the condition of scaling at constant shape in two rather different ways. The first, more basic way, is to require that the ratio of two lengths (e.g., length and diameter or girth) be independent of organ or body size ($b = 0.0$) (Table 19.1, column b, row four). That is, the given (non-dimensional) ratio scales as the *zero* power of volume or mass. Scaling at constant shape represents a special case of absolute invariance (see Table 19.1, column b, row four).

For example, if the ratio of length to diameter for a long bone is independent of bone or body volume (or mass) (i.e., $b = 0.0$), then we have evidence consistent with scaling at constant shape (Table 19.1, column b, row four, and Table 19.3, rows seven and nine). However, this evidence is not beyond dispute; a long bone that scales at or near constant shape by the ratio criterion may nonetheless show varying degrees of *curvature* with increasing body size (see Chapter 12).

A second, more common approach to scaling at constant shape is to express the scaling of organ diameter or length as a *power* of BM or BV. The reference exponent for scaling of organ length or diameter as a function of BM or BV at *constant* shape is 1/3 (Table 19.1, column d, row one). Similarly scaling of *surface area* at constant shape expressed in terms of BM or BV has a reference exponent of 2/3 (Table 19.1, column d, row two).

Keep in mind that the exponents shown in Table 19.1 (column d, rows one and two) apply only to organs or tissues whose volume or mass scales at or near the *first* power of body mass ($b = 1.0$). See footnote to Table 19.1. This is true to a reasonable approximation in a number of cases (e.g., Table 19.5, rows 22 to 26), but is not true for most

Table 19.3 Empirical scaling exponents for diameter, girth, length, thickness, and width (all in cm) as a function of body mass

	Exponents (b)			Intercepts (a)			Structural	Source
	RO	Exponent	99% CL	RO	log (a)	99% CL	y-variable	(Table no.)
Column	a	b	c	d	e	f	g	h
	1	−0.004	0.030	2	−3.198	0.141	RBC diameter	9.3
	2	0.056	0.028	1	−4.554	0.105	Barrier thickness [a]	15.1
	3	0.109	0.035	4	−2.209	0.035	Glomerular diameter	16.2
	4	0.248	0.048	11	−1.142	0.182	Tooth width	10.1
	5	0.284	0.060	3	−3.189	0.211	Epidermal thickness	11.1
	6	0.298	0.042	10	−1.347	0.159	Tooth length	10.1
	7	0.346	0.030	12	−0.266	0.116	Femur length	12.7
	8	0.354	0.003	14	0.446	0.012	Body length	6.1
	9	0.358	0.024	9	−1.434	0.093	Femur diameter	12.7
	10	0.365	0.013	13	0.205	0.061	Body girth	6.2
	11	0.391	0.042	5	−2.177	0.145	Dermal thickness	11.1
	12	0.408	0.047	6	−2.148	0.163	Skin thickness	11.1
	13	0.411	0.037	8	−1.653	0.146	Tracheal diameter	15.7
	14	0.418	0.028	16	1.045	0.107	S.I. length	10.3
	15	0.418	0.071	7	−1.672	0.349	Aortic diameter	9.14
	16	0.439	0.047	15	0.623	0.184	L.I. length	10.5

RO = rank order; RBC = red blood cell; S.I. = small intestine; L.I. = large intestine. Note that "source" in column h refers to tables in the chapters comprising Part II (e.g., '9.3' refers to Table 9.3 in Chapter 9). [a] i.e., thickness of the pulmonary barrier (see Chapter 15).

organs and tissues (see Chapters 3, 24). This somewhat arbitrary restriction was adopted in order to simplify Table 19.1 (compare with Table 3.5).

At the same time, *whole body* (rather than organ or tissue) parameters, such as body length, girth, or surface area scaling at constant shape will, by definition, have *reference* exponents of body volume (or mass) of 1/3 for lengths and 2/3 for surface area (Chapters 3, 24).

Invariance as a function of component size

How the physical dimensions of length, area, or volume scale in adult mammals depends in part on the *level of resolution*. The levels of potential interest span the size range from molecular to microscopic to macroscopic. We see in Table 19.2 (column a, row one) that absolute invariance pertains to diameters, surface areas, and volumes (or masses) of components at the molecular level and to at least some components at the microscopic level (column b, row one), as well as to component number at the organ level (column c, row one). The inverse tends to be true for relative invariance. Relative invariance is applicable to component size (expressed in terms of volume or mass) at the organ or tissue level (Table 19.2, column a, row two), and in some important instances to component number at the molecular and microscopic levels (column c, row two). Relative invariance of component *number* (with a dimension of zero) applies, as far as is known, only at the molecular and microscopic levels (see Tables 19.1 and 19.2, column c, bottom rows).

Repeating units

As another simplification, this chapter is limited to the study of exponents expressed in terms of body volume or mass. For a discussion of the scaling of the various exponents for repeating units in terms of *organ* or *tissue* volume see Chapters 3 and 24.

Rank order

As in most of Part II, the essential analytical instrument employed in the following exposition of the distribution of empirical exponents is *rank order*. We found in Table 19.1 that one or more reference exponents are possible for a given y-variable, as a function of dimensionality and/or invariance (absolute or relative). The aim is to examine the distributions of empirical exponents that emerge when we consider the exponents for each y-variable (for length, surface area, volume) as a function of rank order (smaller to larger exponents). Further study of these patterns may, over time, provide insight into the overall nature of the scaling adjustments made in adult mammals in response to changes in body size.

It is likely that some of these changes are *primary* responses to increased size, and others are *secondary*, made in response to primary changes. For example, it was remarked in Chapter 3 that if some organs and tissues scale with exponents greater than 1, then certain other organs and tissues must, by way of compensation, scale with exponents less than 1. Differentiating between these two classes of responses to change in body size is a task for the future.

Structural exponents for length by rank order

Scaling exponents (and log (intercepts)) for 16 structural y-variables pertaining to the dimension of *length* (e.g., diameter, girth, length) are brought together and enumerated in rank order in Table 19.3 (columns b and e). The mean value of pWR for these 16 y-variables is 5.4 ± 1.6 (STD). See triangles in Figure 19.1, panel a, lower arc.

Observe that the exponents in Figure 19.1, panels a and b, read left to right, are in the same (rank) order as those given in Tables 19.3 (lengths), 19.4 (surface areas), and 19.5 (volumes or masses). For example, the first point on the left-hand side of the lower arc in Figure 19.1, panel a, corresponds to red blood cell (RBC) diameter (see Table 19.3, column g, row one).

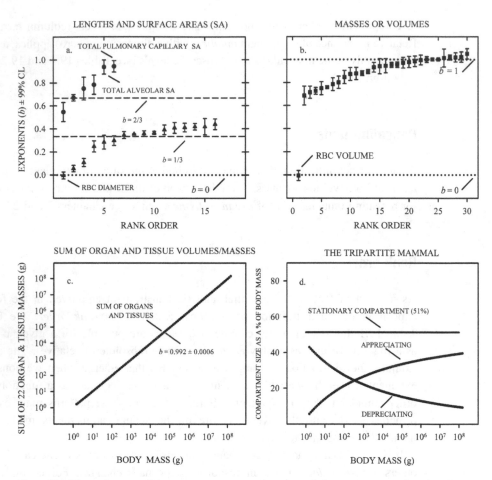

Figure 19.1 Plots of structural exponents, aggregated organ and tissue masses, and a tripartite (three-compartment) model.

The reference exponent for absolute invariance ($b = 0.0$) is shown by dotted lines in Figure 19.1, panels a and b. The reference exponent ($b = 1/3$) for the dimension of length (see Table 19.1, column d, first row) is shown in Figure 19.1 as the lower dashed line and that for surface area ($b = 2/3$) as the upper dashed line. In panels a and b the empirical exponents are plotted in terms of rank order, whereas in panels c and d the ordinates are plotted against body mass.

The first three empirical exponents (for lengths) listed in Table 19.3 (column b) are for *microscopic* structures. They each have exponents closer to 0 than 1/3. The magnitudes of the next 13 *macroscopic* exponents (see Table 19.3, rows 4–16) are clustered around a reference exponent of 1/3 (Figure 19.1, panel a, lower arc (triangles)). The mean exponent for these 13 entries is 0.364 ± 0.056 (STD). This result supports the contention that dimensionality (here length) strongly influences, directly or indirectly, the respective exponents, but does not, as a rule, govern them precisely. Other constraints must be at play.

Table 19.4 Empirical scaling exponents for surface area (cm^2) as a function of body mass

	Exponents (b)			Intercepts (a)			Structural	Source
	RO	Exponent	99% CL	RO	log (intercept)	99% CL	y-variable	(Table)
Col./Row	a	b	c	d	e	f	g	h
1	1	0.546	0.083	1	−2.489	0.310	Tooth area	10.1
2	2	0.670	0.027	4	0.984	0.109	Body SA	6.2
3	3	0.749	0.099	3	0.374	0.395	Basal area of S.I. + L.I.	10.7
4	4	0.782	0.083	2	0.028	0.331	Basal area of S.I.	10.7
5	5	0.936	0.061	6	1.766	0.225	Total alveolar SA	15.1
6	6	0.941	0.050	5	1.661	0.183	Total pulmonary capillary SA	15.1

See Table 19.2. SA = surface area.

Structural exponents for surface area by rank order

The exponents (and log (intercepts)) for the scaling of the dimension of *surface area* as represented by six structural y-variables are shown in Table 19.4 and Figure 19.1, panel a, upper arc (solid circles). The mean value of pWR for these six y-variables is 4.6 ± 1.0 (STD). The reference exponent is 2/3 for surface area (see Table 19.1, column d and Figure 19.1, panel a, upper dashed line). Only one of the six empirical exponents (that for *body* surface area – see Table 19.4, row two) is in good agreement with the reference exponent for constant shape ($b = 2/3$).

The mean of the six exponents for surface area is 0.771 ± 0.140 (STD). The two rather high exponents of 0.936 and 0.941 (see Table 19.4, rows five and six) are for structures consisting of microscopic *repeating units* (Chapters 3, 24). Omitting these two outliers, the arithmetic mean for the first four empirical exponents is 0.687 ± 0.091 (STD) (Table 19.4, rows one to four). These four points are grouped roughly around a reference exponent of 2/3.

Structural exponents for volume or mass by rank order

The empirical structural exponents (and log (intercepts)) for the scaling of organ or tissue volume or mass as represented by 30 y-variables are summarized in Table 19.5 (columns b and e). See also Figure 19.1b. The relevant reference exponents are given in Table 19.1, row three. The mean value of pWR for these 30 y-variables is 6.4 ± 1.1 (STD). Two reference exponents, one for absolute ($b = 0.0$) and one for relative ($b = 1.0$) invariance, or for dimensionality, are shown as dotted lines in Figure 19.1b. One empirical exponent, that for red blood cell volume, is in good agreement with absolute invariance ($b = 0.0$) (see Table 19.5, first row).

Table 19.5 Empirical scaling exponents for cell, organ, and tissue volumes or masses (cm^3 or g)

| | Exponents (b) | | | Intercepts (a) | | | | Source |
	RO	Exponent	99% CL	RO	log (intercept)	99% CL	Structural y-variable	(Table/Figure/ Chapter)
Col./Row	a	b	c	d	e	f	g	h
1	1	−0.005	0.044	1	−10.180	0.253	Red blood cell	9.4
2	2	0.686	0.082	12	−1.903	0.366	Eyeballs	13.4
3	3	0.715	0.053	13	−1.900	0.202	Spinal cord	13.5
4	4	0.727	0.030	2	−4.159	0.083	Pituitary	13.9
5	5	0.750	0.057	16	−1.622	0.206	Testes	14.6
6	6	0.759	0.013	20	−1.271	0.037	Brain	13.2
7	7	0.785	0.055	9	−2.197	0.161	Cerebellum	13.3
8	8	0.801	0.053	4	−2.980	0.224	Adrenals	13.7
9	9	0.847	0.057	8	−2.250	0.273	Pancreas	10.9
10	10	0.873	0.056	18	−1.455	0.215	L.I.	10.5
11	11	0.878	0.053	19	−1.295	0.205	S.I.	10.4
12	12	0.880	0.021	14	−1.786	0.086	Kidneys	16.4
13	13	0.895	0.018	22	−1.187	0.070	Liver	10.8
14	14	0.932	0.042	23	−1.119	0.185	Gut	10.6
15	15	0.941	0.066	15	−1.688	0.290	Stomach	10.2
16	16	0.945	0.062	3	−3.724	0.264	Thyroid	13.10
17	17	0.953	0.022	10	−2.053	0.086	Heart	9.15
18	18	0.954	0.024	27	−0.845	0.086	Skin	11.2
19	19	0.972	0.047	5	−2.563	0.174	Spleen	9.19
20	20	0.978	0.049	21	−1.258	0.192	Plasma volume	9.8
21	21	0.985	0.046	28	−0.733	0.172	Protein	Chapter 8
22	22	0.996	0.043	24	−1.098	0.179	Blood	9.8
23	23	1.000	0.026	11	−1.996	0.109	Lung	15.4
24	24	1.001	0.012	29	−0.373	0.050	Muscle	12.2
25	25	1.002	0.011	30	−0.197	0.036	TBW	8.5
26	26	1.015	0.079	7	−2.440	0.439	Diaphragm	15.2
27	27	1.019	0.022	25	−1.092	0.088	Skeleton	12.5
28	28	1.020	0.040	17	−1.543	0.128	Ash	Figure 8.1
29	29	1.023	0.045	6	−2.461	0.166	Pulmonary capillary	15.1
30	30	1.051	0.045	26	−1.079	0.175	Fat	8.4

See Table 19.3. L.I. and S.I. = large and small intestine, respectively; TBW = total body water; vol. = volume.

Six of the exponents in Table 19.5 (i.e., those for blood volume (row 22), mass of diaphragm (26), lung (23), skeletal muscle (24), skeleton (27), and total body water (25) are not sensibly different from a reference exponent of 1 (*relative invariance*)). Only one exponent in Table 19.5 (for fat) falls significantly, but only just, above the reference exponent of 1. It may be that the exponents for brain, kidney, and liver (see Table 19.5, rows 6, 12, and 13, and Figure 19.1b) reflect a relative reduction in organ size "aimed" at conserving energy (see Chapter 17).

Empirical exponents and dimensionality

It is clear from Figure 19.1, panels a and b, that the empirical scaling exponents for lengths (e.g., diameter) are mostly distinct from those for surface areas (panel a, solid circles) or volumes (panel b, squares). However, the distribution of empirical exponents for surface areas substantially overlaps the empirical distribution for volumes (masses). This may be due, at least in part, to the presence of microscopic repeating units, which substantially increase the numerical value of the applicable scaling exponents (from 2/3 to 8/9) (see Chapters 3, 24). Overall, the evidence is that dimensionality and/or invariance (including in some instances scaling at or near constant shape) *influence* the empirical exponents for lengths (especially) and for surface area and volume (to a lesser degree), but do not precisely define them.

Scaling of aggregate organs and tissues

We seek to discover whether the exponents and intercepts presented in Table 19.5 are reasonably self-consistent (see Chapter 24; the "Haldane restriction" states that a *sum* of different power functions is not itself, as a rule, a power function). More directly, how closely does a calculated sum of known organ and tissue masses approach total body mass over the whole size range? To test for this form of self-consistency I carried out the following experiment. A list of 100 hypothetical body masses was generated, evenly spaced on a logarithmic scale, extending from 1.5 to 150 million grams (pWR = 8.0). The data for 22 organs and tissues were employed in the calculations described below. In constructing this dataset, I omitted data from seven rows (7, 14, 20, 21, 25, 28, and 29) out of 30 in Table 19.5 in order to reduce double-counting (e.g, total body water was omitted, as all organs and tissues contain some water). In addition, row one was omitted as being inconsequential to the outcome. These deletions reduce the number of y-variables to 22 (30 − 8 = 22).

The intercept (calculated as the antilog of the values given in Table 19.5, column e) together with the cognate exponent (Table 19.5, column b) for each specific organ or tissue were employed to generate individual organ and tissue masses at each of the 100 given body masses (see above). One then sums the 22 calculated organ and tissue masses at each body mass. In effect, we aggregate all the organs and tissues of the body into one compartment for each level of body size. This calculated sum of organs and tissues is plotted against body mass in Figure 19.1c. The LSQ exponent of body mass for this sum is 0.992. The sum of the masses of the 22 organs and tissues differs from the given body masses by a mean deviation of 4.7 ± 3.2 (STD) percent. That is, the sum of the organ masses derived as just described is a reasonably good approximation to total body mass over the whole mammalian size range. In this analysis we treat the empirical exponents and intercepts as if they are valid for pWR = 8.0.

In part the above findings may be fortuitous. It is likely that brain mass does not increase with increasing body mass beyond its size in a large elephant (see Chapter 13). I have included blood volume as a separate entry (Table 19.5, row 22), but most organs probably contained some blood when weighed. The mass of the diaphragm (Table 19.5, row 26) may be contained in the estimate of skeletal muscle. So there is no doubt that some double-counting has occurred, and some errors (as for brain mass at large body sizes) are certainly present in the calculations. Moreover, many small tissues and organs are not included in the computations. Nevertheless, taken together, the results (see Figure 19.1c) suggest that these exponents and intercepts for the various organs and tissues are reasonably self-consistent and representative of the scaling of organs and tissues generally, notwithstanding the Haldane restriction (Chapter 24).

The tripartite mammal

The next simplest (see above discussion of the Haldane restriction) representation of scaling in adult mammals at the organ and tissue levels of resolution is based upon aggregating organs and tissues into three compartments. One is termed the *stationary* compartment; it consists of those organs and tissues whose volume or mass scale in accord, or nearly in accord, with relative invariance ($b = 1.0$) (see Table 19.5, rows 22, 23, and 24). A second compartment, here termed *depreciating*, consists of those organs and tissues that scale with exponents of less than 1 ($b < 1.0$) (see 16 entries in Table 19.5; rows 2–6, 8–13, 15–19). The third compartment, termed *appreciating*, comprises those organs and tissues having exponents greater than 1 ($b > 1.0$).

The database used to construct a scaling model for tripartite mammals is much the same as the one employed to construct Figure 19.1c. As before, exponents and intercepts are assumed to be valid for pWR = 8.0. The appreciating compartment was derived by subtraction. That is, the appreciating compartment is computed as the difference between body mass and the sum of the stationary and depreciating compartments. The results of these various calculations are summarized in Figure 19.1d. We see that the stationary compartment (lung and muscle mass plus blood volume or mass) makes up 51% of adult mammals at all body sizes, leaving 49% for the remaining organs and tissues. The depreciating compartment, comprising most major organs and tissues in the body, decreases on an arc from a high of about 43% of body mass in the smallest mammals to a low of about 9% in the largest mammals. Conversely, as required for consistency, the appreciating compartment starts out at nearly 6% (49 – 43) of body mass in the smallest mammals and rises on an arc to 40% (49 – 9) in the largest. The above construction provides a global view of organ and tissue scaling in adult mammals.

It is safe to acknowledge that the detailed results discussed above will change with further work. In particular the cross-over point (where the appreciating and depreciating compartments are equal; here at a body mass of about 330 g and a y-variable of close to 24%) is quite sensitive to the details of the computation and may change significantly in the future. On the other hand, the overall picture presented in Figure 19.1d will likely persist. As pointed out previously (Chapter 3, Figure 3.2), the organs and tissues scaling with empirical exponents for volume or mass greater than 1 must be compensated for by exponents scaling with exponents less than 1 and vice versa. This logical necessity is reflected in the construction of Figure 19.1d.

Table 19.6 Summary statistics for empirical structural exponents

| Dimension | Reference exponents | Number of instances | Exponents (b) | | | |
			Mean ± STD	Median	Range	Delta
Column	a	b	c	d	e	f
Diameter/length	0 or 1/3	16	0.306 ± 0.133	0.356	−0.004 to 0.439	0.443
Surface area	0 or 2/3	6	0.771 ± 0.140	0.766	0.546 to 0.941	0.395
Volume/mass	0 or 3/3	30	0.879 ± 0.194	0.943	−0.005 to 1.051	1.056

STD = standard deviation; Delta = maximum of range less minimum (e.g., in the first row, column f we find 0.439 − (−0.004) = 0.443); for row one see Table 19.3, for row two Table 19.4, and for row three Table 19.5.

Summary statistics

It is convenient for the discussion to follow in Chapter 20 to give a statistical summary of the results presented in Tables 19.3, 19.4, and 19.5. In Table 19.6 we characterize the spectrum of structural exponents in terms of *central tendency* (mean, median) and *range*. This is not meant to mask the fact that the actual structural and functional accommodations to increased body size are likely to be complex. Indeed, wherever we "drill" down into biology, we tend to find novel terrains, each with its own unique characteristics, often including new shapes. Column a of Table 19.6 furnishes several reference exponents (see Table 19.1). Column b gives the number of instances considered for each dimension; columns c and d give the arithmetic mean and the median of the exponents for each y-variable. The range and the difference between the end-points of the range in exponents (termed delta) are shown in columns e and f.

Observe in Table 19.6, columns c and d, that the median exponents are not significantly different from the arithmetic mean exponents. However, the standard deviations (STD) in the mean exponents are large relative to the exponents (Table 19.5, column c). The parameter "delta" (Table 19.6, column f) is about the same for the first two rows but is much larger for row three.

Summary

It is helpful here to refer again to Figure 19.1, panels a and b. We see in panel a that three empirical exponents for length are in the neighborhood of a reference exponent of 0 (absolute invariance) and 13 exponents accord moderately well with a reference exponent of 1/3 (dimensionality for length). Of six exponents for surface area, four are roughly in agreement with a reference exponent of 2/3 (dimensionality for surface areas). In panel b, one exponent agrees with absolute invariance ($b = 0.0$) and at least another six agree with a reference exponent for relative invariance ($b = 1.0$). Thus a total of 27 exponents out of 52 (i.e., 52%) are in reasonable agreement with one or another reference exponent (Table 19.1). At the same time, 25 empirical exponents are not well accounted for by either dimensionality or invariance (52 − 27 = 25). No negative empirical structural exponents were found (see above quote from Stephan and Andy).

A glance at Figure 19.1, panels a and b, shows that physical *dimensionality* may condition, to varying degrees, the numerical values of many of the exponents in these samples. There is little overlap between the 1-D exponents for length (triangles) and 2-D exponents for surface area (circles) or between the 1-D exponents and the 3-D exponents for volume (panel b, squares). However, there is a substantial overlap in the numerical values for some 2-D surface area exponents (circles) with those for 3-D volume (compare the upper arc of panel a (circles) with panel b (squares)).

A reasonable inference from these examples is that dimensionality does influence the distribution of some empirical exponents, but does not rigidly determine it. Physics is decisive in determining the dimensions of small molecules but is probably less important at higher levels of resolution (macromolecules, cells, tissues, and organs). Overall, it is likely that strictly "physical considerations" play a significant but limited role in adult mammalian scaling (see above quote from D'Arcy Thompson).

It is argued (see Figure 19.1, panel c) that the Haldane restriction (see Chapter 24) does not seriously affect our ability to represent body size as a sum of all organ and tissue masses. A simple three-compartment (or tripartite) model of organ and tissue scaling in adult mammals is introduced. These three compartments comprise one that is stationary ($b = 1.0$), one that is appreciating ($b > 1.0$), and one that is depreciating ($b < 1.0$). It is noteworthy that the stationary compartment accounts for about 50% of the mass or volume of adult mammals. The depreciating compartment varies between a high of 43% of body mass and a low of 9%. Correspondingly, the appreciating compartment varies between 6 and 40% of body mass. These various percentages will no doubt change somewhat with further work.

In conclusion, we have learned that certain reference scaling exponents are likely to be important to a first-order interpretation of structural scaling in adult mammals. In particular, the constraint of *dimensionality* is correlated with the empirical exponents for scaling of length and to a lesser degree with the empirical exponents for scaling of surface areas and volume. However, the exponents for surface area and volume mostly overlap. Scaling in accord or near accord with invariance (absolute and relative) is not uncommon. Other constraints, some no doubt yet to be identified, may play a role in bringing about second-order effects seen in the departure of empirical scaling exponents from reference exponents. Looking ahead, identifying such alternate constraints (primary or secondary) will be a challenge for scaling studies yet to come.

20 Functional summary

Natural selection produces systems that function no better than necessary... The product of natural selection is not perfection but adequacy, not final answers but limited short-term solutions.

Bartholomew 1986

With important exceptions (see below), physiological functions are *time-dependent*. Some functions, such as lifespan, represent *durations* and others, such as heart beat, represent *frequencies* or *rates*. We are interested in how such functional quantities vary with body size. In physics we are more accustomed to thinking of time as an independent variable, as in the definition of velocity (distance per unit time). But we can easily construct a physical scaling model where time (duration) is a dependent variable. Consider a family of similar Galilean pendulums, each consisting of a thin rod with a "bob" at one end. For small angular displacements of the pendulum the period (P) of oscillation will vary as the square root of pendulum length (L). For geometrically similar pendulums the length may be assumed to scale as the one-third power of pendulum mass (M). It follows that the pendulum period will scale as the one-sixth power of pendulum mass.

$$P \propto L^{1/2} \text{ and } L \propto M^{1/3} \tag{20.1}$$

$$\text{Hence } P \propto M^{1/6} \approx M^{0.17} \tag{20.2}$$

Thus we have a simple physical system in which a functional period (duration) scales as a power function of mass with an exponent of 0.17 [1]. Nonetheless, physical reasoning has more direct pertinence to scaling of functional rates in mammals than this simple example suggests. Think about mass-specific energy metabolism. It has been argued that if a large mammal such as a steer had the same mass-specific energy metabolism as a mouse its skin temperature would have to exceed the boiling point of water in order for the steer to unload the heat it would generate [2].

Again, small mammals have heart rates of 500–1,000 beats per minute or even higher (see Chapter 9) which are physically impossible in mammals with much larger hearts. Thus there are structural *limits* on maximal functional rates, given constant "design," even though it is currently difficult to express these limitations in terms of *a priori* reference exponents (e.g., as seen for structural exponents in Tables 3.5 and 19.1).

From the standpoint of scaling at constant design, the most directly relevant functional y-variables are those associated with homeostasis. In the adult these include maintenance of mean body temperature, and numerous quantities associated with blood, such as blood pressure, calcium and glucose levels, oxygen capacity, and pH. These and other homeostatic quantities may scale in accord with absolute invariance ($b \approx 0$) (see below).

The following material, drawn from Part II, is based on the LSQ analysis of 20 functional y-variables, with a *mean sample size* (N_{sp}) of 126 ± 177 (STD) and a *range* in sample size from 13 to 586. The total number of records represented by these 20 y-variables is 2,517 ($20 \cdot 126 = 2{,}520$), expressed in tabular form. Each of these 2,517 records is unique. The mean size range for these 20 functional datasets is given by pWR = 5.2 ± 0.6 (STD). The comparable value from Chapter 19 is 6.8 ± 1.6 (STD).

Possible reference exponents for functional *y*-variables

In Chapter 19 we discussed *structural* y-variables in terms of four physical dimensions: number, length, surface area, and volume (or mass). Functional y-variables exhibit many different dimensions, often including time. We are currently unable to assign hypothetical reference exponents based on dimensionality for most functional y-variables.

Absolute invariance ($b = 0.0$) is reasonable for at least some homeostatic y-variables (see below). No example is currently known of a functional y-variable clearly exhibiting *relative invariance*, although two functions (namely VO_2max and pulmonary diffusion capacity) exhibit exponents of 0.916 ± 0.097 and 0.965 ± 0.046 (99% CL), respectively, suggesting that relative invariance for some functional y-variables is not out of the question (see Tables 20.4 and 20.5, bottom rows).

It has often been suggested that exponents for numerous functional exponents may be expected to scale as multiples of $\pm 1/4$ [3, 4]. Five types or classes of y-variables (column a) and possible reference exponents (column b) for four of them are set out in Table 20.1. See also Figure 20.1, panels a, b, and c (dashed and dotted lines), where exponents, but not intercepts, are shown.

Table 20.1 Hypothetical reference exponents for functional *y*-variables

Types of *y*-variables	Possible reference exponents	Number of *y*-variables	Table	Figure 20.1
a	b	c	d	e
Durations	1/4	4	20.2	Panel a, triangles
Rates	−1/4	3	20.3	Panel a, circles
Flows	3/4	7	20.4	Panel b, squares
Absolute invariance (homeostatic)	0	4	20.5	Panel c, circles
Miscellaneous	-	2	20.5	Panel c, diamonds

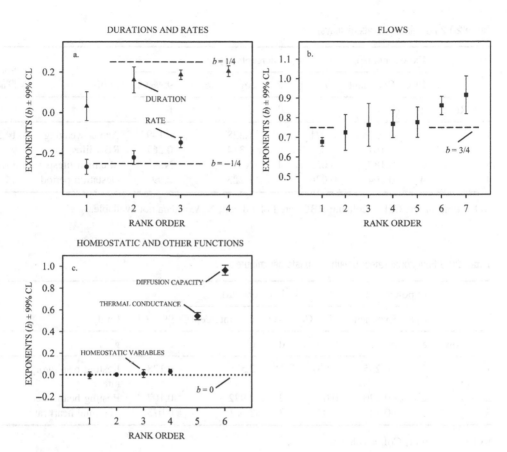

Figure 20.1 Summary of functional exponents.

Durations

The clearest examples of functional dimensions expressed either directly or inversely in terms of time are given by durations and rates (or frequencies). Exponents and log (intercepts) for four durations are summarized in rank order in Table 20.2 (columns b and e) and in Figure 20.1, panel a (triangles). As noted above, some functional y-variables for durations may scale as the one-quarter power of body mass (Table 20.1, column b, row one). The first report of this kind known to me is by Jackson [5], who in 1909 found that the prenatal weight of the human embryo/fetus varies as the 4th power of age (in days), implying that prenatal age scales as the one-quarter power of prenatal weight. From his data [5], taking a lunar month to be 29.5 days, one finds that prenatal age in the human scales as the 0.26 ±0.05 (99% CL) power of prenatal body weight. Take note that the 99% CI (Chapter 5) for this exponent is from 0.21 to 0.31.

Keep in mind that the rank order in which the empirical exponents appear in each panel of Figure 20.1, reading left to right, is the same as in column b of Table 20.2 (upper panel a); Table 20.3 (lower panel a), Table 20.4 (panel b), and Table 20.5 (panel c). For example, the first triangle in the upper left-hand side of Figure 20.1, panel a,

Table 20.2 Functional durations (days)

| Col./Row | Exponents (*b*) | | | Intercepts (*a*) | | | | Source |
| | RO | Exponent | 99% CL | RO | log (intercept) | 99% CL | Entity | (Table) |
	a	b	c	d	e	f	g	h
1	1	0.032	0.071	3	1.328	0.149	Age at weaning	14.3
2	2	0.161	0.064	2	1.314	0.267	RBC lifespan	9.6
3	3	0.187	0.023	4	N.A.	N.A.	Maximal lifespan	17.14
4	4	0.204	0.026	1	1.223	0.091	Gestation period	14.2

RO = rank order; Col. = column; RBC = red blood cell; N.A. = data not available.

Table 20.3 Functional rates: breaths or beats per minute

| Col./Row | Exponents (b) | | | Intercepts (a) | | | | Source |
| | RO | Exponent | 99% CL | RO | log (intercept) | 99% CL | Entity | (Table) |
	a	b	c	d	e	f	g	h
1	1	−0.265	0.037	1	2.536	0.128	Resting respiratory rate	15.6
2	2	−0.220	0.033	2	2.972	0.107	Resting heart rate	9.17
3	3	−0.146	0.025	3	3.061	0.073	Maximal heart rate	9.17

RO = rank order; Col. = column.

is taken from Table 20.2, row one. As remarked in Chapter 14, it is uncertain whether the results for age at weaning (Table 20.2, row one, and Figure 20.1, upper panel a) apply to all mammals.

We see in Figure 20.1, panel a, that none of the four empirical exponents for the duration *y*-variables (triangles) is precisely consistent with a reference exponent of 1/4 (upper dashed line) (Table 20.1, column b, row one).

Lindstedt and Calder [6], in a well-known paper, extracted from the literature exponents for duration for 16 *y*-variables. From their collation one finds a mean scaling exponent pertaining to mammals of 0.25 ± 0.05 (STD). No confidence levels were reported.

Rates

Exponents (and log intercepts) for three examples of the scaling of functional rates are given in Table 20.3; see also Figure 20.1, panel a (solid circles). Notice that all three empirical exponents are negative. Two of the three exponents are potentially consistent with a reference exponent of −1/4 (Table 20.3, rows one, two). The exponent for maximal heart rate (-0.146 ± 0.025) (99% CL) in row three is inconsistent with a −1/4

Table 20.4 Functional flows (cm^3/min)

	Exponents (*b*)			Intercepts (*a*)				Source
	RO	Exponent	99% CL	RO	log (intercept)	99% CL	Entity	(Table)
Col./Row	a	b	c	d	e	f	g	h
1	1	0.677	0.022	3	−1.138	0.054	SMR (VO$_2$)	17.6
2	2	0.739	0.085	7	0.375	0.337	Respiratory minute volume	15.5
3	3	0.764	0.109	2	−3.568	0.454	Daily urine production	16.5
4	4	0.769	0.070	6	0.055	0.286	Resting cardiac output	9.16
5	5	0.777	0.077	5	−3.536	0.330	Daily milk production	14.3
6	6	0.862	0.047	1	−3.683	0.146	Daily water intake	10.10
7	7	0.916	0.097	4	−0.635	0.351	VO$_2$max	17.13

RO = rank order; Col. = column; SMR = standard metabolic rate.

power rule. Overall, of seven exponents reported in Figure 20.1, panel a, two are in reasonable agreement with a ±1/4 rule (see Table 20.3, rows one, two); another five are in moderate to serious disagreement.

Flows

We now consider functional volume-flows, with dimensions of volume per unit time. The exponents and intercepts for seven *y*-variables are given in Table 20.4, column b. In Figure 20.1b, we see that four of the exponents for volume-flows are in reasonable agreement with a three-quarters power rule (dashed lines) (Table 20.4, rows two to five). The mean of these four exponents is 0.762 ± 0.0214 (STD). Three exponents are not in agreement with a three-quarters power rule (Table 20.4, rows one, six, seven). We see that all of the seven empirical flow exponents lie between 0.677 and 0.916, with a mean of 0.786 ± 0.073 (STD).

Note that the 99% CL (±0.109) on the empirical exponent for daily urine production are uncomfortably wide (Table 20.4, column c, row three); the 99% confidence interval extends from 0.655 to 0.873. Recall from Chapter 17 that an exponent of 0.677 for SMR (Table 20.4, row one; see Table 17.6) is not valid for mammals generally, but may apply to some small mammals.

Homeostatic and other functional *y*-variables

Exponents and log intercepts for six functional *y*-variables with diverse dimensions are summarized in Table 20.5, columns b and e; exponents are shown in Figure 20.1, panel c. It is clear from Figure 20.1c that three of the six exponents – all concerned with

Table 20.5 Homeostatic and other functional *y*-variables

	Exponents (*b*)			Intercepts (*a*)				Source
	RO	Exponent	99% CL	RO	log (intercept)	99% CL	Entity	(Table)
Col./Row	a	b	c	d	e	f	g	h
1	1	−0.006	0.031	4	1.318	0.128	Oxygen capacity	9.7
2	2	0.001	0.003	5	1.556	0.008	Body temperature	17.2
3	3	0.010	0.036	6	2.017	0.148	Blood pressure *a*	9.18
4	4	0.032	0.020	2	−0.627	0.082	Bohr effect	9.7
5	5	0.544	0.032	3	−0.072	0.079	Thermal conductance	17.15
6	6	0.965	0.046	1	−4.153	0.157	Diffusion capacity	15.3

RO = rank order; Col. = column; *a* only the arithmetic mean is given in Chapter 10; units are: (volume%) for row one; (°C) for row two; (mm Hg) for row three; (difference in $\log(PO_2)$/(difference in pH)) for row four; ($cm^3 O_2$/(hr·°C)) for row five; ($cm^3 O_2$/(sec·mbar)) for row six.

homeostasis – are consistent with an exponent of 0 (absolute invariance) (solid circles); another exponent (for the Bohr effect) differs from zero by 0.012 (0.032 − 0.020 = 0.012); two exponents (for thermal conductance and diffusion capacity) are far removed from 0 (Figure 20.1c, diamonds). Thus, four exponents are in reasonable agreement with absolute invariance.

The exponent (0.965 ± 0.046) (99% CL) for diffusion capacity (Table 20.5, row six) might (or might not) prove to be in agreement with relative invariance. The rates of *fast* (relative to, say, blood flow) processes such as electron and proton transfer, and photon capture in the retina are likely to be independent of body size. Thus invariance ($b = 0.0$) of such rates may be common in mammals, even though it is difficult at present to cite empirical evidence.

Summary of exponent distributions for four systems

For ease of comparison, four plots from Part II are assembled in Figure 20.2 showing the distribution of structural and functional exponents for the circulatory (panel a), digestive (panel b), musculoskeletal (panel c), and respiratory systems (panel d). These four systems have the largest number of empirical exponents appearing in the various chapters of Part II. See Figures 9.11, 10.5, 12.3, and 15.4.

The distribution of exponents shown in Figure 20.2a is roughly S-shaped; those in panels b and d are both concave downwards, whereas the distribution in panel c suggest a cascade. Observe that the range of exponents is much wider in panel d than in panel b. It will be interesting in the future to see whether the addition of exponents for *y*-variables different from those employed here will alter the above distributions, or reinforce them.

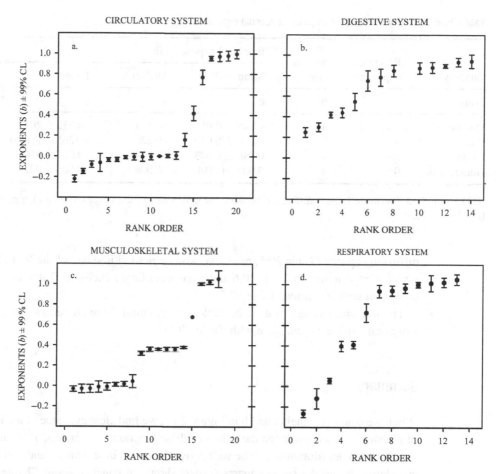

Figure 20.2 Summary of structural and functional exponent distributions for four systems.

A trifling chore

The reader may wish to carry out the following exercise. Use a pseudo-random number generator to generate 20 or more numbers (slopes) between 0 and 1. Then rank-order the slopes. Plot the rank-ordered slopes against their rank order. It is found, as expected, that as the number of slopes increases beyond 20, the distribution of slopes tends to become more and more linear. None of the distributions shown in Figure 20.2 or in Part II is linear overall.

Summary statistics

The distribution of 18 (out of 20) exponents discussed in this chapter (Tables 20.2–20.5) is summarized in Table 20.6. Note that the means and medians (columns c and d) are in reasonably good agreement. Exponents for thermal conductance and perhaps for

Table 20.6 Summary statistics for empirical functional exponents

| Quantity | Reference exponent | Sample size | Empirical exponents (b) | | | |
			Mean ± STD	Median	Range	Delta
Column	a	b	c	d	e	f
Duration	1/4	4	0.146 ± 0.068	0.174	0.032–0.204	0.172
Rates	−1/4	3	−0.210 ± 0.049	−0.220	−0.265–(−0.146)	0.119
Flows	3/4	7	0.786 ± 0.073	0.769	0.677–0.916	0.239
Homeostatic	0	4	0.009 ±0.014	0.006	−0.006–0.032	0.038

STD = standard deviation; Delta = difference between the end-points of the range (column e), e.g., 0.204 − 0.032 = 0.172.

diffusion capacity (Table 20.5, rows five, six) do not fall into any of the four classes of y-variables identified in Table 20.6 and were accordingly excluded. Table 20.6 may be compared with the similar Table 19.5.

The numerical values of delta in Table 20.6 (column f) are all narrower than for the structural exponents (compare with Table 19.5).

Summary

When we compare panels a and b in Figure 20.1, we find clear evidence of the influence of physical *dimensionality* on the values of these functional exponents. Thus in panel a. the four triangles (durations) all lie well above the three circles (rates), and both datasets are plainly below the seven squares (flows) shown in panel b. As in Chapter 19, the appropriate inference seems to be that dimensionality conditions the distribution of certain functional exponents, but does not uniquely determine it. Hence, in developing explanatory models for functional scaling, it will be necessary to take dimensionality into account in most cases; however dimensionality by itself is unlikely to provide a full account. Nature's "selection" of functional exponents may owe more to makeshift expedience than to any sophisticated form of optimization (see above quote from Bartholomew).

In Tables 20.2 and 20.5 we find a total of four (or possibly five if the exponent for age at weaning is included) (out of 20) exponents which accord reasonably well with absolute invariance ($b = 0.0$). In Table 20.3 we find two exponents in agreement with a reference exponent of −1/4; in Table 20.4 another four exponents (all with rather wide CL) concur with a reference exponent of 3/4. Thus, we find that 10 or 11 exponents (out of 20) scale approximately in accord with one or another reference slope.

Despite the remarkable size range spanned by adult mammals and the great variations in external shape (e.g., consider bats and whales), one outcome of this work is to show that mammals are quantitatively alike (i.e., scale in accord with given reference slopes) in more respects than has usually been admitted.

References

1. Günther, B. (1975). On theories of biological similarity. *Fortschritte der experimentellen und theoretischen Biophysik*, **19**:1–111.
2. Schmidt-Nielsen, K. (1975). Scaling in biology: the consequences of size. *Journal of Experimental Zoology*, **194**:287–307.
3. Savage, V.M., Gillooly, J.F., Woodruff, W.H. *et al.* (2004). The predominance of quarter-power scaling in biology. *Functional Ecology*, **18**:257–282.
4. da Silva, J.K.L., Garcia, G.J.M. and Barbosa, L.A. (2006). Allometric scaling laws of metabolism. *Physics of Life Reviews*, **3**:229–261.
5. Jackson, C.M. (1909). On the prenatal growth of the human body and the relative growth of the various organs and parts. *American Journal of Anatomy*, **9**:119–165.
6. Lindstedt, S.L. and Calder, W.A. (1981). Body size, physiological time, and longevity of homeothermic animals. *Quarterly Review Biology*, **56**:1–16.

21 End-sample, mid-sample, and FDS

> *...the size range of the data has a much greater influence on the form of the regression equation than does the sample size.*
>
> Calder 1984

As proposed in Chapter 5, in cases where datasets of 100 or more species became available in Part II, an LSQ analysis was performed on both the *full dataset* (FDS) and the *end-sample*, consisting of the ten largest and the ten smallest species, as determined by body size. A LSQ analysis was also carried out on the *mid-sample*, consisting of the FDS less the end-sample (see Figure 5.3). Thus, the end-sample contains 20% or less of the species present in the FDS; the mid-sample contains 80% or more. Recall that the end-sample has the same value of pWR as does the FDS. The mid-sample is somewhat smaller than the FDS with respect to both pWR and sample size. The purpose of these analyses is to examine empirically the sensitivity of the exponents for the end- and mid-samples to variations in sample size and size range. We are especially interested in how the exponents for the end-samples compare with those for the FDS. See also Chapter 6, Figure 6.2, for seven additional examples (all relating to adult body length).

Structural exponents

Eighteen structural variables studied in this fashion are displayed in Figure 21.1a, and listed in Table 21.1, rank-ordered by the exponent for the FDS (column d).

Observe that in Figure 21.1, each triplicate dataset (end-, mid-sample, FDS) is to be read from left to right (as squares, triangles, circles). The error bars represent 99% CL Keep in mind that the number assigned to each triplet in Figure 21.1 (e.g., 1, 2, 3...) corresponds to a row number in either Table 21.1 (panel a) or Table 21.2 (panel b). For example, the triplet labeled three in Figure 21.1a corresponds to the third row of Table 21.1 (femur length). The exponents and 99% CL for each triplicate dataset (end-sample, mid-sample, FDS) shown in Figure 21.1a are also listed in Table 21.1 (columns b, c, d). The quantity *delta* (Table 21.1, column e) is the absolute value of the difference between the exponent for the FDS and the exponent for the end-sample. The value of delta is to be compared with the 99% CL for the end-sample (column b).

Thus, in Table 21.1, row one, the absolute difference between the end-sample exponent (0.032, column b) and the FDS exponent (–0.005, column d) is 0.037 (column e).

Table 21.1 Summary of structural exponents for end- and mid-samples and FDS

		Exponents ± 99% CL [a]				r^2 for	Source
	Entity	End-sample	Mid-sample	FDS	Delta	FDS	(Table)
Col./ Row	a	b	c	d	e	f	g
1	RBC volume	0.032 ± 0.052	−0.069 ± 0.072	−0.005 ± 0.044	0.037	0.0007	9.4
2	RBC concent.	−0.036 ± 0.050	0.057 ± 0.069	−0.001 ± 0.042	0.035	0.000	9.5
3	Femur length	0.363 ± 0.0335	0.323 ± 0.049	0.346 ± 0.030	0.017	0.900	12.7
4	Body length	0.357 ± 0.017	0.353 ± 0.004	0.354 ± 0.003	0.003	0.982	6.1
5	Femur diameter	0.380 ± 0.0325	0.3265 ± 0.036	0.358 ± 0.024	0.022	0.938	12.7
6	Skin thickness	0.394 ± 0.059	0.423 ± 0.069	0.408 ± 0.047	0.014	0.841	11.1
7	Pituitary mass	0.730 ± 0.050	0.724 ± 0.042	0.727 ± 0.030	0.003	0.961	13.9
8	Testes mass	0.773 ± 0.071	0.732 ± 0.084	0.750 ± 0.057	0.023	0.900	14.6
9	Brain mass	0.713 ± 0.0625	0.769 ± 0.013	0.759 ± 0.013	0.046	0.967	13.2
10	Cerebellar mass	0.774 ± 0.091	0.802 ± 0.0835	0.785 ± 0.055	0.011	0.935	13.3
11	Neonatal mass [b]	0.838 ± 0.034	0.791 ± 0.025	0.800 ± 0.023	0.038	0.952	14.5
12	Adrenal mass	0.813 ± 0.050	0.779 ± 0.092	0.801 ± 0.053	0.012	0.935	13.7
13	Kidney mass	0.894 ± 0.033	0.866 ± 0.028	0.880 ± 0.021	0.014	0.987	16.4
14	Liver mass	0.909 ± 0.031	0.884 ± 0.025	0.895 ± 0.018	0.014	0.992	10.8
15	Thyroid mass	0.932 ± 0.086	0.969 ± 0.108	0.945 ± 0.062	0.013	0.941	13.10
16	Heart mass	0.940 ± 0.021	0.962 ± 0.032	0.953 ± 0.022	0.013	0.991	9.15
17	Spleen mass	0.974 ± 0.082	0.965 ± 0.062	0.972 ± 0.047	0.002	0.967	9.19
18	Lung mass	0.987 ± 0.035	1.019 ± 0.041	1.000 ± 0.026	0.013	0.989	15.4

r^2 is the coefficient of determination; Col. = column; RBC = red blood cell; concent. = concentration;
[a] rank-ordered by exponent for the FDS (column d); [b] litter size greater than or equal to one.

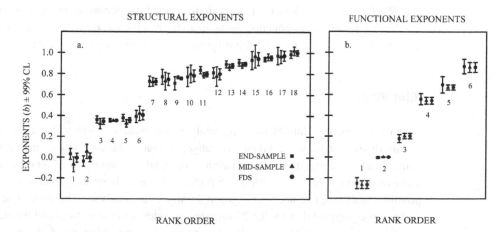

Figure 21.1 Spectrum of exponents for end-samples, mid-samples, and FDS.

This difference is less than the 99% CL (± 0.052) associated with the end-sample (column b), implying that these two exponents may not be significantly different, as judged by the given criterion. Note that in only one case, that for neonatal mass (see Figure 21.1, entry 11, and Table 21.1, row 11, column e), does the exponent for an end-sample, including the 99% CI (confidence interval) *not* include the exponent for the FDS. The discrepancy in this instance ($0.038 - 0.034 = 0.004$) is unlikely to be biologically significant (for pWR = 8.0; $N_d = 26.6 \cdot 0.004 = 0.1064$; $f = 2^{Nd} = 2^{0.1064} = 1.08$), implying an 8% difference over the whole mammalian size range (see Chapter 3).

Note that in 15 out of 18 cases in Table 21.1 (i.e., omitting rows 4, 9, 11), delta is also less than the 99% CL on the exponents for the FDS, notwithstanding the large differences in sample size as between an end-sample and a FDS. In most instances, the exponent for the end-sample agrees with that for the FDS to well within the 99% CL associated with the end-sample. As expected, the 99% CL for an end-sample are usually somewhat wider than for the corresponding 99% CL on a FDS (on average by a factor of 1.15 for structural exponents). The agreement between the end-sample exponents and FDS exponents is good even when the exponents and coefficients of determination are near zero (see Table 21.1, rows one and two). The source table (by chapter) for each entity is given in Table 21.1, column g. For example, the data for row 1, Table 21.1, come from Chapter 9, Table 9.4.

Functional exponents

The results of a similar analysis for six *functional* exponents are displayed in Figure 21.1, panel b. and listed in Table 21.2. As for the structural exponents, the data are rank-ordered by the FDS exponents (column d). Observe that in each of these six cases delta is less than the 99% CL on the end-exponent, consistent with the hypothesis that the end-sample exponents are not significantly different (at the 99% CL) from their respective FDS exponents. On average, the ratio of the 99% CL for the end-samples to the 99% CL for the FDS is 1.9 for the six functional exponents, as compared with 1.15 for the 18 structural exponents. Note that the results for maximal lifespan are not included in Table 21.2 because the original dataset is no longer available.

Inferences

When we prepare a dataset for LSQ analysis we usually prefer to have the x-values evenly distributed over the given size range, so that as far as possible each point in a dataset makes a comparable contribution to the final determination of an exponent (see however Chapter 23). In addition, we prefer to have as large and diverse a sample as possible, so that no one taxonomic group dominates the analysis. The above findings for end-samples suggest that for the 24 samples considered (18 structural and 6 functional), the exponents are quite insensitive to sample size, provided the full size range is respected. (It is interesting that in 11 (out of 24) cases shown in Tables 21.1 and 21.2, the 99% CL are wider for the mid-samples than for the end-samples, even though the mid-samples contain many more species.) The mean values of pWR for Tables 21.1

Table 21.2 Summary of functional exponents for end-samples, mid-samples, and FDS

Col./Row	Entity	Exponents ± 99% CL			Delta	r^2 for FDS	Source (Table)
		End-sample	Mid-sample	FDS			
a		b	c	d	e	f	g
1	RRest	−0.251 ± 0.053	−0.285 ± 0.058	−0.265 ± 0.037	0.014	0.773	15.6
2	Body temperature	−0.003 ± 0.007	0.003 ± 0.003	0.001 ± 0.003	0.004	0.0045	17.2
3	Gestation period	0.180 ± 0.038	0.212 ± 0.031	0.204 ± 0.026	0.024	0.549	14.2
4	Thermal conduct.	0.559 ± 0.053	0.520 ± 0.047	0.544 ± 0.032	0.015	0.953	17.15
5	SMR	0.696 ± 0.078	0.668 ± 0.024	0.677 ± 0.022	0.025	0.912	17.6
6	Water intake	0.871 ± 0.069	0.844 ± 0.067	0.862 ± 0.047	0.009	0.957	10.10

See Table 21.1. RRrest = resting respiratory rate; conduct. = conductance; SMR = standard metabolic rate.

and 21.2 are 6.35 ± 1.4 (STD) and 5.6 ± 1.1 (STD), respectively. The variation between exponents for the end-samples and FDS show little dependence on the numerical value of the FDS exponents (−0.265 to 1.000) or on the coefficient of determination (0 to 0.992) (see columns d and f in Tables 21.1 and 21.2).

To put this another way, while ecological and taxonomic effects on determinations of the best-fit structural and functional exponents cannot be eliminated altogether, the present findings suggest that in these instances, at least, such effects may be small if the size range is suitably large (say pWR > 5.5). In Chapter 23, it will be argued that quantitatively similar exponents may be derived by methods that do not involve LSQ analysis or the assumption that errors are confined to the y-variable.

Summary

In the 24 cases reviewed above, with one exception (for neonatal mass: Table 21.1, row 11), the exponents and 99% CL derived from end-samples (consisting of the ten smallest and the ten largest species) are not significantly different from the exponents derived from the FDS spanning the same size range but with at least five times as many species. These empirical outcomes support the contention that in scaling studies of mammalian structure and function, size range may be more important in determining a precise exponent than is sample size (see above quote from Calder).

This finding suggests that for future studies aimed at determining scaling exponents in mammals generally, it may be more efficient to focus on the largest and smallest mammals possible. This inference in no way reduces the importance of collecting scaling data at intermediate body sizes; that is the only way we can learn whether the exponents for particular groups of mammals conform to the exponents determined from end-samples. However, end-samples may be the most cost-effective way to start out a given empirical scaling study.

22 Human scaling

ONTOGENY

...in the pig the heart has begun to beat before the 21st day of fetal life, when the animal still weighs only 0.2 g.

Widdowson and Dickerson 1964

General results for human ontogeny

Recall that every point (species) on an adult interspecific line represents the end-point of an ontogenetic trajectory that begins during embryogenesis and continues throughout development (Chapter 3, Figure 3.3). The underlying relationships between ontogenetic and adult interspecific trajectories are largely unknown. Here we first review the general results for 21 different human structural and functional *ontogenetic y*-variables. It should be noted that four of the ontogenetic trajectories (for fat, spleen, urine production, and standard metabolic rate) were judged to be appreciably non-linear on a log–log plot. The remainder were considered to be linear or nearly linear. Given that growth curves are usually expressed as curvilinear expressions (e.g., a Gompertz or a polynomial function) of *time*, it is perhaps surprising to find that human fetal growth curves stated in terms of body size are often linear, or approximately so, over about three orders of magnitude in fetal size, on a log–log plot. Another surprise is how close the ontogenetic lines often are to their complementary adult interspecific lines, when plotted in log–log coordinates (see numerous examples in Part II).

The 21 human ontogenetic lines considered here are classified into three positional groups, in terms of whether they fall mainly above, are collinear with, or fall mainly below their respective adult interspecific lines (Chapter 3). The results are summarized in Table 22.1 where the various entries are ordered by "source" (see fourth and eighth columns). There are eight instances where the ontogenetic lines are classified as falling clearly "below" the adult interspecific lines; three cases (for brain, cerebellar, and skeletal masses) are clearly above. Only in one case (for heart size) was an ontogenetic *y*-variable classified as collinear (see Figure 9.8) (see above quotes from Widdowson and Dickerson). Nine cases were either non-linear and/or did not fall clearly into one or another of the above three classes. The mean *adult* interspecific exponent for the eight "below" cases is 0.84 ± 0.32 (STD), and for the three "above" cases is 0.85 ± 0.12.

Table 22.1 How 21 human ontogenetic lines approach their matching adult interspecific lines

Entity	Relative approach	Exponent (*b*) *	Source (Figure)	Entity	Relative approach	Exponent (*b*) *	Source (Figure)
Body length	Sl. above	0.354	6.1	Muscle mass	Below	1.001	12.1
Ca level	Below	1.054	TBL 8.3	Skeletal mass	Above	1.019	12.2
Ash mass	Below	1.020	8.1	Brain mass	Above	0.759	13.1
Fat	Below?	1.051	8.1	Cerebel. mass	Above	0.785	13.3
TBW	Sl. above/ collinear	1.002	8.2	Spinal cord mass	Above/ collinear	0.715	TBL 13.5
Protein	Below	0.985	Chap. 8	Testes mass	Below	0.750	14.2
HRT mass	Collinear	0.953	9.8	Alveolar SA	Below	0.936	15.1
SPL mass	Below?	0.972	9.10	Kidney mass	Below/ collinear	0.880	16.1
BP	Below	0.010	TBL 9.18	Urine production	Below/ collinear?	0.764	16.2
Liver mass	Nearly collinear	0.895	10.3	SMR	Collinear/ above?	0.677	17.2
Skin mass	Below	0.954	11.1	-	-	-	-

* = adult interspecific exponent.
Ca = calcium; TBL = table; TBW = total body water; Sl. = slightly; Chap. = chapter; HRT = heart; BP = blood pressure; SPL = spleen; Cerebel. = cerebellum; SA = surface area; SMR = standard metabolic rate (the slope for SMR is nominal; see Chapter 17).

Thus, there does not seem to be any obvious correlation in this small sample between the "above" and "below" classes and their respective adult empirical interspecific exponents.

Keep in mind that all of the ontogenetic lines approaching the adult lines from below have exponents that are steeper than the adult interspecific lines. At the same time, ontogenetic exponents greater than 1 must be balanced by other organ and tissue exponents less than 1 (see Table 22.2). That is, the sum of all fetal organ and tissue masses should scale with an exponent of 1 (see Figure 19.1c and Chapter 3). Also note that in order to conserve space, a number of ontogenetic lines for which data are readily available were not shown in Part II.

The most striking disparity between a human ontogenetic trajectory and an adult interspecific one is that for brain size (see also Figure 13.1). This may be a peculiarity of human development, or it may show up as well in some other primates and cetaceans. A quick perusal of Table 22.1 shows that the *adult interspecific* exponents in this study range between 0.010 (for blood pressure) and 1.054 (for body calcium level).

Comparison of human organ ontogeny with adult organ size

The human ontogenetic results shown in Table 22.1 are drawn from several laboratories. It is useful to look separately at the extensive dataset from human aborted material

Table 22.2 Comparison of human fetal and adult interspecific exponents for seven organs

Human fetal						Adult interspecific		
N_{pnts}	MPD	pWR	r^2	Exponent ±99% CL	log (intercept) ±99% CL	Exponent ±99% CL	Entity	Source (Figure)
a	b	c	d	e	f	g	h	i
471	22	3.1	0.954	0.967 ± 0.025	−1.4615 ± 0.065	1.000 ± 0.026	Lung	15.2
384	9	3.0	0.992	0.975 ± 0.012	−0.777 ± 0.030	0.759 ± 0.013	Brain	13.1
466	16	3.1	0.973	0.986 ± 0.020	−1.254 ± 0.050	0.895 ± 0.018	Liver	10.3
95	105	2.6	0.890	0.993 ± 0.095	−3.422 ± 0.194	0.945 ± 0.062	Thyroid	13.5
477	15	3.1	0.975	1.016 ± 0.015	−2.188 ± 0.049	0.953 ± 0.022	Heart	9.8
479	17	3.1	0.977	1.153 ± 0.021	−2.454 ± 0.010	0.880 ± 0.021	Kidney	16.1
505	28	3.1	0.945	1.467 ± 0.041	−4.145 ± 0.106	0.972 ± 0.047	Spleen	9.10

N_{pnts} = number of points; MPD = mean percent deviation; pWR = $\log(BW_{max}/BW_{min})$ where BW = body weight or mass; r^2 = coefficient of determination; CL = confidence limits.

provided by T. Shepard (personal communication). These data (see Table 22.2), taken from one laboratory, may be more uniform than those represented in Table 22.1.

For this purpose, I computed best-fit LSQ lines to the ontogenetic data (in log–log form) for seven organs (see Table 22.2, column h). Observe that these data are rank-ordered by the ontogenetic exponents (Table 22.2, column e). The total number of records is 2,377 and the mean number of records for each of the seven human organs is 411 ± 133 (STD) (Table 22.2, column a). The mean scatter (as measured by MPD) around the best-fit line for each of the seven organs is 30 ± 31 (STD) percent (column b). The mean ontogenetic size range (pWR) is 3.0 ± 0.2 (STD) (column c); the mean coefficient of determination (r^2) is 0.958 ± 0.031 (STD) (Table 22.2, column d).

It is notable that some statistics (e.g., N_{pnts}, MPD, r^2) for the seven human fetal lines (Table 22.2) are reasonably comparable (where N_{pnts} is associated with N_{sp}) in a number of cases to those reported for adult interspecific lines generally in Part II. However, values for pWR are usually larger in the adult interspecific cases. These findings (Table 22.2) suggest that a log–log representation of fetal organ data may often be as useful as it is for adult interspecific data. The implications of Table 22.2 may be easier to see in Figure 22.1, where each organ is identified by name.

Take note that there is a correlation between the order (top to bottom) in which the organs appear in Figure 22.1 (i.e., brain, liver, lung, kidney, heart, spleen, thyroid) and the order determined by the relative size of the same organs in the adult human (i.e., liver, lung, brain, heart, kidney, spleen, and thyroid). (The coefficient of determination for this relationship is 0.67; that is, the order in which the organs appear in Figure 22.1 accounts for two-thirds of the ordering seen when adult human organs are ranked by organ size.) It is of interest that the fetal lines in Figure 22.1, excepting only kidney and spleen, do not cross. These nearly parallel fetal lines, but for kidney and spleen, have roughly similar exponents (see Table 22.2, column e), with a mean of 0.987 (close to relative invariance) and a STD of 0.017.

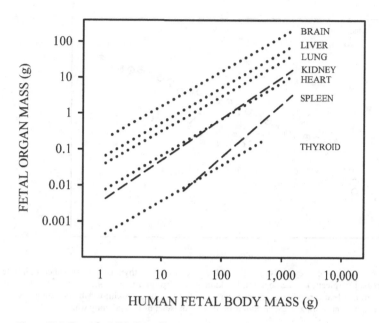

Figure 22.1 Best-fit LSQ lines for seven human fetal organs.

Bear in mind that the comparison of fetal lines with adult interspecific lines involves a (mainly modest) degree of extrapolation of the adult interspecific lines to smaller body sizes. Specifically, for heart, kidney, liver, and lung there is an extrapolation from a whole body mass of about 2.5 g to 1 g. For brain, the extrapolation is from 2 g to 1.5 g and for thyroid from about 16 g to 1 g. No extrapolation is involved for the spleen.

If we do a (lin–lin) LSQ analysis of the adult interspecific exponents (Table 22.2, column g) as a function of the fetal exponents (column e) we obtain a coefficient of determination of 0.07, suggesting there is little or no relation between the two sets of exponents.

Adult *Homo sapiens* as a typical mammal

...Escherichia coli [a gram-negative bacterium] *has essentially the entire repertoire of basic enzyme activities of human cells.*

Campbell 1987

Human beings have always been an unfinished species, a story in the middle, a succession of families, tribes and societies in transition to new awareness.

Hiss 1990

One of the motivations for biological studies generally, and scaling studies in particular, is to contribute to a deeper understanding of our own species. The present material is tendered in this spirit. I reviewed the results drawn from Part II for more than 50 *structural y*-variables. Among these 50 or so *y*-variables, 28 were found to contain data for the adult human. The procedure employed was to remove the human data from these

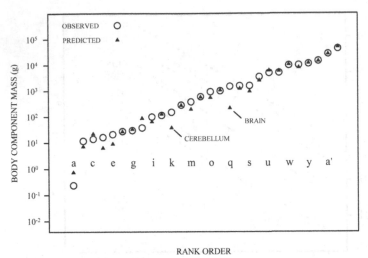

a = lens; b = adrenal; c = eyeballs; d = thyroid; e = ovaries; f = thymus; g = spinal cord; h = testes;
i = pancreas; j = spleen; k = cerebellum; l = kidney; m = pituitary; n = phosphorus; o = lungs;
p = calcium; q = brain; r = liver; s = mammary gland; t = ash; u = skin; v = blood volume; w = fat;
x = skeleton; y = protein; z = extracellular fluid; a' = muscle; b' = total body water

Figure 22.2 Comparison of observed human component size with that predicted from the LSQ adult interspecific lines for 28 structural components.

28 datasets and to then re-compute the best-fit LSQ lines. These best-fit lines (absent human data) were utilized to calculate the size of any given structure at the given body size for an adult human.

For example, for ash, the adult human value of ash content was given for a body mass of 70.55 kg. Hence, the best-fit LSQ line (absent human data) was employed to predict the ash content of a hypothetical adult mammal weighing 70.55 kg (Figure 22.2, item t). This prediction (of 2,500 g) was then compared with the measured value (3,400 g) for the human. The results of carrying out this procedure for 28 structural y-variables are displayed in Figure 22.2.

It is of interest that in 13 of the 28 cases in Figure 22.2, the predicted human component sizes (triangles) lie within the open circles specifying the observed component size. In another eight cases, the predicted and observed component sizes are reasonably close to one another. Hence 75% (100·21/28 = 75%) of the predictions are in fair to good agreement with the observations. The most striking deviations between observed and predicted component sizes are for brain and cerebellum (see Chapter 13). Human brain and cerebellar size exceed the predicted values by factors of 7 and 4, respectively. The MPD between observed and predicted sizes for all 28 structural y-variables is 38%. This MPD is within the range of 30 to 40% usually associated in this study with the scatter around adult interspecific structural LSQ lines.

Note, however, that the data for two atomic species (Ca, P) are confined to a few small and medium-sized mammals such as mouse, rat, guinea pig, and rabbit (Chapter 2). In these instances, a prediction of the component value for the adult human involves a considerable extrapolation. Other structures, associated with the nervous system, including eyeballs, lens, and spinal cord, deviate from the predicted values by 58, 213, and 11%, respectively. In addition, the predicted size of the human testes (item h) exceeds the observed value by 133%.

Summary

About 40% (100·8/21 = 38%) of the human ontogenetic lines in the present sample approach their respective adult interspecific lines from below, implying that these ontogenetic curves are steeper (higher exponents) than the cognate adult interspecific lines. More ontogenetic studies are needed in our own and other species to discover whether the above preliminary results are representative of the relations between ontogenetic and adult interspecific trajectories generally. It is shown in Figure 22.1 that the developmental trajectories for five (out of seven) fetal human organs do not cross, over the given size range.

It was shown (see Figure 22.2) that in 21 cases out of 28, predictions of organ and tissues sizes based from adult interspecific lines (absent human data) were in fair to good agreement with empirical observations. The greatest discrepancy is for the human brain, for which the predicted size is lower than the observed value by a factor of seven.

The great advances in biology since the end of the Second World War predicted, for example, by philosopher Bertrand Russell (whom I once had the privilege of meeting) have made clear the profound similarities among all living organisms on our planet. This gradual but ever-increasing awareness represents a watershed in human thought. However unwelcome it may be to some, the human construct is deeply at one with the natural world. For example, we now know that some 18% of human genes are shared with yeast, 24% with rice, 65% with the chicken, and 90% with the chimpanzee (see also above quote from Campbell). The current chapter aims to extend this picture modestly by showing adult human quantitative similarities to other adult mammals across most of 28 structural y-variables. The most striking differences between us and other mammals relate to the central nervous system (see Figure 13.1). It is apparently our large brain, relative to our body size, that has made this novel and ever-deepening understanding of our place in the scheme of things possible (see above quote from Hiss).

Sad to say, it is also our wayward central nervous system that bears prime responsibility for the accelerated extinction of many species of mammals and birds, as well as other vertebrates, invertebrates, and plants. We seem, as a species, with some notable exceptions, to be remarkably impassive in the face of a mad rush to a likely catastrophe on a global scale (see Chapter 25). Few would dispute that some individual members of our species are remarkably clever, relative to the average, but collectively, it appears that the essential corollaries of forbearance and wisdom are uncommon outcomes of natural selection.

Part IV

Methodology

23 Scaling statistics

It is a grave error to evaluate the "goodness" of an experiment only in terms of the significance level of its results.

Hays 1973

...statistics is a method of investigation used when other methods are of no avail...

Moroney 1982

The discipline of statistics is concerned with the collection, analysis, and display of empirical data. Statistics is also involved with degrees of certainty and uncertainty, calling for an element of subjective judgement (see above quote from Hays). Much of classical statistics relates to idealized distributions of numbers, as the binomial, normal, and Poisson distributions. Each distribution is characterized by certain parameters; for example, a normal distribution is characterized by a mean and standard deviation (STD). The appropriate statistic(s) to employ in any given problem is/are a function of the context in which the problem arises. In the following, I summarize how the various statistics given in Parts II and III are computed and offer a few remarks on the peculiarities of certain statistical practices common to mammalian scaling studies. Emphasis is also put on methods other than LSQ for computing best-fit lines (see above quote from Moroney). The approach is mostly empirical. Other aspects of scaling methodology as applied to mammals are discussed in Chapter 24.

Recall that a primary aim here is to contribute to an understanding of the internal adjustments mammals *generally* make to increased body size (e.g., a decrease in size-specific energy metabolism with increasing body size). These adjustments, or lack of, may be seen in comparisons across adult members of different species, or ontogenetically, within a given species. One knows empirically (see Part II) that many *structural* and *functional* variables (e.g., heart mass or heart rate) are, to a first approximation, a linear function of body size when displayed on a log–log plot. In these cases, one takes linearity to imply that a given variable (y) is a *power function* of body weight or mass (x). That is:

$$y = ax^b \qquad \text{(for } x > 0\text{)} \tag{23.1}$$

We invoke the constraint that x must be greater than zero for biological reasons; there are no mammals of zero body size.

Taking logarithms, one has:

$$\log(y) = \log(a) + b{\cdot}\log(x) \qquad \text{(for } x > 0) \qquad (23.2)$$

In relation (23.2) one invokes the constraint that x must be greater than zero for mathematical reasons, independent of relation (23.1). That is, the logarithm of 0 is mathematically undefined. It is implicitly assumed in relations (23.1) and (23.2) that x and y are *continuous* variables. Relation (23.2) shows that $\log(y)$ is a linear function of $\log(x)$. Note that in the literature the parameter a is usually termed the *intercept* (the value of y in relation (23.1) when $x = 1.0$); b is variously referred to as exponent, power, regression coefficient, or slope. Here the terms *exponent* and *slope* and occasionally *power* are employed interchangeably. In mammalian scaling studies, the principal emphasis is on the parameter b. A long-term aim is to find estimates of exponents that apply to mammals in general; that is, exponents that are influenced only weakly by ecology or phylogeny (see Chapter 21). It is true, of course, that many workers are interested in how specific groups of mammals deviate from the general pattern; this can only be ascertained with confidence when the general pattern is well characterized.

In biology, the x- and y-variables normally exhibit some degree of variation (or scatter). Statistical analysis in mammalian scaling studies is devoted mainly to deriving a best-fit line through a given set of points, points that rarely all lie exactly on a straight line in log–log coordinates. From the standpoint of *applied mathematics*, this task is a special case of fitting a curve to (often non-linear) numerical data. A linear best-fit line affords a numerical value for the exponent (generally lying between $-1/3$ and $10/9$ for adult mammals) as well as a prediction of the value of the y-variable for a given value of the x-variable (see relation (23.1) and below). We denote the value of the y-variable predicted by a best-fit line by y'. The difference, for any given x-value, between the observed y-value (y) and the predicted y-value (y') is termed the residual (i.e., $y - y'$) (see Figure 23.1). Sometimes, residuals are employed for analytical purposes (as to distinguish one group of mammals from another). It would be well in these cases to provide, as far as possible, an estimate of the confidence limits on the residuals.

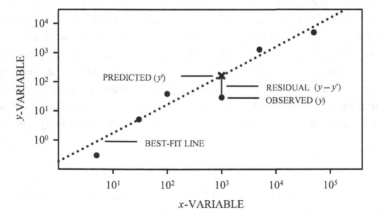

Figure 23.1 Definition of residual.

The solid circles in Figure 23.1 represent the empirical y-values (on the ordinate) associated with x-values (on the abscissa). The dotted line represents a best-fit line determined by one or another suitable algorithm (see below). The short vertical line represents a residual. The predicted y-value ($y' = 167$) in Figure 23.1, for $x = 10^3$, is denoted by a cross. The best-fit line affords a measure of the degree to which variation in y is correlated (or associated) with variation in x (i.e., the "explained" variance in y, where variance is a measure of dispersion). No causal relation is implied by a best-fit line between the x- and y-variables, and none should be inferred in the absence of independent evidence. Moreover, the x- and y-variables may not be independent (consider plotting organ mass against body mass). The residuals collectively provide a measure of the degree to which variation in y is *not* associated with variation in x (i.e., the "unexplained" variance).

An ideal dataset and an ideal best-fit line

An ideal dataset for scaling studies in adult mammals would be linear on a log–log plot. The x-values would be evenly distributed (on a log-scale) on the x-axis, with little or no duplication of x-values. The best-fit line, or better line-segment, would display as many y-values above the line as below, and the residuals would be evenly distributed, on average, above and below the line, over its whole length. Each species would be represented by the same number of points (usually one). The contribution of each data point to the best-fit value of the exponent should be independent of the x-values (see below). In practice, empirical scaling datasets and best-fit lines for mammals only approximate these conditions. For a partial definition of an ideal dataset for mammalian scaling studies generally see Table 5.1, bottom row.

Comparison of means and medians

The arithmetic mean is a measure of central tendency (or center of "gravity") for a set of numbers. Specifically, an arithmetic mean of a set of n numbers is simply the sum of the numbers divided by n (see Table 23.1, where the mean of the five numbers (101 to 105) is 103 (column f, row one)). On the other hand, a *median*, strictly speaking, provides a

Table 23.1 Comparison of the response of mean and median to outliers

Col./Row	Data values					Mean	Median
	a	b	c	d	e	f	g
1	101	102	103	104	105	103	103
2	0	102	103	104	10,000	2,062	103

Col. = column.

measure of central tendency for a frequency distribution of numbers. For an odd number of numbers the median is simply the middle number (e.g., 103) (row one, column g). For an even number of numbers the median is the average of the two middle numbers. For example, given the numbers 1, 3, 4, and 8 the arithmetic mean is 4 and the median is 3.5. The arithmetic mean is much more widely used than is the median. However, the median has the advantage, in at least some applications, of being more robust (less sensitive to outliers or incorrect values) than is an arithmetic mean.

To see this, consider Table 23.1, row two. The first and last entries (columns a and e, row two) have been dramatically altered (from 101 to 0 and from 105 to 10,000, respectively), so as to simulate the effect of outliers. Observe that the arithmetic mean (column f, row two) changes markedly, from 103 (row one) to 2,062 (row two) whereas the median (103) is unaffected by the outliers (row two, column g).

The median is also robust in another sense, for datasets with an odd number of numbers. Consider Table 23.2, row one, columns a, b, and c, where we have three arbitrary data values: 1, 10, and 100. The *arithmetic mean* is 37 (111/3 = 37) (column d) and the *median* is 10 (column f). Row three, columns a, b, and c, provides the logarithms (0, 1, 2) of the three given data values (i.e., 1, 10, 100). The mean and median of the logarithms are both 1 (see Table 23.2, column d, row three). Taking the antilog of 1 we obtain 10 (row three, columns e and g). The key point is that the antilog (see column e, row three) of the arithmetic mean does *not* recover the original arithmetic mean of 37 (row one, column d), whereas the antilog (10) (column g, row three) of the median does recover the original median of 10 (column f, row one).

Put more directly, if we take the median of the logarithms (for an odd number of numbers) and then take the antilog, we recover the median. The same is not true for the arithmetic mean. The restriction to datasets with an odd number of numbers is of minor concern; we are concerned with the principle involved. It should be admitted, however, that both mean and median are *immune* to transformations consisting of multiplication of a set of numbers by a constant. The reader can show that if we multiply a set of numbers by 2 and then calculate a new mean and a new median, the original mean and median may be recovered by dividing the derived mean and median by 2. A similar argument holds for addition of a constant to each member of a set of numbers.

In sum, the median is insensitive to the presence of outliers and is unaffected by a logarithmic transformation followed by an antilogarithmic transformation (see double- and single-median best-fit methods below). Both measures (mean and median) are robust with respect to addition or multiplication by a constant.

Table 23.2 Comparison of the response of mean and median to a logarithmic transformation

Col./Row	Data values			Arithmetic means		Medians	
	a	b	c	d	e	f	g
1	1	10	100	37	–	10	–
2	Logarithms of above data			–	Antilog (of 1)	–	Antilog(of 1)
3	0	1	2	1	10	1	10

Computation of exponent and intercept by LSQ

In most mammalian scaling studies, a best-fit line to a given bivariate dataset is found using *ordinary least squares*; i.e., the best-fit line is that line which minimizes the sum of the squares of the residuals. This least-squares (LSQ) minimization procedure is usually referred to as regression analysis. However, since regression analysis in mammals is almost always carried out using LSQ, the term "regression" is redundant. I have employed the acronym LSQ rather than the longer term regression.

The primary statistical outcome of a LSQ analysis pertinent to the present study is an estimate of the numerical value of the exponent or slope. It may be assumed that a LSQ line will pass through the center of the dataset (i.e., the arithmetic means of the $\log(x)$ and the $\log(y)$ values). (Recall that an arithmetic mean of a given dataset is also a minimum in the LSQ sense.) We denote the arithmetic mean of the logarithms of x and y by ML(x) and ML(y), respectively. The center point (ML(x), ML(y)) may be computed directly from a logarithmically transformed dataset, without deriving a best-fit line. Given the *exponent* (or slope) associated with a best-fit line, and the *coordinates of the center*, one may then compute the numerical value of the *intercept* from the relation $\log(a) = \text{ML}(y) - b \cdot \text{ML}(x)$; the intercept ($a$) in native coordinates is then obtained by taking the antilogarithm of $\log(a)$. (For example, the antilog of 2 is 100.)

It has been argued on occasion that the intercept is of limited biological significance. A reason sometimes given is that the intercept often lies well outside the body size range. But this point is not, by itself, convincing. By choosing different (possibly arbitrary) units for the x-variable, one could generally derive an intercept that lies within any given body size range. Nonetheless, the fact that the numerical value of the intercept depends on the units employed is a significant disadvantage (see Chapter 24).

One is also interested in other statistical parameters, including the standard error of the exponent, the 99% CL on the exponent, and the coefficient of determination. A synoptic discussion of these and other topics is given below. More detailed accounts of these (and many other) statistical matters are readily available in classical works [1,2] as well as in many more contemporary studies (see below). There are two particular reservations about the use of LSQ in mammalian scaling studies: one is the implicit assumption that the variations (errors) are confined to the y-variable, and the second is the possibly malign influence of outliers on the value of the exponent (Table 23.1). See below.

End-sample, mid-sample, and FDS

In order to test whether the results of the *end-sample*, *mid-sample*, and *FDS* experiments are reasonable (see Chapters 6, 21) the following numerical exercise was performed. First, one generates a set of 100 x-values, evenly distributed (on a log scale) over, say, four orders of magnitude. Then a corresponding set of 100 y-values is generated as (for example) the 0.900 power of the x-values (relation (23.1)). Next, the y-values are modified using a *pseudo-random* number generator. Finally, LSQ analyses are carried out for the end-sample ($N_{\text{pnts}} = 20$) and mid-samples ($N_{\text{pnts}} = 80$) and for the FDS ($N_{\text{pnts}} = 100$).

Table 23.3 Simulation of LSQ analyses of end- and mid-samples in relation to a FDS

	N_{pnts}	MPD	pWR	r^2	Exponent (b)	99% CL	Delta	Entity
Column/Row	a	b	c	d	e	f	g	h
1	20	28.5	4	0.993	0.912	0.051	0.010	End-sample
2	80	18.5	3.2	0.987	0.893	0.031	0.009	Mid-sample
3	100	20.8	4	0.990	0.902	0.024	0.000	FDS

N_{pnts} = number of points; MPD = mean percent deviation; pWR = $log(BW_{max}/BW_{min})$; r^2 = coefficient of determination; CL = confidence limit(s); Delta = absolute difference between rows one, two and row three in column e.

The outcomes of such an exercise are shown in Table 23.3. The exponents and 99% CL are given in columns e and f. *Delta* (column g) gives the absolute difference between the exponents in rows one and two and that in row three. For example, in column g, row one, the figure 0.010 is calculated as 0.912 (column e, row one) less 0.902 (column e, row three). Observe that in column g, rows one and two, the absolute values of delta are less than the 99% CL (column f), implying that the exponents in column e, rows one and two are *not* significantly different from that for the FDS (column e, row three). Note the high coefficients of determination (r^2) (column d).

The main conclusion from Table 23.3 is that exponents determined for the end- and mid-samples are not significantly different from the exponents for the FDS, as determined by the 99% CL (0.051 and 0.031, column f, rows one and two). In particular, the exponent for the end-sample (0.912) is not significantly different from the exponent of 0.902 $\pm$ 0.024 (99% CL) for the FDS. This result is consistent with the empirical findings (Chapters 6 and 21) and with the general hypothesis that *size range* (pWR) is likely to be more important than *sample size* in determining the various statistics of primary interest in mammalian scaling studies. Note that introducing an arbitrary MPD (or scatter) of 21% (column b, row three) in the y-values only changed the best-fit exponent from an assumed ideal value of 0.900 to an empirical value of 0.902 (Table 23.3, column e, row three).

Extrapolation

By extrapolation we mean the projection of a best-fit LSQ line beyond the smallest or largest body sizes present in the underlying dataset. Implicit and sometimes explicit extrapolation is quite common in mammalian scaling studies. For example, it arises when the ratio of two variables is taken (see "Invariant ratios" and "Special ratios" below). Such extrapolations may be dubious. The reason is that the power function, so widely used in scaling studies, is strictly empirical. One often cannot be certain for any given y-variable that a simple power function will apply across the entire mammalian body size range (see also Chapter 24). Recall an apparent failure in the case of brain mass expressed in terms of body mass (Chapter 13), where two lines with significantly differing intercepts and exponents are required for an adequate empirical description.

If, nevertheless, one resolves on extrapolation, this will be more plausible when the extrapolation is modest, and restricted to mammals taxonomically, structurally, and functionally similar to those represented in a given dataset. For example, extrapolation of skeletal mass from a group of rodents to somewhat larger or smaller rodents is less objectionable than an extrapolation to elephants. It would be good if the nature and degree of any extrapolation were to be stated.

Of course, in a *modeling* environment, one is entitled to say: "if so-and-so is true, then such-and-such follows," where so-and-so may include explicit extrapolation of an exponent, potentially to all mammals. (In evaluating the impact of differences in exponents for pWR = 8.0, I have routinely made such extrapolations.) See Chapter 3. This type of modeling, possibly reaching well beyond the available empirical data, may furnish pathways to an understanding of scaling in mammals not otherwise available.

Sensitivity of LSQ calculations to outliers

The exponent (b) of a best-fit straight line (on a log–log plot), as determined by LSQ, may be expressed as the ratio of two terms: a numerator (N) and a denominator (D). Thus:

$$b = N/D \tag{23.3}$$

where $N = \sum (x_i - \bar{x})(y_i - \Delta)$ and $D = \sum (x_i - \bar{x})(x_i - \bar{x})$ $\qquad$ (23.4)

where "$\sum$" stands for summation. The two terms ($x_i - \bar{x}$) and ($y_i - \bar{y}$) are termed "deviations from the mean" for the general point (x_i, y_i). In relations (23.4), and in what follows, $\bar{x}$ and $\bar{y}$ denote the means of the x- and y-values. Unless otherwise stated, the x- and y-values and their means are expressed as *logarithms*. One may evaluate the numerator and denominator in relation (23.4) point-by-point. The findings from an analysis of numerator terms are shown in Figure 23.2.

In Figure 23.2b, we have an empirical dataset with 11 points ($N_{pnts} = 11$) and a dotted line showing the best-fit straight line as determined by LSQ ($b = 0.985 \pm 0.014$ (s.e.)). Panel a shows the contribution each point (x_i, y_i) in the dataset makes to the numerator (i.e., to ($x_i - \bar{x}$)($y_i - \bar{y}$)), expressed as a *percent* of the total value of the numerator (see relation (23.4)). If each point made the same contribution to the numerator that would amount to 9% per point ($100/11 \approx 9$); see horizontal dotted line in panel a. Note that the two points, one at each end of the distribution, actually contribute 16.5% and 20.2%, for a total of almost 37%. Alternatively, the five points lying above the dotted line (panel a) together account for 75% of the total numerator whereas the desired amount for equality of influence on an exponent would be 45% ($5 \cdot 9 = 45$).

We see that the end-points of the distribution (which are far from the respective x- and y-means) make a much larger contribution to the numerator than do points near the middle (Figure 23.2a). Actually, if there were a point at the exact middle ($\bar{x}, \bar{y}$) it would make no contribution to the sum. The exponent calculated from just the two end-points of the dataset is 0.986 (rather than 0.985) (relation (23.4)). The points for the

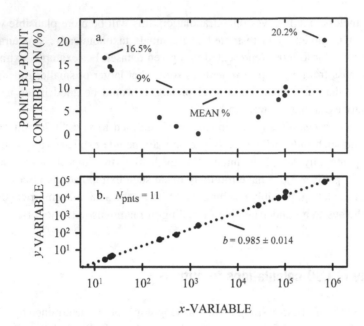

Figure 23.2 Point-by-point determination of LSQ exponent.

denominator (not shown) are almost superimposable on the numerator points. Thus the end-points also dominate the expression for the denominator.

The fact that the end-points make a large contribution to both the numerator and denominator implies that outliers at the two extremes may have a disproportionate influence on the respective sums and hence on the derived exponent. This is a weakness of the LSQ algorithm. A preferable algorithm for adult interspecific studies in mammals would give each species represented in the calculations the same "weight" (see below). It may be noted that the assertion sometimes made that LSQ depends on a normal distribution of the errors (residuals) is doubtful [3]. Finally, observe that the exact value of the *exponent* of the best-fit line for adult mammals for any given *y*-variable as a function of body size is usually *unknowable*, as there will almost always be some uncertainty, however small, due to biological variation. In the following, the *x*- and *y*-variables are assumed to be in logarithmic form.

Standard error of estimate in *y*

The *standard error of estimate* in *y* is denoted by $(S_{y.x})$. This parameter provides a measure of the mean variation between the observed *y*-values (*y*) and those values predicted (*y′*) by a best-fit straight line. It is computed from the relation:

$$S_{y.x} = \sqrt{(\Sigma(y_i - y')^2 / (n - 2))} \tag{23.5}$$

Standard deviations in *x* and *y*

The *standard deviation* is a measure of the dispersion around a mean. The standard deviations in x (STD_x) or y (STD_y) are computed from the relations:

$$\mathrm{STD}_x = \sqrt{\Sigma(x_i - \bar{x})^2/(n - 1)} \text{ and } \mathrm{STD}_y = \sqrt{\Sigma(y_i - \bar{y})^2/(n - 1)} \qquad (23.6)$$

where n is the number of observations (some software packages may use n rather than $(n - 1)$ in the denominator when calculating a standard deviation). In physics and engineering, where $\bar{x}$ or $\bar{y}$ may be zero, the standard deviation is often referred to as the root-mean-square (r.m.s.).

Standard errors in exponent and intercept

The *standard error* (s.e.) in the exponent (s_b) is used (see below) to compute confidence limits on an exponent or intercept. The standard error in an exponent is computed from the relation:

$$s_b = S_{y.x}/\sqrt{\left(\Sigma(x_i - \bar{x})^2\right)} \qquad (23.7)$$

where $S_{y.x}$ is given by (23.5) and $\bar{x}$ is the mean of the x-values. The standard error of the exponent for the data shown in Figure 23.2 (panel b) is ± 0.014. (Throughout this work the standard error (s.e.) in the exponent (or slope) is given in plots, and 99% CL are usually given in tables.)

The standard error for the intercept (s_a) is computed as follows:

$$s_a = S_{y.x}\cdot\sqrt{\Sigma x_i^2/n\Sigma(x_i - \bar{x})^2} \qquad (23.8)$$

Testing for ecological or phylogenetic influences on exponent

As a rule, whenever one computes an exponent for the scaling of some y-variable on body size, there is a possibility that the computed exponent is not representative of mammals at large; the exponent may be confounded by *ecological* or *phylogenetic* effects. That is more likely to be the case when sample sizes are small. When the sample sizes are large (say $N_{\mathrm{sp}} \geq 100$) and the size range appreciable (say pWR ≥ 4.0) one can test for ecological or phylogenetic effects, to a degree, by omitting specific groups of mammals. I have carried out numerous experiments of this type for many y-variables; in a few instances, the products of these experiments are reported in Part II. Overall, for sufficiently large samples, it was found that omitting individual taxonomic or functional groups, one-by-one, usually fails to alter the revised exponents outside the 99% CL of the parent exponent (Chapter 21).

A different approach, for samples of 100 or more, is to omit 80% or more of the points, retaining only 10 points at each end (termed the end-samples). The consequences of these experiments (Chapters 6 and 21) suggest that in large samples the possible effects of ecology or phylogeny do not significantly bias the empirical exponents. It is true, however, that any adult interspecific exponent depends on data derived from varied species, and usually from two or more distinct taxonomic orders. Datasets extending over the entire mammalian size range will necessarily include cetaceans and species drawn from bats and/or insectivores and/or rodents. Thus, ecological and taxonomic effects cannot be entirely eliminated by partitioning the data into subsets.

The *t*-table

Values of the "*t*" statistic are used, together with standard errors (see above), to compute confidence limits (CL) on, say, an exponent or intercept. The first step is to choose a preferred significance (or probability) level. In this study, I have generally used a 1% significance level (0.01). The second step is to derive the number of degrees of freedom (df), given by the number of observations *minus* 2. One then enters a *t*-table (available in many statistics texts and on the internet) with the given significance level and df in hand. Consider the dataset shown in Figure 23.2b. There are 11 observations, so df is 9. Since exponents may ultimately prove to be either larger or smaller than the current best-fit value, one employs a two-tailed *t*-test. That requires entering a *t*-table with a significance level of 0.01/2 = 0.005. For the data shown in Figure 23.2, the value found for *t*, at the 99% CL, given df = 9 and a significance level of 0.005, is 3.250. A similar procedure applies to intercepts. See below.

The reader can easily show that, on average, for any given standard error in the exponent, the 99% CL tend to be about 25% wider than for 95% CL, for degrees of freedom between, say 5, and 1,000.

99% confidence limits and intervals

...statistically narrow confidence limits do not necessarily imply biologically equally narrow confidence limits.

Schmidt-Nielsen 1984

When confidence limits are reported in mammalian scaling studies they are usually at the 95% level. However, 99% CL are more rigorous. Although the use of 99% CL has been uncommon in the mammalian scaling literature, Stahl employed them as far back as 1965 [4]. The 99% CL on an exponent are given by:

$$99\% \text{ CL} = t \cdot s_b \tag{23.9}$$

where the standard error in the exponent (s_b) is given by relation (23.7). For the data shown in Figure 23.2b (where s.e. = 0.014), the 99% CL (see above, where t = 3.25) are $\pm 0.014 \cdot 3.25 = \pm 0.046$. Since the best-fit LSQ exponent is 0.985, it follows that the 99% confidence interval (CI) lies between 0.939 and 1.031 (i.e., 0.985 ± 0.046). This means that if one repeated the calculations 100 times, using similar yet not identical data samples, the 99% CI on the respective exponents would include the mean (0.985) 99 times. On the other hand, 95% CL imply only 19 times out of 20. More generally, confidence limits give us a sense of how much trust to place in any given empirical exponent. Although the standard error is not considered to be a confidence limit per se, it is conducive to brevity to include it along with the CL in some tables, as needed.

It is worth reminding ourselves that differences in exponents such as $3/4 - 2/3 = 1/3 - 1/4 = 0.08$ are almost universally considered to be potentially significant with respect to variables like energy metabolism. However, Calder [5] at one point refers to the "slight difference in exponents (0.76 vs 0.68)." If these distinctions (of ± 0.08) were indeed *not* statistically and biologically significant, much of the mammalian bioenergetics literature would be of doubtful relevance. It is true, however, that if the 99% CL on these exponents were ± 0.041 or greater one might very well doubt that the given pairs of exponents were significantly different, statistically speaking (i.e., $0.76 - 0.041 = 0.719$; $0.68 + 0.041 = 0.721$, showing that the two confidence intervals overlap). It follows that important inferences may depend on the relevant values of the CL (see also "Special pleading" below).

Consider hypothetical 99% CL of ± 0.02. This means that the 99% confidence interval (CI) is 0.04. It follows that the values of the y-variable (for pWR = 8.0) could differ by as much as one doubling (N_d), or a factor (f) of 2 ($N_d = 26.6 \cdot 0.04 \approx$ one doubling, equivalent to a factor of 2; see Chapter 3). Taking 99% CL of ± 0.02 to be a critical threshold may prove to be conservative in some instances. The issue fundamentally comes down to what level of change in any given y-variable is "seen" by natural selection as increasing or decreasing fitness. No doubt, population geneticists will report that impacts on fitness vary with the nature of the y-variable. Perhaps in some cases the critical 99% CL on empirical exponents will be less than ± 0.02, and in other cases larger. Ultimately, the decision as to what constitutes a critical 99% CI should be made by the wider community. For the present, it may be observed that perhaps the most common shortcoming in mammalian scaling studies has been the widespread failure to report confidence limits on exponents. In the absence of CL the biological significance of any given exponent is uncertain (see below). Wide CL may mean that a given difference in slope is not statistically significant. But as Schmidt-Nielsen implied, narrow CL do not ensure biological significance (see above quote).

Coefficient of determination

The *coefficient of determination* (r^2) tells us what fraction of the variation in y-values for a given dataset is associated with variation in the x-values. It is numerically equal to the correlation coefficient (r) squared. The parameter r may be calculated most simply from the relation:

$$r = b \cdot (STD_x / STD_y) \qquad (23.10)$$

where b is the exponent and STD_x and STD_y are the standard deviations of the x- and y-variables respectively (see relation (23.6) above). For example (see Figure 23.2), if the exponent (b) is 0.985 and the standard deviations in the logarithms of x and y are 1.626 and 1.603, respectively, then the correlation coefficient (r) is 0.999 (i.e., $0.985 \cdot 1.626 / 1.603 = 0.999$). Hence the coefficient of determination (r^2) is 0.998 ($0.999^2 = 0.998$). Note from relation (23.10) that if $r = 1$, then $b = STD_y / STD_x$.

Also observe that when $STD_x = STD_y$ the correlation coefficient is numerically identical to the exponent or slope (i.e., $r = b$) (relation (23.10)). I have suggested tentatively that for scaling studies in mammals at large, coefficients of determination (r^2) of 0.99 or higher may be regarded as *excellent*, and those less than 0.95 as *poor* (Chapter 5, Table 5.4). Whether the data are in native form or transformed logarithmically, the correlation coefficient, and the coefficient of determination are *non-dimensional* quantities. The correlation coefficient takes the same sign ($\pm$) as the exponent; the coefficient of determination (r^2) may be 0 or greater than 0, but is never negative and is never greater than 1.

Percent, mean percent, and absolute deviations

It often happens in a study such as this that one would like to compare the *dispersion* across variables having different dimensions. For this purpose, the usual statistical measures of dispersion (e.g., standard deviation), expressed in native coordinates, are unsuitable because they incorporate the dimensions of the variables in question. For example, we may *not* wish to compare the dispersion in a sample of lengths (say measured in cm) with the dispersion in a sample of areas (say in cm^2) without first removing the influence of dimensions. To this end, one may invoke *percent deviation* (PD) between two instances of a sample (say x_1, x_2), as defined by the relation:

$$PD = 100 \cdot |(x_1 - x_2) / x_1| \qquad (23.11)$$

where the vertical bars denote absolute value. If, for example, x_1 is 8 and x_2 is 12, then PD is given by $100 \cdot (8 - 12)/8 = -50\%$. Taking the absolute value, PD becomes 50%. If one has 10 estimates of PD, then the *mean percent deviation* (MPD) is simply the sum of the ten PDs divided by 10:

$$MPD = (1/n) \Sigma PD \qquad (23.12)$$

where Σ implies summation and n is the number of estimates.

The parameters PD and MPD, both non-dimensional, work reasonably well in most applications to scaling in mammals. There is one situation, at least, where these measures can be misleading. Consider the following pairs of samples of (x_1, x_2): (10, 12), (15, 13), (31, 25), (45, 49), (0.01, 0.001), with PD of 20, 13, 19, 9, and 90, respectively, and a MPD of 30 (i.e., $(20 + 13 + 19 + 9 + 90)/5 = 30$). It is evident that the pair (0.01, 0.001), having a *small* absolute deviation (0.009) (relative to two or more), and a relatively *large* percent deviation (90%), in effect bias the estimate of MPD. That is, the MPD for five instances is 30%, while the MPD of the first four samples is only 15%. In these instances a better alternative may be to employ the *absolute deviation* (AD) and the *mean absolute deviation* (MAD), defined by:

$$AD = |x_1 - x_2| \text{ and } MAD = (1/n)\sum AD \qquad (23.13)$$

Using the same five pairs as above, one finds that MAD for the five estimates is 3.5, whereas that for the first four estimates is 2.8 (differing by a factor of only 1.25). It can be shown that MPD is apparently measuring much the same thing as the *standard error of estimate*. For example, if we compute a LSQ best-fit line for the standard error of estimate (in logarithmic form) expressed as a function of MPD for 42 structural variables, we obtain a coefficient of determination (r^2) of 0.931. Thus, variation in MPD in a moderate sample accounts for 93% of the variation in the standard error of estimate.

It may be noted in passing that we often encountered datasets in Part II, having MPD from the best-fit LSQ line in the range of 30 to 40%. The implication of this fact is that predictions of y-values for given x-values will often differ from observed values by similar percentages.

On the invariance of alleged "invariant" ratios

Adolph [6] referred to what he considered to be several invariant physiological ratios in mammals, including, for example, the ratio of oxygen consumption to nitrogen output. Günther and Guerra [7] argued that in homeotherms the duration of the respiratory cycle is four times that of the cardiac cycle. The same argument has been applied by other workers to these and other parameters [8, 9]. There are two different types of invariance involved. First, we have those cases where a single y-variable scales very nearly as the first power of body size, implying that the ratio of the y-variable to body mass is a constant (i.e., *relative invariance*). We saw in Part II that this form of invariance is likely to be true, for example, for blood volume, total mass of skeletal muscle, total body water, and lung mass (Chapters 9, 12, 8, 15, respectively).

The second type of invariance occurs when two variables (y_1, y_2) apparently scale with body mass to the same (or nearly the same) exponents (b_1, b_2). A possible example is the ratio of resting respiratory rate to resting heart rate; this ratio is often put at 1:4. (But it is doubtful that this ratio is invariant across species; see Table 20.3.) In any case, we have:

$$y_1 = a_1 x^{b_1} \text{ and } y_2 = a_2 x^{b_2} \qquad (23.14)$$

$$\text{hence } y_1/y_2 = (a_1/a_2) \cdot x^{b_1 - b_2} \qquad (23.15)$$

When $b_1 = b_2$ the exponent in relation (23.15) is 0, consistent with *absolute invariance*. Stahl [10, 11] referred to the quantity $(b_1 - b_2)$ as the *residual mass exponent* (r.m.e.) or the *residual mass index* (r.m.i.). At various times he considered an r.m.e. of 0.03, or 0.05, or 0.08 to be significant, in that larger values of r.m.e. would produce "considerable error with larger animals" [10–12]. Stahl gave no explicit criterion for making this assessment, nor did he explain the rather wide variation in critical values of r.m.e. (0.03 to 0.08) which he considered at different times to be significant. If two lines sharing a common intercept differ in exponents by 0.03, then the end-points (for pWR = 8.0) differ by a factor (f) of 1.7 (i.e., $N_d = 26.6 \cdot 0.03 = 0.798$; hence $f = 2.^{0.798} = 1.7$) (Chapter 3). The same argument applied to a difference in exponents of 0.08 gives a factor (f) of 4.4.

In accord with the above discussion (see "99% confidence limits" above), it is suggested as a rule of thumb that r.m.e. of 0.04 or greater is likely to be biologically significant when the implications are evaluated (i.e., often extrapolated) over the whole mammalian size range (pWR = 8.0) ($0.04 \cdot 26.6 = 1.1$ or factor (f) $= 2^{1.1} \approx 2.1$).

When *ratio invariances* are considered in the literature, the topic of confidence limits (CL) is rarely mentioned. Or when invoked, it may be implied or stated that the CL are sufficiently wide that the r.m.e. can be safely ignored; part of the reason for this stance is that CL on exponents have seldom been given [5]. In making a judgement as to whether a given ratio may be invariant, CL are of the essence. In many cases, it may turn out that an argument for invariance is undermined by the span of the 99% CI (see below).

Special ratios

This [calculation] *gives circulation time as a function of body mass for mammals. The mass exponent is 0.25...*

Schmidt-Nielsen 1984

Certain exponents arising in scaling studies in mammals are assumed to have special significance. For example, the exponents (1/4, 3/4) are given pride of place because of their presumed relationship to standard metabolic rate (see, however, Chapter 17). A good example is provided by circulation time, calculated as the *ratio* of total blood volume to resting cardiac output [13]. It is reported that blood volume scales empirically as the first power ($b = 1.0$) of body mass and resting cardiac output as the three-quarters power ($b = 3/4$). Hence it is inferred that circulation time scales as the one-quarter power of body mass ($1 - 3/4 = 1/4$) (see above quote from Schmidt-Nielsen).

When such calculations are presented, CL are seldom given. The respective datasets (here for blood volume (see Table 9.8) and cardiac output (see Table 9.16)) can usually be assumed to have different numbers of species (N_{sp}), to extend over different size ranges (pWR), and to exhibit different diversity indices (DI). The propriety of carrying out simple algebraic manipulations on the exponents when two datasets are of different composition is doubtful.

Table 23.4 Estimates of confidence limits on the exponent and (log) intercept for blood volume (cm^3), resting cardiac output (cm^3/min), and circulation time (min), as a function of adult body mass

	N_{ord}	N_{sp}	MPD	pWR	DI (%)	r^2	Exponent $\pm$ 99% CL	log(a) $\pm$ 99% CL	Entity	Source (Table)
Col./Row	a	b	c	d	e	f	g	h	i	j
1	11	35	20	4.9	13.5	0.992	0.996 ± 0.043	-1.0975 ± 0.080	BV	9.8
2	8	21	23	4.4	7.5	0.981	0.769 ± 0.070	0.0554 ± 0.286	CO	9.16
3	-	-	-	-	-	-	0.218 ± 0.082	-2.53 ± 0.80	CT	-

See Table 23.3. Col. = column; CL = confidence limits; a = intercept; BV = blood volume; CO = resting cardiac output; CT = circulation time.

In order to look into this question more carefully, two empirical datasets were reviewed, one on the scaling of total blood volume in adult mammals, the other pertaining to resting cardiac output. A partial analysis of these two datasets is displayed in Table 23.4.

Carrying out a conventional subtraction of exponents, one would infer that circulation time scales as the 0.23 power of body mass (0.996 – 0.769 = 0.227). In order to obtain a statistically valid estimate of the exponent and the corresponding 99% CL for circulation time, I provided the raw data for the above two datasets (corresponding to Table 23.4, rows one and two) to Professor Paul Sampson (Department of Statistics, University of Washington) and then-graduate student Benjamin Ely. They carried out a bootstrap operation (a computer-intensive procedure for estimating statistical distributions with minimal assumptions) on these data. From their computations they derived an exponent of 0.218 ± 0.082 (99% CL) (see Table 23.4, row three, column g). That is, based on the above data, the best-fit exponent for circulation time would lie between 0.136 and 0.300 at the 99% CL. This result does not support a one-quarter power rule.

Although one cannot be certain that similar findings would obtain for the other exponents of ratios that have been published, that possibility clearly exists. A well-known case in point is the contention that the number of heart beats in a lifetime is the same in all mammals. Until similar calculations to those just described are made in these cases, one should withhold judgement on the implications of the reported exponents. It may be that in some or even most cases the evidence for invariance over the full mammalian size range will prove to be at best a rough approximation [14].

FOUR OTHER ALGORITHMS FOR COMPUTING BEST-FIT EXPONENTS

Computation of best-fit exponent by minimizing absolute residuals

The method of LSQ for finding the exponent (or slope) of a bivariate trajectory assumed to be linear on a log–log plot has the twin advantages of convenience and tradition. The fact that the LSQ method gives greater weight to the values lying near the ends of the

Table 23.5 Determination of best-fit exponent by minimizing the sum of the absolute residuals

N_{sp}	LSQ r^2	Slope (b)	99% CL	99% CI	Slope by LAR method	Entity	Source (Table/Chapter)
a	b	c	d	e	f	g	h
1,266	0.982	0.354	0.003	0.351–0.357	0.352	Body length	6.1
708	0.967	0.759	0.013	0.746–0.772	0.758	Brain mass	13.2
87	0.978	1.051	0.045	1.006–1.096	1.042	Fat mass	8.6
134	0.992	0.895	0.018	0.877–0.913	0.889	Liver mass	10.8
11	0.998	0.985	0.046	0.939–1.031	0.976	Protein mass	Chapter 8
62	0.0007	−0.080	0.030	−0.111–0.050	−0.087	RBC count	9.5
78	0.002	−0.004	0.030	−0.034–0.025	−0.023	RBC diameter	9.3
31	0.979	0.715	0.053	0.662–0.768	0.704	Spinal cord mass	13.5
100	0.967	0.972	0.047	0.925–1.019	0.995	Spleen mass	9.2

CL = confidence limits; CI = confidence interval; LAR = least absolute residuals; RBC = red blood cell.

trajectory is a potential disadvantage (see Figure 23.2). Moreover, the assumptions underlying ordinary LSQ are sometimes called into question (see below). For these reasons, it is useful to know that there are several methods other than LSQ that often give exponents that by-and-large agree with the LSQ method to within the given 99% CL (see below).

As mentioned above, the LSQ method rests upon minimizing the sums of the squares of the residuals (i.e., differences between the *y*-observed and *y*-predicted values; Figure 23.1). An alternative and quite similar method, here termed the "absolute residual" method, rests not on minimizing the sum of the squares of the residuals, but rather on minimizing the sum of the absolute values of the residuals. This procedure (denoted by LAR, for "least absolute residuals") is quite easy, if somewhat tedious, to carry out. The results of an analysis along these lines are shown in Table 23.5, for nine different *structural y*-variables. The best-fit LSQ exponents and 99% CL are given in columns c and d. The 99% confidence intervals (CI) on the LSQ exponent, computed from columns c and d, are given in column e. The best-fit exponents determined by the LAR method are displayed in column f. Note that each exponent determined by the LAR method falls within the 99% LSQ CI (column e). The mean absolute difference between the LSQ and LAR exponents is 0.010 ± 0.007 (STD).

It is relevant to this discussion of minimizing the sum of the absolute residuals that sample sizes (N_{sp}, column a) vary between 11 and 1,266, the LSQ coefficients of determination (r^2) (Table 23.5, column b) vary between 0.0007 and 0.998, and LSQ exponents (column c) vary between −0.004 and 1.051. It follows that the goodness of fit between the LSQ and LAR exponents in the present sample is relatively insensitive to variation in these parameters. If forced to choose between the exponents determined by the two methods, the absolute method might be preferable. However, this is not the main point. The two methods, using somewhat different mathematical processes, give statistically similar answers (as anticipated).

Computation of best-fit exponent by the major axis method

Consider a hypothetical best-fit line to some dataset and one specific point in that dataset. Construct a "drop-line" from the given point to the best-fit line, where the drop-line is perpendicular to the latter (rather than vertical, as in ordinary least squares). Choose as the best-fit line that which minimizes the sum of the squares of the lengths of all the drop-lines. This best-fit line is termed the major (or principal) axis. It was pointed out by Bacon [15] in 1953 that this calculation may be physically meaningless when, as is often the case, the x- and y-variables have different physical dimensions. See also Sokal and Rohlf [2].

Computation of best-fit exponent by reduced major axis method

Consider again some given dataset, one specific point in that dataset, and a hypothetical best-fit line. Construct two drop-lines from the given point to the best-fit line, where the first line is parallel to the x-axis and the second parallel to the y-axis (as in ordinary LSQ analysis). The given data point together with the two points defined by the intersection of the said drop-lines with the hypothetical best-fit line form a triangle. Compute the area of this triangle. Choose as the best-fit line that which minimizes the sum of the areas associated with all the data points. This best-fit line is referred to as the "reduced major axis" (RMA). The main argument in favor of RMA is that it accommodates the possibility of variation (errors) in both the x- and y-variables. It may be shown that the best-fit exponent (b_{RMA}) determined by the RMA method is related to the best-fit LSQ exponent (b_{LSQ}) through the relation:

$$b_{RMA} = b_{LSQ}/r \qquad (23.16)$$

where r is the coefficient of correlation [2]. Observe that the absolute value of b_{RMA} is usually larger than b_{LSQ} (i.e., whenever $r < 1$) and is never smaller.

The LSQ and RMA exponents for 33 *structural* and 17 *functional* variables are shown in Tables 23.6 and 23.7, respectively, columns a and c. The 99% CL on the LSQ exponents together with the RMA exponents are shown in Figure 23.3, panels a (structure) and c (function; solid circles). (For clarity the LSQ exponents themselves are not shown in Figure 23.3; if drawn they would fall in the middle of each error bar.) Note that in one case (for body length, row one, column c, Table 23.6) the *structural* RMA exponent (0.357) falls just at the edge of 99% CL. In the remaining 32 cases the structural RMA exponents fall within (often well within) the 99% CL and hence are judged to be *not* significantly different from the LSQ exponents. The mean absolute deviation between the LSQ and RMA exponents for 33 *structural* variables is 0.010 ± 0.011 (STD).

Six of the 17 functional RMA exponents (Table 23.7, rank orders 4 through 8, and 12) fall outside the 99% CI; of these six exponents perhaps only two deviations are biologically significant (i.e., rank order 7, for age at weaning and rank order 10 for gestation period). Keep in mind that in Figure 23.3c, the coefficients of correlation (r) (not shown)

Table 23.6 Structural exponents as determined by the LSQ, RMA, and double-median (DM) methods

Rank order	Exponents				Entity
	LSQ ± 99% CL	LSQ 99% CI	RMA	DM	
Col./Row	a	b	c	d	e
1	0.354 ± 0.003	0.351–0.357	0.357	0.352	Body length
2	0.365 ± 0.012	0.353–0.377	0.366	0.364	Body girth
3	0.670 ± 0.027	0.643–0.697	0.672	0.655	Body surface area
4	0.686 ± 0.082	0.604–0.768	0.726	0.693	Eyeball mass
5	0.715 ± 0.053	0.662–0.768	0.723	0.728	Spinal cord mass
6	0.727 ± 0.030	0.697–0.757	0.742	0.746	Pituitary mass
7	0.750 ± 0.057	0.693–0.807	0.791	0.687	Testes mass
8	0.759 ± 0.013	0.746–0.772	0.771	0.772	Brain mass
9	0.785 ± 0.055	0.730–0.840	0.811	0.840	Cerebellar mass
10	0.801 ± 0.053	0.748–0.854	0.829	0.830	Adrenal mass
11	0.847 ± 0.057	0.790–0.904	0.857	0.861	Pancreas mass
12	0.873 ± 0.056	0.817–0.929	0.891	0.846	L.I. mass
13	0.878 ± 0.053	0.825–0.931	0.895	0.839	S.I. mass
14	0.880 ± 0.021	0.859–0.901	0.886	0.866	Kidney mass
15	0.895 ± 0.018	0.877–0.913	0.898	0.885	Liver mass
16	0.941 ± 0.066	0.875–1.007	0.959	1.063	Stomach mass
17	0.944 ± 0.059	0.885–1.003	0.944	0.980	Whole body K
18	0.945 ± 0.062	0.883–1.007	0.975	0.938	Thyroid mass
19	0.953 ± 0.022	0.931–0.975	0.958	0.972	Heart mass
20	0.954 ± 0.024	0.930–0.978	0.957	0.941	Skin mass
21	0.962 ± 0.075	0.887–1.037	0.975	0.935	Gut mass
22	0.972 ± 0.047	0.925–1.019	0.988	0.981	Spleen mass
23	0.985 ± 0.046	0.939–1.031	0.986	0.996	Whole body protein
24	0.996 ± 0.075	0.921–1.071	0.997	0.965	Whole body ECF
25	0.996 ± 0.043	0.953–1.039	1.000	1.008	Total blood volume
26	1.000 ± 0.012	0.988–1.012	1.001	1.012	Muscle mass
27	1.000 ± 0.026	0.974–1.026	1.006	1.018	Lung mass
28	1.002 ± 0.011	0.991–1.013	1.003	1.000	Total body water
29	1.019 ± 0.022	0.997–1.041	1.021	1.020	Fresh skeletal mass
30	1.020 ± 0.040	0.980–1.060	1.024	1.019	Whole body ash
31	1.043 ± 0.061	0.982–1.104	1.043	0.998	Whole body P
32	1.051 ± 0.045	1.006–1.096	1.062	1.045	Whole body fat
33	1.054 ± 0.056	0.988–1.166	1.054	1.058	Whole body Ca

See Table 23.5. RMA = reduced major axis; DM = double-median; Ca = calcium; ECF = extracellular fluid; L.I. = large intestine; P = phosphorus; S.I. = small intestine.

for those points with rank orders between 4 and 8, inclusive, lie between 0.10 and 0.70. Thus, the variations seen in the functional RMA exponents relative to 99% CL in the LSQ exponents appear to be largely unrelated to the value of the coefficient of correlation. The mean absolute deviation (panel c) between the LSQ exponents and the RMA exponents for 17 *functional* variables is 0.035 ± 0.043 (STD). The RMA method has been criticized by Jolicoeur [16] (see also Calder [5] and Niklas [17]).

Table 23.7 Functional exponents as determined by the LSQ, RMA, and double-median (DM) methods

Rank order	Exponents				
	LSQ ± 99% CL	LSQ 99% CI	RMA	DM	Entity
Column	a	b	c	d	e
1	−0.265 ± 0.037	−0.302 to −0.228	−0.301	−0.314	Resting respiratory rate
2	−0.220 ± 0.033	−0.253 to − 0.187	−0.242	−0.250	Resting heart rate
3	−0.146 ± 0.025	−0.171 to −0.121	−0.153	−0.146	Maximal heart rate
4	−0.006 ± 0.031	−0.037–0.025	−0.060	−0.052	Oxygen capacity
5	0.0015 ± 0.0032	−0.0017–0.0047	0.022	0.020	Body temperature
6	0.010 ± 0.036	−0.026–0.046	0.055	0.079	Blood pressure
7	0.0315 ± 0.071	−0.0395–0.1025	0.224	0.181	Age at weaning
8	0.032 ± 0.020	0.012–0.052	0.045	0.047	Bohr effect
9	0.161 ± 0.064	0.097–0.225	0.178	0.153	RBC lifespan
10	0.204 ± 0.026	0.178–0.230	0.275	0.288	Gestation period
11	0.544 ± 0.032	0.512–0.576	0.557	0.534	Thermal conductance
12	0.677 ± 0.022	0.655–0.699	0.709	0.672	Standard metabolic rate
13	0.725 ± 0.091	0.634–0.816	0.743	0.753	Respiratory minute volume
14	0.769 ± 0.070	0.699–0.839	0.776	0.783	Resting cardiac output
15	0.777 ± 0.077	0.700–0.854	0.782	0.748	Daily milk production
16	0.862 ± 0.047	0.815–0.909	0.881	0.865	Daily water intake
17	0.916 ± 0.097	0.819–1.013	0.931	0.853	VO_2max

See Table 23.6.

Computation of best-fit exponent by a double-median method

Our aim is to compare exponents computed by an adaptation of Sen's method, often applied to temporal data (see Gilbert [18]), with those computed by ordinary LSQ. For convenience, one may characterize this approach as the *double-median* (DM) method and the exponents derived as DM exponents. It works as follows. First, one sorts the x-values (together with their complementary y-values) from smallest to largest. Then one takes the logarithms of the x- and y-values. Next, calculate two medians, one (denoted by xx) for the column of $\log(x)$-values and one for the column of $\log(y)$-values (denoted by yy). For each logarithmic data point (x_i, y_i) one calculates an absolute value of the exponent or slope (b_i) of a line connecting the given point (x_i, y_i) to the median (xx, yy), from the relation:

$$b_i = |(y_i - yy)/(x_i - xx)| \qquad (23.17)$$

where the vertical bars denote absolute value. If there are n points in the dataset, then relation (23.17) generates n estimates of the exponent. The next to last step is to take the *median* of the set of computed exponents. Note that the sign (±) for the exponent is lost when one takes the absolute value (relation (23.17)). The last step is to assign a sign (±) to the best-fit median exponent by hand (as by inspection of a plot). Because one takes the median (twice), the DM method may be considered more robust than the LSQ

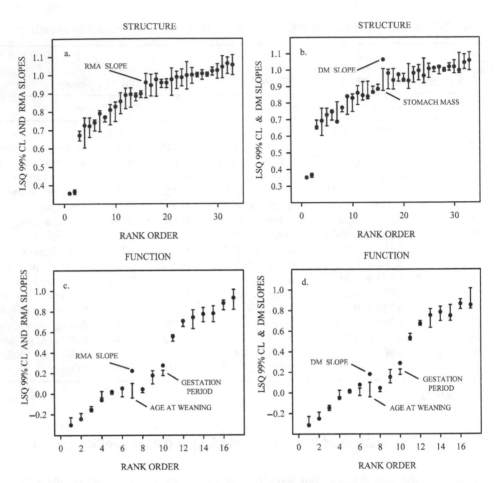

Figure 23.3 Comparison of exponents determined by the RMA and DM (double-median) methods with LSQ 99% CL.

method, in that it is less sensitive to outliers. It is not, however, wholly insensitive to perturbations. Significant disturbing values arise when x_i equals or very nearly equals xx.

The exponents for 33 *structural* variables determined by the DM method are summarized in Table 23.6, column d, and in Figure 23.3b, where the solid circles denote DM exponents. The data used to construct Figure 23.3b are summarized in Table 23.6, column d. In only one case (Table 23.6, row 16) does the double-median exponent fall well outside the 99% CL on the LSQ exponent (i.e., for stomach mass). The mean absolute deviation between the 33 structural LSQ exponents and the corresponding double-median exponents is 0.021 ± 0.023 (STD).

In Figure 23.3d we see that for 17 *functional* exponents determined by the DM method, six exponents fall outside the 99% CL; of these, perhaps only three deviations are of biological significance (those for blood pressure, age at weaning, and gestation period). The data on which Figure 23.3d are based are shown in Table 23.7, columns a and d. The mean absolute deviation between the 17 LSQ exponents and DM exponents is 0.037 ± 0.037 (STD).

Computation of best-fit exponent by a single-median method

It is useful to know that there are multiple methods for fitting a straight line to log–log transformed data. A few of these methods were discussed briefly above. Each method has its advantages and shortcomings. The last method to be discussed here is the simplest, both conceptually and in practice. It may be termed the *single-median* (SM) method. Suppose one is given a dataset in native coordinates consisting of n pairs of x- and y-values. One sorts the data from small to large x-values as described above, and transforms the data to logarithmic form. We next compute the absolute value of the exponent (b_i) of the straight line joining each and *every* pair of points (x_i, y_i) and (x_j, y_j) in the dataset:

$$b_i = |(y_i - y_j)/(x_i - x_j)| \qquad (23.18)$$

Note that given n points, the total number (N_{pr}) of pairs of points for any given dataset is given by:

$$N_{pr} = n(n-1)/2 \qquad (23.19)$$

Observe that if $n = 10$, then $N_{pr} = 45$. For $n \geq 100$, $N_{pr} \approx n^2/2$; hence if $n = 100$ then $N_{pr} \approx 5{,}000$. The last step is to take the median of the N_{pr} exponents and then, as needed, to assign a sign ($\pm$), say by inspection of a plot.

In computing exponents by the SM method I selected datasets with positive exponents ($b > 0$) as determined by LSQ, taking into account the 99% CL. In practice, in computing exponents based on every pair of points, it may happen that two y-values (y_i, y_j) share the same x-value (x_i, x_i), giving rise to an infinite exponent (see relation (23.18)). This was found to be the case for a number of datasets in Part II. In those cases where there was only a single infinite exponent, I replaced the infinite value with a numerical exponent slightly larger than the prior largest calculated exponent.

For example, if the largest finite exponent in a dataset was 7, and the next exponent was infinite, I substituted an exponent of 8 for the infinite value and then computed a median exponent. Keep in mind that the median exponent will not be affected if one substitutes 800 for the value of infinity rather than 8. The number substituted for infinity simply serves as a "place-holder" (see Table 23.1). Those datasets that generated more than one infinite exponent were discarded for this part of the study. (An alternative would be to modify the respective datasets so that each y-value corresponds to a unique x-value.) As it turned out, single-median (SM) exponents were determined for 28 *structural* variables and 7 *functional* variables.

It will be seen from Figure 23.4 that 28 *structural* exponents (panel a) and 7 *functional* exponents (panel b) all agree with the LSQ respective exponents to within the 99% CL. The mean absolute deviation between the LSQ exponents and the 28 SM

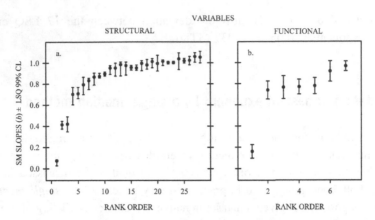

Figure 23.4 Evaluation of exponents as determined by the single-median method with 99% CL determined by the LSQ method.

structural exponents is 0.0115 ± 0.011 (STD); for 7 functional exponents the corresponding figures are 0.0142 ± 0.008. The mean number of species for the 28 structural variables is 59 ± 43 (STD) with a range of 6 to 159. The comparable figures for the seven functional variables are 21 ± 6 (STD) with a range from 13 to 29.

Note that in Figure 23.4 the solid circles represent exponents as determined by the SM method. The "error" bars reflect the 99% CL on the LSQ exponents. For clarity the LSQ exponents themselves are not shown, but would lie in the middle of the confidence interval.

Comments on different methods of computation of best-fit lines

In the above, we discussed briefly a number of methods for fitting a linear function to log–log transformed data. These include LSQ, LAR, and RMA methods, as well as DM and SM methods. The aim of these exercises is to show that different algorithms, based on different assumptions, may give exponents that are statistically equivalent. Neither of the two median methods involves minimization; both proceed by finding representative exponents that are implicit in the data. In particular, it is argued that none of the RMA exponents differs from the corresponding LSQ exponents by an amount that is likely to be biologically significant (see Figure 23.3a). A possible virtue of the RMA method is that it takes into account errors (variations) in the x-values. On the whole, however, the case for using the RMA method for the purposes of the present study appears to be weak (see also [17]).

It is not intended here to discourage use of the LSQ method. Although not a robust method, it has the virtues of convenience and long usage. Software packages to compute LSQ statistics are readily available, and the behavior of the LSQ method is well known. Nevertheless, the DM and SM methods described above, among others, may merit further study for potential applications to mammalian scaling studies. The main inference to be drawn from the above analyses is that different algorithms, based on different mathematical processes, may give rise to statistically similar exponents. This finding implies that the assumptions of most importance may be those – such as the assumption of linearity on a log–log plot – that are common to the different methods. Other methods of robust analysis have been described and the interested reader is advised to consult them (e.g., Rousseeuw and Leroy [19]). What is lacking is an in-depth analysis of the relative merits of the different methods in a collaborative effort between biologists and statisticians.

As a final point, one finds that LSQ exponents (not shown) are insensitive (relative to the 99% CL) to pseudo-random changes in the x-values of several datasets for mean deviations of 10–15%. The test process is to multiply all the x-values in a given dataset by pseudo-random numbers (e.g., 1.05, 0.87, 1.01, 1.13, etc.) to produce a desired mean percent deviation (MPD). Then one re-computes the LSQ statistics, including the exponent and 99% CL (see also Calder [5]). Having carried out several analyses along these lines, I find that LSQ exponents are relatively insensitive to moderate (mean of 10 to 15%) pseudo-random changes in the x-values as judged by the 99% CL.

Special pleading

...the numerical size of the difference [between two numbers] *has little or no meaning without an adequate frame of reference...*

Minium and Clarke 1982

The term "special pleading" refers, for example, to the situation where certain evidence is cited in favor of a given model and other information from the same source that does not support the model is, perhaps by oversight, not cited. The phrase "cherry-picking" is sometimes used in a similar sense. The concept is illustrated in Figure 23.5. The heavier dashed line indicates a hypothetical model exponent of 1. We assume that the model exponent is without error. One asks whether the points P_1 to P_6 (with exponents of 0.98, 1.0, 1.02, 0.965, 0.985, and 1.005, respectively) differ significantly from the model exponent. (Here, the points P_1 and P_3 represent the CL of the point P_2. By assumption,

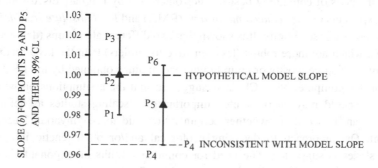

Figure 23.5 A concept of special pleading.

the points P_1 and P_3 are as probable as the point P_2. In the same way the points P_4 and P_6 represents the CL on the point P_5 and are, by assumption, as probable as the point P_5.)

Let us take (the arbitrary figure of) ± 0.02, for convenience, as the criterion of significance. By this criterion, the points P_1, P_2, P_3, P_5, and P_6 may be said to support the model exponent of 1 (that is, they differ from an exponent of 1 by ± 0.02 or less).

On the other hand, the point P_4 ($b = 0.965$), which is as probable as P_6 ($b = 1.005$), differs from an exponent of 1 by 0.035, and hence does not support the model exponent, by the stated criterion. To say that the point P_5 supports the model, without acknowledging that the point P_4 (exponent = 0.965) does not (say at the 99% CL), may be regarded as a form of special pleading. A more common example of special pleading is the failure to give any CL at all.

Special pleading in these two senses is not uncommon in works on mammalian scaling. When exponents are published, CL (perhaps at the 99% level) should be given as a matter of course. In order to reach agreement on what constitutes special pleading, the community of scaling biologists needs to define what constitutes a "critical" difference in exponents. In the above numerical example, an absolute difference in exponents of 0.02 or more is taken to be statistically and biological significant (see above quote from Minium and Clarke). This is meant to be a possible *starting point* for further discussion.

Summary

Going through this chapter, one first encounters the notion of a residual. Then aspects of an ideal dataset and best-fit line for mammalian scaling studies are described. The use of means and medians is discussed briefly. The least squares method of deriving exponents and intercepts from bivariate datasets is discussed. The sensitivity of slopes to sample size and size range is stated in terms of end- and mid-samples as these relate to full datasets (FDS). The risks of extrapolation are noted. The potential sensitivity of LSQ estimates of slopes to outliers is emphasized. The methods used to compute numerous statistical parameters cited in Parts II and III are reviewed briefly.

Four methods of computing best-fit lines other than by LSQ are discussed. Calculation of exponents by the *reduced major axis* (RMA) and *least absolute residuals* (LAR) methods are discussed briefly. It is shown that two different algorithms (denoted by DM and SM), which are more robust (less sensitive to outliers) than the LSQ method, give exponents that agree mostly or entirely with the findings obtained by the LSQ method, to within the computed 99% CL. It is suggested that the assumptions on which LSQ analysis is based may be of limited importance to scaling studies as defined here. Questions are raised as to whether certain ratios said to be "invariant" actually are invariant. One approach to detecting ecological and/or phylogenetic influences on overall slopes is suggested. The need for confidence limits on exponents is stressed. The concept of *special pleading* is considered briefly.

References

1. Hays, H.L. (1973). *Statistics for the Social Sciences,* 2nd edn. New York: Holt, Rinehart and Winston.
2. Sokal, R.R. and Rohlf, F.J. (1980). *Biometry: The Principles and Practice of Statistics in Biological Research.* New York: W.H. Freeman.
3. Hamilton, W.C. (1964). *Statistics in Physical Science: Estimation, Hypothesis Testing, and Least Squares.* New York: Ronald Press.
4. Stahl, W.R. (1965). Organ weights in primates and other mammals. *Science,* **150**:1039–1041.
5. Calder, W.A. (1984). *Size, Function, and Life History.* Cambridge, MA: Harvard University Press.
6. Adolph, E.F. (1949). Quantitative relations in the physiological constitutions of mammals. *Science,* **109**:579–585.
7. Günther, B. and Guerra, E. (1955). Biological similarities. *Acta Physiological LatinoAmericana,* **5**:169–186.
8. Boggs, D.F. and Tenney, S.M. (1984). Scaling respiratory pattern and respiratory "drive". *Respiratory Physiology,* **58**:245–251.
9. Li, J.K.-J. (1983). A new similarity principle for cardiac energetics. *Bulletin Mathematical Biology,* **45**:1005–1011.
10. Stahl, W.R. (1962). Similarity and dimensional methods in biology. *Science,* **137**:205–212.
11. Stahl, W.R. (1967). Scaling of respiratory variables in mammals. *Journal of Applied Physiology,* **22**:453–460.
12. Stahl, W.R. (1963). The analysis of biological similarity. *Advances in Biological Medical Physics,* **9**:355–489.
13. Schmidt-Nielsen, K. (1984). *Scaling: Why is Animal Size so Important?* Cambridge: Cambridge University Press.
14. Prothero, J. (1993). Adult lifespan as a function of age at maturity. *Experimental. Gerontology,* **28**:529–536.
15. Bacon, R.H. (1953). The "best" straight line among the points. *American Journal of Physics,* **21**:428–446.
16. Jolicoeur, P. (1975). Linear regressions in fishery research: some comments. *Journal of Fisheries Research Board Canada,* **32**:1491–1494.
17. Niklas, K.J. (1994). *Plant Allometry: The Scaling of Form and Process.* Chicago, IL: University of Chicago Press.
18. Gilbert, R.O. (1987). *Statistical Methods for Environmental Pollution Monitoring.* New York: John Wiley & Sons.
19. Rousseeuw, P.J. and Leroy, A.M. (1987). *Robust Regression Analysis and Outlier Detection.* New York: John Wiley & Sons.

24   Scaling sums

...numerical precision is the very soul of science...

Thompson 1943

The real world... can be viewed as comprising sets of interlinked systems at various scales and of varying complexity, which are nested into each other to form a... hierarchy.

Chorley and Kennedy 1971

This chapter provides an *informal* derivation of various mathematical statements made in Parts I, II, and III. In addition, a few topics not considered earlier are touched upon. We begin with a discussion of the nature of those mathematical functions that have been or might be employed in mammalian scaling studies.

Properties of three simple mathematical functions

For present purposes it is sufficient to restrict ourselves to *functions* involving only two variables. The term "function" here implies a correspondence (or mapping) between two quantities, say x and y. In this study we are concerned with *single-valued* functions (each x-value implies a unique y-value) and (for the most part) continuous functions. (The notion of a discontinuous function, for example a square wave, which has multiple y-values for certain x-values, is more intuitive than is a formal mathematical definition of continuity.) *Bivariate* functions (relations between two variables) are denoted by:

$$y = f(x) \tag{24.1}$$

where for each value of x the function $f(x)$ establishes a one-to-one correspondence with a value of y. Commonly x is termed the *independent* and y the *dependent variable*, although sometimes the order may be reversed. No causal relationship between x and y is implied. Moreover, it often happens in adult mammalian scaling analyses that x and y are not strictly independent. For example, when we look at how organ mass scales with body mass, we should keep in mind that body mass includes organ mass. Again, since many y-variables are highly correlated with body size, it follows that they are correlated to varying degrees with each other [1, 2].

The linear function

The simplest bivariate function is that for a straight line:

$$y = a + b \cdot x \text{ and } \Delta y = b \cdot \Delta x \qquad (24.2)$$

In relation (24.2), (a and b) are adjustable "constants," termed *intercept* (a) and *slope* (b), respectively. The intercept gives the y-value when $x = 0$. Each constant (a, b) and each variable (x, y) may, in principle, take on any finite value. A change (Δy) in y is directly proportional to a change (Δx) in the x-value. The values of x and y may fall into any of the four Cartesian quadrants. In relation (24.2) there is no necessary or implied mathematical relationship between a and b. It is true, however, that for a *family* of straight lines there may be a relation between a and b, as when the straight lines *intersect* at or near a common point [3]. For example, depending on where the point of intersection falls, there may be an inverse hyperbolic relation between a and b. Recall also that the numerical values of the intercept and slope, expressed in native coordinates, depend on the units employed (see Table 24.5). Generally empirical relations between intercept and slope are of limited interest in *structural* and *functional* mammalian scaling studies (however, for a specific theoretical relationship see "Inherent limits . . ." below).

The exponential function

One of the simplest bivariate functions is the *exponential*. It describes a process akin to cell doubling in the absence of cell death. An exponential function is given by:

$$y = ae^{bx} \text{ and } dy = by dx \qquad (24.3)$$

where e is a constant (classically 2.718..., but it could just as well be 10 or another value) and dx and dy are small (infinitesimal) changes in x and y, respectively. As in relation (24.2), the terms a and b are independent adjustable constants. The exponential function has the important property that a small change (dy) in y is directly proportional to y for a given dx. The greater is y, the greater is dy.

The power function

The *power function* is by far the most commonly used function in adult interspecific mammalian scaling studies. It is defined by the relation:

$$y = ax^b \text{ and } dy/y = b(dx/x) \text{ where } x > 0 \qquad (24.4)$$

The constant a (often suppressed in mathematical discussions) is called the intercept (or sometimes the weight coefficient), and b the exponent, power, or slope. Like the exponential function, a small change (dy) in the y-value is proportional to the value of

y; unlike the exponential function a change (dy) in y is inversely proportional to x. Taking logarithms of both sides of relation (24.4) one finds:

$$\log(y) = \log(a) + b \cdot \log(x) \text{ where } a > 0, \quad x > 0, \quad y > 0 \qquad (24.5)$$

Relation (24.5) implies that a power function is linear on a log–log plot. Note that the x- and y-values are confined to the first Cartesian quadrant ($x > 0$, $y > 0$). Given two discrete points (x_1, y_1) and (x_2, y_2), we infer from relation (24.5) that the quantity [log (y_2) – log(y_1)] is equal to the quantity $b \cdot$[log(x_2) – log(x_1)]. That is to say, log(y_2/y_1) = $b \cdot$log(x_2/x_1). Relation (24.5) is the simplest representation of a power function. As far as is known, the empirical exponents (b) for adult mammals *generally* lie between –1/3 and 10/9. See above quote from D'Arcy Thompson.

A comparison of linear, exponential, and power functions

Certain properties of these three functions (linear, exponential, power) are summarized in Table 24.1. Columns c and d show the conditions under which these three functions are *linear*. For example, the exponential function is linear on a scale that is linear in x and logarithmic in y (i.e., x, log(y)). Table 24.1, column e, shows that the intercept is evaluated at $x = 0$ for the linear and exponential functions and at $x = 1.0$ for the power function (recall that the logarithm of zero is undefined).

How the slope may be calculated given two valid points ((x_1, y_1), (x_2, y_2)) for each function is shown in Table 24.1, column f. The scales given in Table 24.1 (columns c and d) are selected so as to give linearity. Another comparison among these three functions is made in Figure 24.1. Be aware that in Figure 24.1 the three given functions share the same numerical values for intercept (0.5); two functions (panels a and c) share the same slope or exponent (0.75).

Figure 24.1, panel a, is a linear–linear plot; panel b, a semi-log plot; and panel c, a log–log plot (see Table 24.1, columns c and d). The most conspicuous feature of Figure 24.1 is the huge difference in the rate at which y increases with increasing x for the exponential function (panel b) relative to either the linear or power functions. Provided the exponent (b) is positive, an exponential function will always "swamp" a linear or a power function for some value of x. In the bioscaling literature a power

Table 24.1 Linear, exponential, and power functions compared

Function	Equation	Linearity as a function of linear or logarithmic x–y scales		Intercept (a) value of x is:	Slope (b) 2-point slope is:
a	b	c	d	e	f
Linear	$y = a + bx$	x	y	0	$(y_2 - y_1)/(x_2 - x_1)$
Exponential	$y = a \cdot 10^{bx}$	x	$\log(y)$	0	$\log(y_2/y_1)/(x_2 - x_1)$
Power	$y = ax^b$	$\log(x)$	$\log(y)$	1	$\log(y_2/y_1)/\log(x_2/x_1)$

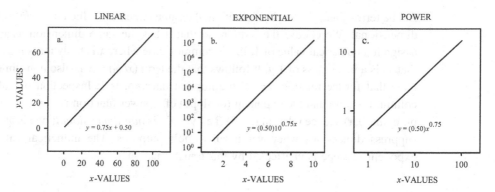

Figure 24.1 Plots of linear, exponential, and power functions.

function is often referred to as an exponential function; but the two functions behave quite differently over large size ranges (see, however, "A discrete representation..." below).

Numerical equality between a linear and a power function

It is easily confirmed that if we plot a linear function (see relation (24.2)) in log–log coordinates, then the resultant plot is *non-linear* in the region between $bx < a$ and $bx > a$ where $a > 0$. Another way to compare a *linear* function (x, y_1) and a *power* function (x, y_2) is as follows. Suppose we have:

$$y_1 = a_1 + b_1 x \text{ and } y_2 = a_2 x^{b_2} \qquad (24.6)$$

We ask: under what conditions are y_1 and y_2 numerically identical? It is evident by inspection of relations (24.6) that identity of y_1 and y_2 requires:

$$a_1 = 0, \quad a_2 = b_1, \quad \text{and } b_2 = 1 \qquad (24.7)$$

Observe the *peculiarity* that identity between a linear function and a power function requires that the intercept (a_2) of the power function be equal to the slope (b_1) of the corresponding linear function. On occasion, a linear function has been advocated for use in scaling studies; generally, however, a *power function* is preferable for adult interspecific studies in mammals [4].

A power function in non-dimensional form

It is useful to be reminded that a power function can always be written in *non-dimensional* form. To see this we rewrite relation (24.4) above in the form:

$$(y/a) = (x/x_0)^b \text{ or } Y = X^b \text{ where } Y = (y/a) \text{ and } X = (x/x_0) \qquad (24.8)$$

As the terms y and a have, by definition, the same dimensions, the ratio $(y/a) = Y$ is non-dimensional. We choose the term x_0 so that it has the same dimension as x, and we assign it a numerical value of 1. In the common case where x is body mass in g, we have that x_0 is a body mass of 1 g. It follows that the term $(x/x_0) = X$ is also non-dimensional. Note that for the relation $y = X^b$ the implicit intercept is 1. Inspection of Table 24.1, column f, shows that the exponent (or slope) of a power function is non-dimensional (a pure number). Hence relation (24.8) (i.e., $y = X^b$) is *non-dimensional*; one is at liberty to suppress dimensions whenever it suits one's purposes. The numerical value of the exponent or slope is unaffected by this transformation.

An empirical power function in terms of size range

As one progressively narrows the size range over which empirical scaling data extend, one progressively weakens the case that a power function is the best available two-parameter descriptor. We may demonstrate this with the following example. We generate an exact dataset using a sinusoidal function:

$$y = \sin(\theta) \tag{24.9}$$

In Figure 24.2a, theta (θ) plays the role of the x-variable over half of a sinusoidal cycle. Panel b shows the same data plotted in log–log form. Observe the substantial distortion of the sinusoidal form. A LSQ line was fitted to the transformed sinusoidal data from theta = 2 (P_1) to 40 (P_2) degrees (a 20-fold range). The best-fit line is shown as dots in panel b (this line is *displaced* upwards by a factor of 2 from the sinusoidal curve for clarity). One finds that the best-fit line deviates on average from the sinusoidal curve (both expressed in native coordinates) over the given range (P_1 to P_2) by 0.4%. However, if we were to infer from the dotted line that the underlying function is a power function we would be grossly mistaken. A two-fold or greater extrapolation of the best-fit line (to 80 degrees or more) predicts y-values systematically wide of the mark.

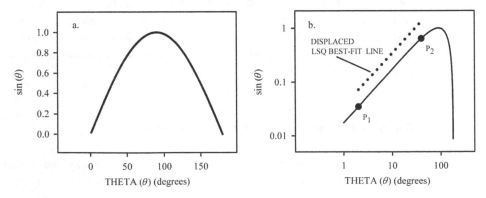

Figure 24.2 A power function and size range.

Experience with (log–log transformed) empirical scaling data for mammals shows that a power function often gives a quite reasonable fit (see Part II). Still, the use of a power function is strictly empirical. Over narrower size ranges (say pWR < 2.0) it is possible that some other two-parameter function (e.g., $y = a + bx$) would fit just as well. As the size range increases (say to pWR $\geq$ 4), *assuming linearity*, it becomes less and less likely, as a generalization, that any other two-parameter function will give a better fit than does a power function (for $b \neq 1$).

Percent change in *y* in terms of slope and percent change in *x*

... the equation $y = 70.5x^{0.73}$ merely states that a body weight increase of 1 per cent is associated with a basal metabolism increase of 0.73 per cent; increasing body weight 100 per cent increases basal metabolism 73 per cent.

Brody 1945

The above quote from Brody [5] is an approximation, as he stated elsewhere in his book. However, others (e.g., [6]) have not always pointed out that the above statement is an approximation.

We now derive an *exact* relation (for a power function) expressing a *percent change* in *y* (PD_y) as a function of a *percent change* in *x* (PD_x), and a given slope (*b*). We evaluate a given power function at two points: P_1 (x_1, y_1) and P_2 (x_2, y_2). That is:

$$y_1 = ax_1{}^b \text{ and } y_2 = ax_2{}^b \qquad (24.10)$$

$$\text{Thus } y_2/y_1 = (x_2/x_1)^b \qquad (24.11)$$

Next one defines PD_x and PD_y by the relations:

$$PD_x = 100 \cdot (x_2/x_1 - 1) \text{ and } PD_y = 100 \cdot (y_2/y_1 - 1) \qquad (24.12)$$

Rewriting the above expression (24.12) for PD_x, we have:

$$x_2/x_1 = PD_x/100 + 1 \qquad (24.13)$$

Hence, from relations (24.11), (24.12), and (24.13) one finds:

$$PD_y = 100 \cdot (y_2/y_1 - 1) = 100 \cdot \left((x_2/x_1)^b - 1\right) = 100 \cdot \left((PD_x/100 + 1)^b - 1\right) \qquad (24.14)$$

Brody [5] gives relation (24.14) for the special case when $PD_x = 100\%$. If $b = 1.0$ then relation (24.14) reduces to $PD_y = PD_x$. On the other hand, if in relation (24.14) one assumes that $PD_x/100$ is much less than 1, then one may invoke the binomial approximation (i.e., $(1 + x)^b \approx 1 + bx$ for $x \ll 1.0$), leading to the relation:

$$PD_y = 100 \cdot \left((PD_x/100 + 1)^b - 1\right) \approx 100 \cdot (1 + bPD_x/100 - 1) = b \cdot PD_x \qquad (24.15)$$

Table 24.2 Approximate and exact values of PD$_y$ for given values of PD$_x$ and slope

PD$_x$ (%)	Slope (b) = 0.1		Slope (b) = 0.5		Slope (b) = 1.1	
	Approximate	Exact	Approximate	Exact	Approximate	Exact
a	b	c	d	e	f	g
1	0.1	0.1	0.5	0.5	1.1	1.1
10	1.0	1.0	5	4.9	11	11
50	5.0	4.1	25	22.5	55	56
100	10.0	7.2	50	41.4	110	114
1,000	100.0	27	500	232	1,100	1,298

Relation (24.15) (PD$_y \approx b \cdot$PD$_x$) is the basis of the above quote from Brody [5] to the effect that if PD$_x$ is 1% and b is 0.73, then PD$_y$ is 0.73%. The nature of the approximation is illustrated in Table 24.2. The *exact* values of PD$_y$ for the given values of PD$_x$ (column a) and slope (b) are based on relation (24.14). (See Table 24.2, columns c, e, and g.) Relation (24.15) is exact when $b = 1.0$. *Approximate* values of PD$_y$ (calculated from PD$_y \approx b \cdot$PD$_x$) are shown in Table 24.2, columns b, d, and f.

We see from Table 24.2 that the approximation (PD$_y \approx$ b·PD$_x$) is perhaps much better than might be expected (e.g., for b = 0.1, 0.5, and 1.1). For instance, if the slope (b) is 0.1 and PD$_x$ (column a) is 100%, then the approximate value of PD$_y$ (column b) is 10% (row four) while the exact value is 7.2% (column c, row four). Generally, if b $\neq$ 1 and PD$_x$ is less than 100%, then the approximation (PD$_y = b \cdot$PD$_x$) is satisfactory for most practical purposes. In the example Brody [5] gives (see above quote) a 100% increase in body weight with a slope of 0.73 furnishes an approximate increase in basal metabolism of 73%, whereas the exact value (from relation (24.14)) is 66% (i.e., 100· $(2^b - 1) = 100 \cdot (2^{0.73} - 1) = 66$). For PD$_x$ greater than 100% and b significantly different from one, the approximation PD$_y \approx b \cdot$PD$_x$ may be poor (e.g., columns b and c, bottom row). For a shorter derivation of relation (24.15), (PD$_y \approx b \cdot$PD$_x$) start from relation (24.4), i.e., $\Delta y/y \approx b(\Delta x/x)$).

The relation PD$_y \approx b \cdot$PD$_x$ is worth remembering for values of PD$_x$ less than 100%. It affords a useful insight into how a power function works under the given approximation.

Inherent limits on organ or tissue slopes greater than 1

We now study the scaling of organs and tissues as a function of body size. It is self-evident that if a structure scales with an exponent greater than 1, it will tend with increasing body size to make up the whole body [7]. One may derive an expression for the *maximum* slope (b_{max}), as a function of the intercept (a) and size range (pWR) at which organ size hypothetically comes to *exactly* equal body size. In Parts I and II (e.g., see Chapter 2) pWR was routinely employed as a measure of size range:

$$\text{pWR} = \log(x_2/x_1) \tag{24.16}$$

where x_1 and x_2 are two different body masses ($x_2 > x_1$). Here it is convenient to take x_1 to be 1. Then we may write:

$$\text{pWR} = \log(x_2 / x_1) = \log(x_2). \tag{24.17}$$

We now ask at what body size does organ or tissue mass (y_{max}) come to equal body mass (i.e., x_2). This requirement is satisfied by the relation:

$$y_{max} = ax_2^{b_{max}} = x_2 \tag{24.18}$$

Taking logarithms in relation (24.18), and drawing on relation (24.17), we find:

$$b_{max} = 1 - \log(a)/\log(x_2) = 1 - \log(a)/\text{pWR} \tag{24.19}$$

Relation (24.19) is the same as that given in Chapter 3 (see relation (3.3) and Figure 3.2). This relation says that for any given intercept (a) and value of pWR, the exponent (assumed to be >1) associated with the scaling of any tissue or organ must be less than b_{max}. In mammals, and in vertebrates generally, any organs or tissues scaling with exponents greater than 1 must be *compensated* for by other organs and tissues scaling with exponents less than 1 [1]. On average, these compensating organs and tissues may be found to scale on or above the reflected curve given by $b_{reflected} = 2 - b_{max}$ (see Figure 3.2). Thus the confined space between the upper (b_{max}) and lower ($b_{reflected}$) curves may largely specify the crucial domain within which *natural selection* can alter structural exponents in making internal structural adjustments to changes in adult body size. See above quote from Chorley and Kennedy.

A discrete representation of a power function

One may wish on occasion to represent a given power function at a *discrete* number of points [8]. Assume that the two end-points of a best-fit LSQ line-segment are given by (x_{min}, y_{min}) and (x_{max}, y_{max}). We aim to display the best-fit line at n discrete points, rather than continuously. This may be achieved using two *geometric series*, one for the x-values and one for the y-values. The *common factor* in each series is represented by k_x and k_y. It may be shown that k_x and k_y are given by:

$$k_x = (x_{max}/x_{min})^{\left(1/(n-1)\right)} \text{ and } k_y = (y_{max}/y_{min})^{\left(1/(n-1)\right)} \tag{24.20}$$

A specific example is shown in Figure 24.3, where (x_{min}, y_{min}) = (1.5, 0.072) and (x_{max}, y_{max}) = (10^8, 792,000) and $n = 10$. For these data, $k_x = 7.40$ and $k_y = 6.06$. As a take-home exercise, the reader may show that the slope (b) of the best-fit LSQ line-segment in Figure 24.3 is given by:

$$b = \log(k_y)/\log(k_x) = \log(y_{max}/y_{min})/\log(x_{max}/x_{min}) = \log(6.06)/\log(7.40) = 0.9 \tag{24.21}$$

Table 24.3 A discrete representation of a power function

Point	x-values	y-values
a	b	c
1	1.5	0.072
2	11.1	0.436
3	82.2	2.643
4	608	16.02
5	4,502	97.05
Common factor	7.40	6.06

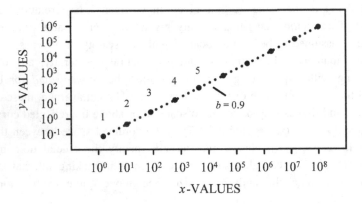

Figure 24.3 A discrete representation of a power function.

The numerical data for the first five points plotted in Figure 24.3 are given in Table 24.3. The reader may easily extend both series to include all 10 points and also to show that $k_y = k_x^b$. Keep in mind that all empirical datasets are discrete.

Observe that the successive x-values and y-values shown in Table 24.3 (columns b and c) are given by:

$$x_{\min} \cdot (1, \ k_x, \ k_x^2, \ k_x^3, \ k_x^4 \ \dots) \text{ and } y_{\min} \cdot (1, \ k_y, \ k_y^2, \ k_y^3, \ k_y^4 \ \dots) \tag{24.22}$$

In both series (relations (24.22)), each term after the first is expressed as an *exponential*, showing that when we construct a *discrete* representation of a power function there is a close relationship to two series of *exponential* terms. (See also "Testing for non-linearity" below.)

Divergence between two end-points in terms of change in slope

The situation we consider next (see also Chapter 3) is illustrated in Figure 24.4, panel a. Two best-fit lines with different slopes diverge from a common intercept ($x = 1$). The upper end-points of the two lines, where pWR = 8.0 (i.e., $x = 10^8$), are denoted by P_1

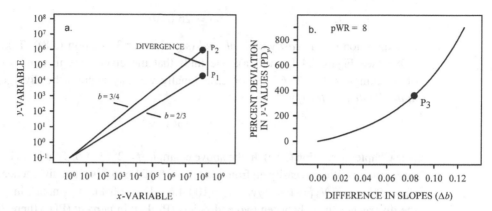

Figure 24.4 Divergence between end-points of paired best-fit lines having different slopes.

and P_2, for $P_2 > P_1$. We aim to determine the magnitude of the divergence between these two end-points as a function of differences in slope and size range (pWR). The equations for these two lines are given by:

$$y_1 = ax^{b_1} \text{ and } y_2 = ax^{b_2}, \ b_2 > b_1 \tag{24.23}$$

$$y_2/y_1 = x^{b_2-b_1} = x^{\Delta b} \tag{24.24}$$

We denote the difference $(b_2 - b_1)$ between two slopes by Δb. We evaluate PD_y where pWR = 8.0 or $x = 10^8$. The percent deviation (PD_y) between P_1 and P_2 (Figure 24.4, panel a) is given by:

$$PD_y = 100 \cdot (y_2 - y_1)/y_1 = 100(x^{\Delta b} - 1) = 100 \cdot (10^{8\Delta b} - 1) \tag{24.25}$$

Relation (24.25) is illustrated in Figure 24.4b, for pWR = 8. The point P_3 corresponds to a *difference* in slope (Δb) of 0.083 (e.g., 3/4 – 2/3 or 1/3 – 1/4 = 0.083). The reader can confirm that the value of PD_y when $\Delta b = 0.083$ is 360%. Thus two best-fit lines, intersecting at $x = 1$, differing in slope by 0.083, and projected over the whole mammalian size range (pWR = 8.0), differ in y-values at the upper end-points by 360%. One may express the *divergence* in y-values between the two end-points by the *number of doublings* (N_d) involved (Chapter 3). We use the same convention as above, measuring pWR (relation (24.17)) from $x = 1$. Then:

$$y_2/y_1 = (x_2/x_1)^{\Delta b} = 2^{N_d} \tag{24.26}$$

Taking logarithms we have:

$$\Delta b \cdot \log(x_2/x_1) = \Delta b \cdot pWR = N_d \cdot \log(2) \tag{24.27}$$

Thus we find that:

$$N_d = \Delta b \cdot pWR/\log(2) \tag{24.28}$$

If we set pWR = 8.0 then:

$$N_d = 26.6 \cdot \Delta b \tag{24.29}$$

an expression that is *easily remembered* (see Chapter 3, relation (3.8)). Taking $\Delta b = 0.083$ (see Figure 24.4, panel a), one finds that the divergence in terms of number of doublings is 2.2 ($26.6 \cdot 0.083 = 2.2$). Finally we may express the divergence as a *multiplicative factor* (f):

$$f = 2^{N_d} \tag{24.30}$$

(see Chapter 3, relation (3.9)). In the above example, $f = 2^{2.2} = 4.6$. That is, y_2 is 4.6 times larger than y_1. We can easily go from the factor (f) of 4.6 to a percent divergence of 360% (i.e., $100 \cdot (y_2 - y_1)/y_1 = 100(y_2/y_1 - 1) = 100(4.6 - 1) = 360\%$). In sum, one may describe the divergence (in y) between two end-points (P_1, P_2): in *percent* (PD_y) (here 360%); in *number of doublings* (N_d) (here 2.2); or as a *multiplicative factor* (f) (here 4.6).

The Haldane restriction

Haldane (cited in Huxley [9]) pointed out that a sum of power functions is not a power function:

$$a_1 x^{b_1} + a_2 x^{b_2} + a_3 x^{b_3} + \cdots \neq a x^b \qquad b_1 \neq b_2 \neq b_3 \ldots \tag{24.31}$$

For example, if we had empirical power functions describing the structural scaling of all organ and tissues in the body, these could not, as a matter of principle, always sum exactly to body mass (see also Turner [10]). This principle is of some conceptual importance; it shows that a self-consistent description of the scaling of organs and tissues exclusively in terms of power functions is mathematically impossible. Nevertheless, the Haldane principle is of limited practical importance; most organs and tissues studied here scale as a function of body mass with structural exponents lying between about 0.85 and 1.05 (see Table 20.5). As a practical matter, deviations from a slope of 1 in the range 0.05 to 0.15 (and somewhat greater) do not necessarily have a significant impact on the summation process (see Figure 19.1c). Furthermore, biological variation in individual organ and tissue masses *and* in adult body masses confers a degree of uncertainty in the calculation of total mass from variable organs and tissues. Finally, the Haldane restriction does not apply to *functional* variables, which are not usually additive in any event.

It should also be recorded that there is a simple exception to the Haldane restriction. We can sum any number of power functions *exactly* at two values of the x-variable (say x_1, x_2) where $x_1 = 1.0$. We denote the two respective sums by y_1 and y_2. Then we may write:

$$\text{For } x_1 = 1 \qquad y_1 = a_1 x_1^{b_1} + a_2 x_1^{b_2} + a_3 x_1^{b_3} + \cdots = a_1 + a_2 + a_3 = \Sigma a_i \tag{24.32}$$

$$\text{For } x_2 \qquad y_2 = a_1 x_2^{b_1} + a_2 x_2^{b_2} + a_3 x_2^{b_3} + \cdots = \sum a_i x_2^{b_i} \tag{24.33}$$

where $\sum$ denotes summation. By assumption, the various (a_i, b_i) and (x_1, x_2) values are known. Given numerical values for (x_1, y_1) (see relation (24.32)) and (x_2, y_2) (see relation (24.33)) one can derive the correct *slope* (b) from the two-point relation for a straight line in log–log form (see Table 24.1, column f, bottom row). The intercept is the same as y_1 (relation (24.32)). The reader may confirm that the above formulation is exact at two points.

Moreover, if the slopes are not too disparate, and the size range not too great, the power function defined by the given intercept and slope will produce y-values at intermediate x-values ($x_{\min} < x < x_{\max}$) that agree with the sum of the given set of power functions to within a few percent. (See Figure 19.1, panel c.) The Haldane restriction is worth knowing about; at the same time, it is perhaps not a cause for concern in most adult mammalian scaling studies.

A potentially misleading plot

. . .the curve goes up steeply at the small-animal end [here hummingbirds] *and it indicates that at 2.5 g the rate of metabolism would be infinitely high.*

Pearson 1953

Suppose one has (x, y) data that are apparently linear on a log–log plot. We express the y-values in x-specific form by dividing them by the respective x-values, giving a series of ratios of the form (y/x). We plot this ratio (y/x) on a linear scale as a function of the logarithm of the x-values (i.e., a semi-log plot, here in $\log(x)$ rather than $\log(y)$; see Figure 24.5a). We find that the ratio (y/x) appears to increase rapidly with decreasing x-values. Schmidt-Nielsen [11] gives an example (with data) of this phenomenon for mass-specific resting VO_2 expressed as a function of body size [11]. At the same time, a log–log plot of the same data (in absolute form) is apparently linear (see Figure 24.5d; Schmidt-Nielson [11] shows both plots). These two different plots (Figure 24.5, panels a and d) imply quite different behaviors. An attempt to resolve the contradiction between an apparent exponential change (in mass-specific form) on a semi-log plot (panel a) and a linear change (in absolute form) on a log–log plot (panel d) is made in Figure 24.5, panels b and c.

Panel a shows *mass-specific* resting VO_2 on a linear scale (ordinate) plotted against body mass (abscissa) on a logarithmic scale. Note the very large apparent increase in mass-specific VO_2 as between a mouse and shrew (an increase by a factor of 3). In panel b, we have a rather similar plot, but for larger mammals. Again, we see a substantial increase between the dog and cat (by a factor of 2) that is much less apparent in panel a. (However, the above comparison is somewhat misleading, as the dog weighs almost five times as much as the cat, whereas the mouse is less than twice as heavy as the shrew.) Panel c shows the same mass-specific data as panel a, plotted on a log–log plot. Except for the shrew, the data appear to be roughly linearly distributed on a log–log plot. Finally, in panel d we have converted *mass-specific* resting VO_2 $(cm^3/g \cdot min)$ to *absolute* resting VO_2 (cm^3/min). Note that the ordinal data in panel c extend over about

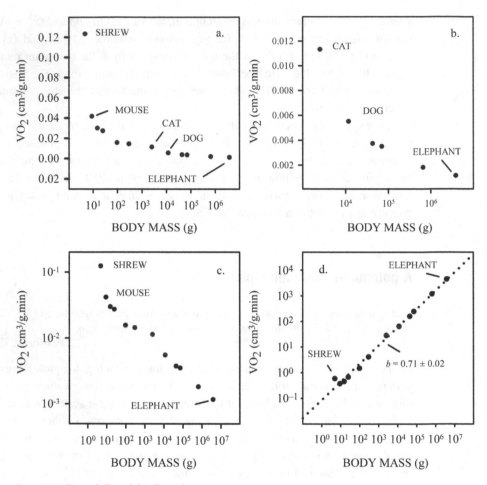

Figure 24.5 Potentially misleading plots.

two orders of magnitude, whereas those in panel d extend over slightly less than four orders of magnitude.

In panel d, we see that the data are more nearly linear in distribution. The shrew appears to be an outlier. That is plausible, as an alert shrew seldom rests. It is also true that some species of shrews exhibit remarkable adaptations for high oxygen transport (e.g., Bartels [12]). Plots akin to Figure 24.5a should not be taken to imply that there is something like a "metabolic explosion" in small mammals generally. Panel d avoids the possible fallacy associated with plotting a ratio against its denominator (recall that such plots may introduce spurious correlations) [4]. Finally, panel d is consistent with the assumption that resting VO_2 in mammals is, to a first approximation, governed by a power function. This is a more secure assumption than that drawn from panel a, which implies a metabolic rate in small mammals that is "infinitely high". There is no compelling physiological reason to expect that an adult mammal with a body mass of 2.5 or even 1.5 g will have an indefinitely high metabolic rate (see above quote from Pearson).

Table 24.4 Two forms of structural invariance expressed as a function of body size

Invariance	Scaling exponents	Scaling parameters	Examples
a	b	c	d
Absolute	0	Size, shape	Actin, myosin, hemoglobin, water molecules, RBC
Relative	1	Number	As above

RBC = red blood cells.

Structural invariance: absolute and relative

In Chapter 3 we defined *absolute* and *relative invariance*, where the exponents or slopes (expressed as a function of body size) are 0 or 1, respectively. Examples of absolute and relative invariance for several structural variables are brought together in Table 24.4.

The exponents (0 or 1) shown in Table 24.4, column b, are expressed as a function of body size. The invariant parameters (size and shape; number) are given in column c. Examples of structures that may exhibit absolute invariance (with respect to size and shape) or relative invariance (with respect to component number) are given in column d. Thus the *size* and *shape* of red blood cells are independent of body size, on average, whereas their *number* apparently increases directly with body size. The same is likely to be true for actin, myosin, and water molecules, which together account for a large fraction of body size in adult mammals. All of these examples, no doubt among many others, are molecular, macromolecular, or cellular. Whether there are any *macroscopic* entities that exhibit both absolute (size, shape) *and* relative (number) invariance is an open question. In addition, most organs of an adult mammal, such as the heart (one) or brain (one) or kidneys (two), show *absolute invariance in number*. Thus absolute invariance is not confined exclusively to the atomic, molecular, and cellular levels of resolution at which it is usually found.

Functional invariance: absolute and relative

Absolute invariance is characteristic of *homeostatic* variables such as body temperature, blood pressure, blood pH, and the like. Keep in mind that these functional examples of *y*-variables are *intensive* (non-additive), in contrast to structural *y*-variables involving, for example, tissue or organ volume or mass, which are *extensive* (additive). (Note that the ratio of two extensive quantities is also intensive.) There are currently no known examples of functional variables showing relative invariance in mammals generally, although maximal oxygen consumption (VO_2max) ($b = 0.916 \pm 0.097$) (99% CL) and diffusion capacity ($b = 0.965 \pm 0.046$) (99% CL) might conceivably be found, with further research, to closely approach relative invariance (see Tables 20.4, 20.5).

Testing for non-linearity

When datasets are large (say $N_{sp} > 300$) and span considerable size ranges (say pWR > 4), it may happen that the data are in fact *non-linear* (do not fall on a straight line on a log–log plot), yet this *non-linearity* may be obscured by the scatter in the data (e.g., see Chapters 3, 17). There are two straightforward methods for revealing potential non-linearity. The first method is by "binning." Here we divide the x-variable for some dataset into a number of bins. Each bin will span (as nearly as possible) the same fraction of the logarithm of total (body) size range. For example, we might choose to divide the data for a given variable into 21 bins. If pWR is 5.1, then each bin will span 0.24 logarithmic units in x (5.1/21 = 0.24). The first bin includes the first point and the last bin includes the last point. For each bin we calculate the *mean* of the logarithms of the x-values and the *mean* of the logarithms of the y-values. If the data are significantly non-linear, this may be apparent when the array of mean logarithmic values is plotted (e.g., see Chapter 17, Figure 17.3). The above approach is conceptually similar to the topic "A discrete representation..." discussed above.

A second method to reveal possible non-linearity, given sufficiently large samples, is to *partition* the data into two or more disjoint groups, based on body size. For example, we might compute a best-fit LSQ line for all mammals in a dataset with body masses *less than* 100 g, and a second for body masses *greater than* 1,000 g (see Chapter 17). If the slopes for the two groups are significantly different (say at the 99% CL) we have reason to suspect the data are non-linear. When data are suspected of being non-linear, inferences based on the outcomes of ordinary LSQ analysis (e.g., correlation coefficients) should be treated with caution. In some cases non-linear data can be linearized by an appropriate transformation (e.g., see Chapter 17). It is, however, not within the scope of this work to consider non-linear LSQ analysis any further.

Variation in scaling parameters with units

Here we look at the dependence of the intercept (a), slope (b), and their standard errors (s.e.), together with the coefficient of determination (r^2), on change of units (here in the y-variable) for linear and power functions. We assume that the dimension of the x-variable is unchanged. To test whether a parameter (intercept or slope) and the s.e. in the parameters including the coefficient of determination change, we simply replace the variable y with the variable ky, where k is a constant, representing a change in units (if we change kg to g, then $k = 1,000$). If a change in units does not affect a parameter, it means that the constant k cancels out in the expression for the given parameter. This test is applicable to those scaling variables in mammals measured on a *ratio* scale (see below).

The findings of this analysis for five parameters are shown in Table 24.5 (columns c, d, e, f, and g). Note that for the linear function ($y = a + bx$) of the five parameters we take into account, only the coefficient of determination (r^2) is independent of a change in units (Table 24.5, column g, row one). On the other hand, for a power

Table 24.5 Do selected scaling parameters change with a switch in the units for the *y*-variable?

Function	Plot	Intercept (*a*)		Slope (*b*)		r^2
		a	s.e.	*b*	s.e.	
a	b	c	d	e	f	g
$y = a + bx$	lin–lin	yes	yes	yes	yes	no
$y = ax^b$	log–log	yes	no	no	no	no

lin–lin = linear–linear.

Table 24.6 A characterization of idealized spatial dimensions

Entity	Dimension	Exponent	Minimum number of coordinates
a	b	c	d
Point	$[L^0]$	0	0
Line	$[L^1]$	1	1 (*X*)
Surface	$[L^2]$	2	2 (*X,Y*)
Volume	$[L^3]$	3	3 (*X,Y,Z*)

function ($y = ax^b$, $b \neq 1$), only the *intercept* changes with a change in units (Table 24.5, row two). The findings in Table 24.5 also show one of the advantages of log–log plots vs lin–lin plots in scaling studies in that, with the exception of the intercept, the several log–log parameters are invariant under a change in units. The fact that the numerical value of the intercept varies with the units employed severely limits its usefulness in scaling studies.

Spatial dimensions

The abstract concept of spatial dimension employed in mathematics refers to the *minimum* number of coordinates required to specify the location of a point within a given space (see also Chapter 19). For example, if we have a point lying on a straight line, only one coordinate (say *x*) is required to specify where it is located (see Table 24.6, row two, column d). For surfaces and volumes we require two and three coordinates, respectively. By convention the symbol for dimension is written in upper case in square brackets. For example, the commonly used symbol for the dimension of length is [L]. There are no units associated with these dimensions. We speak of a line as being one-dimensional, a surface two-dimensional, and a volume three-dimensional (see Table 24.6, column c). Note that the above conception of dimension is *independent* of the concepts of shape or size. Thus we judge this abstract concept of dimension to be more fundamental, and logically prior to the concepts of shape or size (see below).

Physical dimensions

In everyday usage, as distinct from that in mathematics, the term dimension refers to a physical entity such as length (l), mass (m), or time (t). These are often written in *lower case*. Most physical quantities are measured on a ratio scale. For example, a page in a book may measure about 22 cm by 28 cm. This means that the width of the page is 22 times a standard unit (here 1 cm). Thus a measured dimension (e.g., length or width) is a pure number (22) associated with a unit (e.g., cm). The units in which physical quantities are measured are arbitrary. It was argued in Chapters 19 and 20 that physical dimensions significantly influence the numerical values of their associated scaling exponents, while not, at least in most cases, uniquely *delimiting* them.

An abstract concept of shape

The shape or form of an object refers to certain geometrical properties of its external surface. *Shape* is unlike any other *y*-variable considered in this study. Shape can be specified quantitatively for regular objects such as a cube or rhombohedron. But this is not the case for most organs nor for the external surface of mammals. Recall that there exists no simple scale for measuring shape, and there are no units for shape. The statement that one shape is twice another is meaningless; likewise the term "shapeless" has no clear meaning. No object has zero shape. In scaling studies, biologists usually ask not what a given shape is, but whether a given shape changes with body size.

Three dichotomies

In the following discussion the reader may find it useful to keep three dichotomies in mind. The first dichotomy is concerned with *forward* and *inverse* reasoning. Forward reasoning occurs when, for example, we derive the common geometrical properties shared by two objects assumed to have the same shape. Inverse reasoning occurs when, for example, we have a log–log plot of object length as a function of object volume. If a best-fit line has a slope of 1/3, we may then infer (perhaps incorrectly) that the objects in question have the same shape (see below).

The second dichotomy arises when different volumes are used for the *independent* variable. For example, we may elect to plot body length against *body volume* (the independent variable) or organ diameter against *organ volume* (the independent variable). This usage is straightforward and does not present any inherent problems. A difficulty arises when, for example, we express the diameter of an organ in terms of body volume rather than organ volume. This dichotomy has in some cases given rise to a simple but common logical error (see below).

A third dichotomy concerns the distinction between a *model* and a *prototype*. In applied physics and in engineering a (usually small) scale-model is studied (say in a water tank) in order to make predictions of the behavior of a full-scale prototype. In mammals, there is no naturally occurring species that can serve as a model for all other adult species. A long-term goal for scaling studies may be to construct a mathematical model of an idealized 1 gram mammal which together with cognate slopes can be used as a basis for predicting the structure and function of larger mammals. In all likelihood this will be possible only in part.

Dimensionality and shape

Therefore, if biological organisms were geometrically similar, we would expect long-bone allometric exponents to be multiples of one-third.

Garcia and da Silva 2006

Variations in the external shape of bats, land mammals, and cetaceans are striking and biologically important [13]. These variations no doubt reflect, in part, adaptations towards efficient forms of locomotion in air, on land, and in water. Equally important are the shapes of the long bones, eyes, heart, and teeth. Indeed, it seems likely that natural selection has molded the shapes of most organs and tissues to make them more efficient.

The generally accepted criteria for scaling at constant shape (also termed geometric similarity, isometry, and sometimes constant proportions) are that object diameters and lengths scale as the one-third power of object volume (or mass) and areas as the two-thirds power (see above quote). In the following it is suggested that in some instances the above criteria are too stringent, in the sense that exponents need not, in principle, be precisely 1/3 or 2/3 in order to reflect constant shape. The converse situation also arises. That is, empirical slopes of nearly 1/3 or 2/3, derived from log–log plots, need not be indicative of scaling *precisely* at constant shape. In the following remarks, involving a modicum of mathematics, an effort is made to justify these remarks. Before restating the criteria used to define constant shape it is helpful to define a few technical terms first. We start with the concepts of model and prototype.

Model and prototype

When the word *model* is used in physical scaling studies, one usually has in mind a scale-model, some of the empirical properties of which are to be scaled to predict the properties of a (usually larger) *prototype* (Chapter 3). For example, measurements of drag made in a water tank using a scale-model (e.g., of a ship) may be employed to predict mathematically the drag of a full-size prototype. Here one is usually concerned with one model and one prototype; in bioscaling studies we may construct a hypothetical idealized mammal (say of 1 g body mass) as a model and any number of other adult mammals (of varying body size) as prototypes (see Figure 24.6).

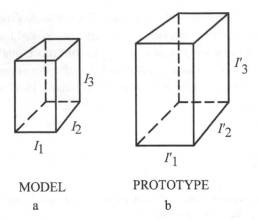

MODEL

a

PROTOTYPE

b

Figure 24.6 Definition of model and prototype.

Degrees of freedom for shape

It is a common practice to use *cube*s to derive the mathematical relations consonant with scaling at constant shape and varying size. This makes for simplicity, but for the logical problem that all cubes are the same shape, independent of size, and adult mammals are not. It is convenient to think of cubes as having zero *degrees of freedom* with respect to shape. The simplest regular 3-D objects that may or may not exhibit constant shape at the same or different sizes are right prisms (cuboids). By degrees of freedom we mean the number of parameters needed to define the size of an object, less one. Thus, the size of a cube is defined by one parameter (side-length), so it has *zero* degrees of freedom with respect to shape. A right prism requires three parameters to define its size and therefore has two degrees of freedom with respect to shape.

Homologous lengths

With reference to Figure 24.6, we define three ordered pairs of *homologous* (or corresponding) lengths as: (l_1, l'_1), (l_2, l'_2), (l_3, l'_3). For each pair of homologues, one length (e.g., l_1) represents a side-length of the model and one (e.g., l'_1) that of the length of the homologous side of a prototype. Note that the lengths l_1, l_2, l_3 are assumed to share a common origin and to be *mutually perpendicular*. This conception of homology plays a key role in the following derivation of criteria for scaling at constant shape.

Magnification factors

We are at liberty to generate from a given scale-model a prototype of *arbitrary* shape by multiplying each side-length of the model by a separate *magnification* (or *scale*) factor. Thus one writes:

$$l'_1 = m_1 l_1 \qquad l'_2 = m_2 l_2 \qquad l'_3 = m_3 l_3 \tag{24.34}$$

where (m_1, m_2, m_3) are, by assumption, different magnification factors. It is convenient to rewrite relations (24.34) in the equivalent form:

$$m_1 = l'_1/l_1 \qquad m_2 = l'_2/l_2 \qquad m_3 = l'_3/l_3 \tag{24.35}$$

Thus each magnification factor is simply the ratio of a pair of homologous lengths. Hence magnification (or scale) factors are pure numbers.

Aspect ratios

Aspect ratios (AR) (or *similarity criteria*) may be employed to define the shape of a model or a prototype. Aspect ratios are computed as the ratio of two lengths [14]. For example, in the US and UK, the aspect ratios of TV screens are calculated as the ratio of the width to the height of the display and are typically 4:3 or 16:9. The aspect ratio for wings (animate or inanimate) is defined as wingspan over wing width (measured front to back). Gliding birds tend to have wings with high aspect ratios. With respect to Figure 24.6, panel a, we may define three aspect ratios for a *model* by the relations:

$$AR_1 = l_1/l_2 \qquad AR_2 = l_1/l_3 \qquad AR_3 = l_2/l_3 \tag{24.36}$$

As a specific example, the length of a long bone divided by its diameter (measured, say, at mid-shaft) may be termed an aspect ratio. Observe that each aspect ratio (or similarity criterion) is a pure number (as was the case for magnification factors above).

The reader may easily show from relation (24.36) that:

$$AR_3 = AR_2/AR_1 \tag{24.37}$$

That is, a generalized shape, for example a right prism, has only two independent aspect ratios. A cube has aspect ratios of $(1, 1, 1)$; a right cylinder or a square prism has aspect ratios of $(1, AR_2, AR_3 = AR_2)$, and a right prism has aspect ratios of $(AR_1, AR_2, AR_3 = AR_2/AR_1)$.

It is apparent that we may also define *degrees of freedom* for shape as the number of independent (freely varying) aspect ratios. Hence a cube has zero degrees of freedom, a right cylinder has one degree of freedom, and a right prism has two degrees of freedom with respect to shape. Adult land mammals and bats, among others, are likely to require at least three degrees of freedom to adequately describe the scaling of their external shape.

Criteria for scaling at constant shape

To determine objectively whether any two objects have the *same shape*, we need criteria based on measurements. What are those criteria? The simple answer is that two objects (of the same or differing size) have the same shape if pairs of *homologous* lengths have the same *aspect ratios*. We denote the three aspect ratios of any given *prototype* by AR'_1, AR'_2, AR'_3. Then we have:

$$AR'_1 = l'_1/l'_2 \quad AR'_2 = l'_1/l'_3 \quad AR'_3 = l'_2/l'_3 \tag{24.38}$$

See Figure 24.6. To achieve constant shape as between a model and a prototype, we enforce the following conditions:

$$AR'_1 = AR_1 \quad AR'_2 = AR_2 \quad AR'_3 = AR_3 \tag{24.39}$$

Hence two objects have the same shape if homologous aspect ratios are identical. This requirement implies *absolute invariance* for each of the three aspect ratios AR'_1, AR'_2, AR'_3.. That is, the aspect ratios are independent of object size.

From relations (24.34, 24.38) and (24.39) we have that:

$$AR'_1 = l'_1/l'_2 = m_1 l_1/m_2 l_2 = (m_1/m_2)\cdot(l_1/l_2) = (m_1/m_2)AR_1 \tag{24.40}$$

From relation (24.40) we infer that the prototype aspect ratio (AR'_1) will be equal to the model aspect ratio (AR_1) if and only if the two magnification factors (m_1, m_2) are equal:

$$m_1 = m_2 \tag{24.41}$$

If we also require that $AR'_2 = AR_2$ and $AR'_2 = AR_2$ we find that:

$$m_1 = m_2 = m_3 = m \tag{24.42}$$

Thus the condition for scaling at *invariant shape*, namely constant aspect ratios, implies that each side-length in the prototype is equal to the homologous length in the model multiplied by a constant magnification factor (m). That is, the prototype has the same shape as does a model because it is simply a uniformly magnified version of it. This result is intuitively self-evident. In geometry, a magnification factor of 1 is referred to as an *isometry*. More generally in mathematics, an isometry is a transformation wherein lengths are unchanged. The usage is broadly similar in physiology where an *isometric contraction* of muscle is one at constant length.

On the contrary, in *bioscaling* studies the term isometry came to mean scaling in accord with geometric similarity or constant shape (i.e., implying absolute invariance of aspect *ratios*) at varying body volumes. As a corollary, slopes of 1/3 for lengths, 2/3 for surface areas, and one for volumes, as inferred from log–log plots, have often been cited in the mammalian scaling literature as evidence for isometry. As argued below, such inferences may, at least in some cases, be imprecise.

Scaling of model and prototype volume in terms of side-lengths

We denote the volume of a model and a prototype by V_m and V_p, respectively. Then we may write:

$$V_m \propto l_1 \cdot l_2 \cdot l_3 \text{ and } V_p \propto l'_1 \cdot l'_2 \cdot l'_3 \tag{24.43}$$

Relations (24.43) are equalities for right prisms but only proportionalities for more general shapes.

Prototype side-lengths as a function of prototype volume

We have in mind a series of prototypes differing in volume. We may assume that prototype side-lengths (l'_1, l'_2, l'_3) scale as power functions of prototype volume (V_p), with exponents of b_1, b_2, and b_3, respectively. Then we have:

$$l'_1 \propto V_p^{b_1} \quad l'_2 \propto V_p^{b_2} \quad l'_3 \propto V_p^{b_3} \tag{24.44}$$

It follows from relations (24.43 and 24.44) that:

$$l'_1 \cdot l'_2 \cdot l'_3 \propto V_p^{b_1} \cdot V_p^{b_2} \cdot V_p^{b_3} \propto V_p^{b_1+b_2+b_3} \propto V_p \tag{24.45}$$

Relations (24.45) obtains if and only if:

$$b_1 + b_2 + b_3 = 1 \tag{24.46}$$

Relation (24.46) is required for logical consistency; it does not imply scaling at constant shape.

Criteria for whole body scaling at constant shape

We now derive the conditions for scaling at constant shape relative to whole body volume (as opposed to organ volume). This is achieved by requiring that the homologous aspect ratios for any prototype be identical to those for a model. From relations (24.35, 24.39, and 24.44) we find:

$$AR'_1 = l'_1/l'_2 \propto V_p^{b_1}/V_p^{b_2} = V_p^{(b_1-b_2)} = AR_1 = \text{constant} \tag{24.47}$$

In order for the term $V_p^{(b1 - b2)}$ to be constant one must have:

$$b_1 = b_2 \tag{24.48}$$

Exactly the same argument applies to AR'_2 and AR'_3. Thus we have:

$$b_1 = b_2 = b_3 \tag{24.49}$$

Combining relations (24.46) and (24.49) we find:

$$b_1 = b_2 = b_3 = 1/3 \tag{24.50}$$

We have shown that if a series of prototypes and a model all have the same shape, then the side-lengths (or diameters) will scale as the one-third power of body volume. This is the same result as is usually derived from a cube, where one is assuming constant shape from the beginning. Here we have derived this condition, at greater length, without assuming constant shape *ab initio*. By an extension of the above argument, one can show that constant shape implies that external surface area scales as the two-thirds power of volume.

Inverse reasoning applied to whole objects

So far we have discussed certain logical relations associated with constant shape. It is convenient to refer to this as reasoning in the *forward direction*. In actual practice in scaling studies we are usually given a log–log plot of a length or area as a function of body volume or mass. If the slope of the LSQ best-fit line is 1/3 or 2/3, we infer that the results are consistent with constant shape. This may be termed *inverse* reasoning. However, the underlying assumption is open to doubt. Given a set of numbers, we can compute a mean; however, given (only) a mean, one cannot infer the original set of numbers. Or given the coordinates of the center of a straight-line segment and its slope we can infer the intercept. However, given an intercept and a slope and no other information, we cannot infer the coordinates of the mid-point. Amongst the many known mathematical transformations, it is common to find ones for which there is no inverse.

We consider three examples where *inverse reasoning* as applied to scaling of shape may lead to incorrect inferences. We look first at the scaling of *ellipsoid* diameter as a function of volume, where an empirical exponent near 1/3 is usually taken as evidence of scaling at constant shape.

In Table 24.7, columns a, b, and c give the major and minor diameters and volumes, respectively for a series of ellipsoids (rows 1 to 6). The exponents for the scaling of major and minor diameters computed on ellipsoid volume are given in the bottom row (columns a and b). Note that the exponent for the minor diameter (0.320 ± 0.008) (99% CL) (column b) would usually be interpreted as consistent with geometrical similarity. Still, the *aspect ratios* (see Table 24.7, column d) vary from row one to row six by a factor of 2, implying an appreciable change in shape. That is, what might be viewed as small deviations in exponents from 1/3 (Table 24.7, columns a, b, bottom row) may

Table 24.7 Scaling of diameter as a function of volume for six ellipsoids of varying shape

Row/Column	Major semi-axis	Minor semi-axis	Volume	Aspect ratio
	a	b	c	d
1	3	1.74	38.2	1.7
2	10	5.26	$1.2 \cdot 10^3$	1.9
3	30	13.33	$2.2 \cdot 10^4$	2.3
4	100	44.44	$8.3 \cdot 10^5$	2.3
5	300	109.09	$1.5 \cdot 10^7$	2.8
6	1,000	294.2	$3.6 \cdot 10^8$	3.4
Exponent	0.360 ± 0.015	0.320 ± 0.008	1	-

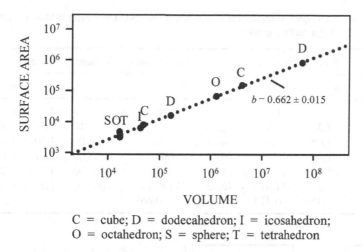

C = cube; D = dodecahedron; I = icosahedron;
O = octahedron; S = sphere; T = tetrahedron

Figure 24.7 Scaling of surface area for regular solids of different shapes.

signify appreciable changes in external shape when applied over a significant size range
(pWR = 7.0) (column c).

Sometimes an empirical slope of 2/3 for the scaling of surface area on volume is
invoked as evidence consistent with scaling at constant shape. A plot is shown in
Figure 24.7 of surface area against volume for regular solids having six different shapes.
The units for surface area and volume in Figure 24.7 are arbitrary. The three points in a
vertical row (labeled S, O, and T) in the lower left-hand corner of Figure 24.7 stand for
sphere, octahedron, and tetrahedron, respectively. These regular solids span 3.6 orders
of magnitude in volume. The LSQ slope for surface area is 0.662 ± 0.052 (99% CL).
Note that the aspect ratios (column d) are computed as the ratio of entries in column a
over those in column b. This illustration raises doubts as to whether a slope of nearly 2/3
necessarily implies scaling at constant shape.

The above argument is a simple extension of one given previously [4]. In a word, the
scaling of surface area may be rather insensitive to moderate changes in shape.

Above we discussed changes in shape in relation to the physical dimensions of either
diameter (length) or surface area. In the next example (see Table 24.8) we discuss
changes in shape in relation to both length and surface area with increasing volume. We
start with a cube (row 1). We increase the side-lengths of the cube in accord with
exponents (on volume) of 0.314, 1/3, and 0.353 to produce right rectangular prisms (or
parallelepipeds) (columns a, b, c, bottom row). Columns a, b, and c show the successive
side-lengths (rows one to six). The surface areas and volume (computed from columns
a, b, and c) are given in columns d and e, respectively. The aspect ratios computed from
columns a, b, and c are given in columns f, g, and h.

The exponents for the scaling of side-lengths (Table 24.8, columns a, b, and c) on
volume are given in columns a, b, and c, bottom row; the exponent for *surface area*
scaling on volume is given in column d, bottom row. Note that the two exponents

Table 24.8 Scaling of a right rectangular prism at different shapes but with exponents on volume near 1/3 (for lengths) and 2/3 (for surface area)

	Side-lengths			Surface area	Volume	Aspect ratios		
Column/Row	a	b	c	d	e	f	g	h
1	1	1	1	6	1	1	1	1
2	2.25	2.06	2.15	27.9	10	1.09	1.05	0.96
3	5.08	4.24	4.64	129.6	100	1.20	1.09	0.91
4	11.46	8.73	10.00	603.7	1,000	1.31	1.15	0.87
5	25.82	17.97	21.54	2,815.5	10,000	1.44	1.20	0.83
6	58.21	37.01	46.42	13,148	100,000	1.57	1.25	0.80
Exponents	0.353	0.314	1/3	0.668	1	-	-	-

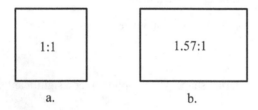

Figure 24.8 Aspect ratios for a cube and a right rectangular prism.

(0.353 and 0.314) (columns a and b, bottom row) for the scaling of side-lengths on volume each differ from an exponent of 1/3 (column c, bottom row) by ±0.02. Observe that surface area scales on volume with an exponent of 0.668 (column d, bottom row). Also note that volume (column e) spans five orders of magnitude. With each increase in volume (rows two to six) there is an increase (columns f, g) or decrease (column h) in the aspect ratios.

The greatest change in aspect ratio is seen in Table 24.8, column f, where there is an overall increase in the aspect ratio by a factor of 1.57. This change in shape is illustrated schematically in Figure 24.8. The results shown in Table 24.8 support the contention that seemingly small changes in scaling exponents (e.g., ±0.02) are compatible with obvious changes in shape over wide size ranges.

The above three examples (Tables 24.7, Figure 24.7, Table 24.8) suggest that inferences as to scaling at constant shape made from LSQ computations of log–log transformed data may be open to doubt. It can be shown, however, that if the exponents for lengths (or diameters) are nearly 1/3 *and* those for areas nearly 2/3 then scaling at constant shape is a reasonable inference.

Criteria for scaling at constant shape for embedded objects

In the above, we examined the relations between physical dimensionality and shape for whole objects (e.g., bodies or organs). The diameters, lengths or surface areas of a given object were expressed in terms of the object's volume. In biology, it often happens that the physical dimensions of an organ (e.g., a long bone) are expressed in terms of body volume (or mass) rather than in terms of organ (e.g., bone) volume or mass. This practice has often led to a simple error in logic.

It is convenient to think of most organs (with some exceptions, such as skin or appendages) as "embedded" within the body. This term is meant to imply that the physical dimensions (diameters, lengths) of an organ are free to vary, within limits, independently of body size. Possible examples are heart, kidney, and lung. Skin may be regarded as semi-embedded, as only skin thickness is potentially free to vary independently of body size. In the following it will be shown with numerical examples that scaling at *constant shape* need not, and usually does not, require scaling of organ lengths or diameters precisely as the one-third power of body volume or mass, as often assumed.

We may define three mutually perpendicular lengths (l_1, l_2, l_3) of an organ in terms of body volume (BV) by the relations:

$$l_1 \propto BV^{b_1} \quad l_2 \propto BV^{b_2} \quad l_3 \propto BV^{b_3} \tag{24.51}$$

where b_1, b_2, and b_3 are exponents of body volume (see relations (24.44)). Organ volume (OV) is related to body volume via the relations:

$$OV \propto l_1 \cdot l_2 \cdot l_3 \propto BV^{(b_1+b_2+b_3)} \propto BV^b \tag{24.52}$$

$$\text{where } b = b_1 + b_2 + b_3 \tag{24.53}$$

It is important to recognize that there is no mathematically necessary relationship between organ and body volume. Within the limits of body volume, scaling of organ size may, in principle, vary greatly. However, if we assume that organ volume scales as a constant fraction of body volume ($b = 1.0$), then the following condition must be satisfied:

$$b_1 + b_2 + b_3 = 1 \tag{24.54}$$

That is, the sum of the exponents must be 1. This condition does not require scaling at constant organ shape; consider $b_1 = 1.05$, $b_2 = 1$, and $b_3 = 0.95$, which sum to 1 but imply varying shape. There is no mathematically necessary relationship between *relative invariance* (here denoting constant volume-fraction) and scaling at *constant shape*. Relative invariance and constant shape are *independent* constraints. In order to ensure scaling at constant organ shape we must impose the condition:

$$b_1 = b_2 = b_3 = b/3 \tag{24.55}$$

Observe that relation (24.55) does not require scaling at constant volume-fraction ($b = 1.0$) [13]. That is, the slopes for an embedded object may be consistent with constant shape (i.e., equal to one another, though differing from 1/3). If we enforce the condition of scaling at constant volume-fraction *and* constant shape, we have:

$$b = b_1 + b_2 + b_3 = 1 \text{ and } b_1 = b_2 = b_3 = b/3 = 1/3 \tag{24.56}$$

Table 24.9 Hypothetical scaling of organ dimensions (lengths) at constant shape for four different scaling regimes, expressed in terms of either organ volume (OV) or body volume (BV)

	Organ				Organ aspect ratios			Body volume		
	Side-lengths			Volume						
	l_1	l_2	l_3	OV	$AR_1 =$ l_1/l_2	$AR_2 =$ l_1/l_3	$AR_3 =$ l_2/l_3	BV_1	BV_2	BV_3
Column/Row	a	b	c	d	e	f	g	h	i	j
1	2	3	4	24	2/3	1/2	3/4	34.2	120	139.4
2	4	6	8	192	2/3	1/2	3/4	344	960	907.7
3	8	12	16	1,536	2/3	1/2	3/4	3,471	7,680	5,909
4	10	15	20	3,000	2/3	1/2	3/4	7,302	15,000	10,800
Exponents of OV	1/3	1/3	1/3	1.000	-	-	-	-	-	-
Exponents of BV_1	0.300	0.300	0.300	0.900	-	-	-	-	-	-
Exponents of BV_2	1/3	1/3	1/3	1.000	-	-	-	-	-	-
Exponents of BV_3	0.370	0.370	0.370	1.110	-	-	-	-	-	-

BV_1, BV_2, and BV_3 represent three hypothetical arrays of body volume selected to give exponents for organ volume (OV) on body volume of 0.900, 1.000, and 1.110 (see column d, bottom three rows).

The above points are illustrated in Table 24.9 – structured in a slightly intricate fashion so as to demonstrate several different but interrelated points. First, we review the overall organization of the table. The hypothetical side-lengths (l_1, l_2, l_3) of an "organ" are given in columns a, b, c, rows one to four. Second, organ volume and the corresponding three aspect ratios for the side-lengths are given in columns d, e, f, and g, respectively, rows one to four. Third, three series of hypothetical "body" volumes (BV) are given in columns h, i, and j, rows one to four.

By construction, the side-lengths (l_1, l_2, l_3) each scale as the one-third power of *organ volume* (OV), implying scaling at constant shape (see columns a, b, and c, fourth row from the bottom). This is confirmed by the fact that the aspect ratios (columns e, f, g, rows one to four) do not change with organ volume (column d, rows one to four).

The way in which *organ volume* scales with *body volume* (BV_1, BV_2, or BV_3) is shown in column d, bottom three rows of Table 24.9. When organ volume scales as the 0.900 power of body volume (BV_1) then the side-lengths each scale with BV_1 as the 0.300 power (0.9/3) (see columns a, b, c, third row from the bottom).

If organ volume scales as the first power of body volume (BV_2), it follows that the side-lengths obey the one-third power rule (see columns a, b, c, second row from the bottom), consistent with scaling at *both* constant volume-fraction and constant shape. Finally, when organ volume scales as the 1.110 power of body volume (BV_3), the side-lengths each scale as the 0.370 power of body volume (columns a, b, c, bottom row) (1.10/3 = 0.370).

In sum, it is shown numerically in Table 24.9 that organ volume scaling as the 0.900, 1, and 1.110 powers of body volume may each be consistent with scaling at constant shape, as shown by *constant aspect ratios* (columns e, f, g, rows one to four). The reader may easily confirm the above results by carrying out LSQ calculations on the logarithmically transformed data shown in Table 24.9 (columns a, b, c, d, and columns h, i, j, rows one to four). It is *not* the case, as a rule, that organs scaling at constant shape must have diameters that scale with body volume or mass with exponents that are exact multiples of 1/3. For the brain the exponent would no doubt be nearer 0.25. (See above quote from Garcia and da Silva.)

The key point in the above analysis is that all organ side-lengths (or diameters) scaling at constant shape have exponents of *organ* volume of 1/3. There is no dispute about this. But when these same exponents are expressed in terms of *body* volume (rather than organ volume) then the exponents for organ side-lengths (or diameters) scaling at constant shape become $b/3$, where b is the exponent by which organ volume scales as a function of body volume. It has been generally assumed, in effect, that organs scale at constant volume-fraction ($b = 1$), but this is true only for a few organs and tissues (e.g., skeletal muscle mass, blood volume). When, as usual, the exponent b $\neq 1$, then side-lengths scale as the $b/3$ power of body volume for constant shape, not the one-third power.

Summary

Linear, exponential, and power functions are compared. It is shown that a power function can always be cast in non-dimensional form. It is argued that the wider the size range (in x) the more likely it is, *assuming linearity*, that a power function is the simplest two-parameter representation of scaling data for adult mammals. An expression is given for the percent change in y-values as a function of percent change in x-values and slope. A simple algorithm is derived relating a maximal *structural* slope (for slopes greater than 1) to cognate values of intercept and slope.

It is shown that the discrete terms of two geometric series may be used to represent a power function at any specified number of points. An equation is derived expressing the divergence at the end-points of two lines diverging from a common intercept and extending over eight orders of magnitude in the x-variable. Evidence is presented that the Haldane restriction is unlikely to be a problem in most mammalian scaling studies. It is argued that some semi-log plots are potentially misleading.

Two types (absolute, relative) of *structural* invariance are briefly reviewed. It is suggested that both forms of structural invariance are common in the interspecific scaling of adult mammals. Some *functional* y-variables scale in accord with absolute invariance, but at present none is known to clearly scale in harmony with relative invariance. Two possible tests for non-linearity in datasets are put forward. These tests involve "binning" and partitioning the data. It is shown that the major scaling parameters associated with a power function, with the exception of the intercept, are independent of a change in units.

The concepts of dimensions, models, prototypes, degrees of freedom for shape, and magnification factors are examined. Forward and inverse reasoning with respect to scaling at constant shape are defined. The classical criteria for scaling at *constant shape* at the whole body and organ levels are reviewed. It is argued that empirical log–log slopes of near to 1/3 and 2/3 need not imply scaling precisely at constant shape. It is also shown that the scaling of an organ diameter on body volume or mass at constant organ shape need not imply an exponent of 1/3. Indeed an exponent for an organ diameter on body volume scaling at constant shape could be as low as 0.25 (e.g., for the adult brain).

References

1. Adolph, E.F. (1949). Quantitative relations in the physiological constitutions of mammals. *Science*, **109**:579–585.
2. Western, D. (1979). Size, life history and ecology in mammals. *African Journal of Ecology*, **17**:185–204.
3. Lumer, H. (1942). On the significance of the constant b in the law of allometry $y = bx^a$. *American Naturalist*, **76**:364–375.
4. Prothero, J. (1986). Methodological aspects of scaling in biology. *Journal of Theoretical Biology*, **118**:259–286.
5. Brody, S. (1945). *Bioenergetics and Growth*. New York: Reinhold.
6. Gould, S.J. (1979). One standard lifespan. *New Scientist*, **81**:388–389.
7. Calder, W.A. (1984). *Size, Function, and Life History*. Cambridge, MA: Harvard University Press.
8. Kleiber, M. (1961). *The Fire of Life: An Introduction to Animal Energetics*. New York: John Wiley & Sons.
9. Huxley, J.S. (1972). *Problems of Relative Growth*. New York: Dover Publications.
10. Turner, J.S. (1988). Body size and thermal energetics: how should thermal conductance scale? *Journal of Thermal Biology*, **13**:103–117.
11. Schmidt-Nielsen, K. (1979). *Animal Physiology: Adaptation and Environment*. Cambridge: Cambridge University Press.
12. Bartels, H., Bartels, R., Baumann, R. *et al.* (1979). Blood oxygen transport and organ weights of two shrew species (*S. etruscus* and *C. russula*). *American Journal of Physiology*, **236**: R221-R224.
13. Prothero, J. (1992). Scaling of bodily proportions in adult terrestrial mammals. *American Journal of Physiology*, **262**:R492–R503.
14. Alexander, R.McN. (1989). *Dynamics of Dinosaurs and Other Extinct Giants*. New York: Columbia University Press.

Part V

A broader view

25 A sense of scale

We have become experimental creatures of our own making. This experiment has never been tried before. And we, its unwitting authors, have never controlled it. The experiment is now moving very quickly and on a colossal scale.

Wright 2004

What shall we do? No one yet knows. Unless we think about fundamentals, our specific measures may produce new backlashes more serious than those they are designed to remedy.

White 1967

Scaling may be defined as the study of change, or lack of change, in the properties of a system with change in size or sometimes duration. A scaling approach was perhaps first consciously applied by the architects who designed the Great Pyramids 4,500 years ago (Chapter 3). Scaling in the modern period follows the work of Galileo (1564–1642) (Chapter 12). Until the twentieth century the scaling perspective was mostly confined to applied mathematics, physics, and engineering. Since then scaling concepts, as variously defined, have been applied, for example, to architecture [1], biology (see Appendix A), human institutions [2, 3], human language [4], landscape architecture [5], the design of organ pipes [6], and polymer physics [7]. As these few examples suggest, the arena of potential applications for scaling concepts is surprisingly wide.

Scaling is mainly concerned with regular (coordinated) changes as well as non-changes in specific parameters of a given system with increasing size or period of time. The approach is inherently quantitative. Thus, in a series of objects of constant shape and varying size, we expect surface area to increase systematically as the two-thirds power of volume (Chapter 24). Scaling is also concerned with *limits*, or *scale effects* (Chapter 3). That is, all scaling phenomena are expected to exhibit deviations from regular scaling at sufficiently large or small sizes, reflecting qualitative rather than strictly quantitative changes in the system under study.

For instance, injecting carbon dioxide into the atmosphere (see below) increases global surface temperatures. One effect of increased mean surface temperature is to melt ice and snow. Ice and snow have high *albedos*, meaning they reflect a higher proportion of incident sunlight back into space than do earth or water, which have lower albedos. Consequently, when ice and snow melt more heat is absorbed on average, thereby further raising the surface temperature, melting even more snow and ice. This self-reinforcing

Table 25.1 An alphabetical assortment of current global challenges

Antibiotic resistance	Jevon's paradox [a]	Soil erosion
Biodiversity loss	Killer bees	Tar sands
Climate change	Logging	Unlimited economic growth
Disease	Malaria	Voter turnout
Environmental degradation	Neoliberal dogma	Water shortages
Famine	Overconsumption	Xylene pollution
Greenhouse gases	Population growth	Yellow river silt
Human health	Quality of water	Zinc pollution
Inequality (wealth & income)	Rising sea-levels	-

[a] Jevon's paradox means that increases in efficiency, such as in fuel efficiency, may (by lowering costs) lead to increased (rather than decreased) consumption. The net environmental improvement may be zero.

process, left unchecked, will eventually result in the melting of all our planet's surface ice and snow. That is, about two-thirds of the earth's stock of fresh water will have vanished into the seas (see Table 25.7, column c, rows three and four) ($100 \cdot 1.6/2.5 = 64$).

A possibly critical emerging domain for scaling studies is the collection of *global* problems now threatening the survival of human civilization. An assortment of these problems is itemized in Table 25.1. That these problems are pervasive can be seen by the fact that the reader can easily construct an independent set of global challenges in less time than it takes to complete a crossword puzzle (e.g., Arctic thaw, boom-and-bust economic cycles, carbon emissions, dirty oil...). These problems are interdependent and often complex.

It is argued here, as seems self-evident, that the majority of difficulties we face as a global civilization are mainly "matters of scale" (see above quote from Wright). When total human numbers were a million or under (some 300,000 years ago) and consumption not much above subsistence, human impact on the environment was no doubt minimal and short-lasting. Now, by mid-century (2050), we will approach a population of ten billion (an increase by a factor of 10,000) ($10^{10}/10^6 = 10^4$) (by some recent accounts as of 2014, the human population will reach 11 billion by the year 2100); human per capita consumption has increased by a factor of 14 or more (see Figure 25.4, panel b). Thus, the total impact of "rogue primates" on our planet has increased by a factor of roughly 140,000 in a "mere" 300,000 years. We are now in a position to substantially destabilize the entire biosphere through widespread carelessness, greed, and indifference.

The major physical components of the *biosphere* are the *atmosphere* (air), the *hydrosphere* (water), and the *pedosphere* (soil). In most published works on the global problems we face, only one of these components is discussed in detail. Here we discuss only some major aspects of our impact on each of these components. The primary impacts will be discussed chiefly in terms of the dimensions of mass, area, and time. A more wide-ranging scaling analysis would require the consideration of numerous other dimensions, examples of which are given in Table 25.2; this table points to the

Table 25.2 Selected dimensions for a scaling analysis of pressing local and global problems

Ecological	Economic	Political	Spatial	Temporal
Organism	Self-employed individual	Family	Local	Years
Population	Family business	Village, city	Regional	Decades
Community	Small corporation	State/province	National	Centuries
Ecosystem	National corporation	Nation	Continental	Millennia
Biosphere	Multinational corporation	United Nations	Global	MY

MY = million years.

Table 25.3 The masses of our universe, our galaxy, and our sun

Entity	Mass (g)	Percent	Source
Visible universe*	2×10^{57}	100	[8]
Milky Way galaxy	3.6×10^{44}	1.8×10^{-11}	[8]
Sun (or solar system)	2×10^{33}	10×18^{-22}	[9]

* The figure given is for the mass of the *visible* universe. Because of "dark matter" and "dark energy," this figure is considered to be only a few percent of the total mass of the observable universe.

fact that lasting solutions to the problems discussed below are likely to be intricate and multidimensional, requiring the marriage of specialized and integrated insights and understandings.

OUR PLACE IN THE UNIVERSE

The earth, our unique home in the vastness of space, is in crisis.
Brady and Weil 2002

We begin with a few remarks aimed at defining where we fit physically into the universe. It will be seen from Table 25.3 that our sun has a mass smaller than that of the visible universe by a factor of 10^{24}. Large numbers such as this are often taken to imply that our sun and our planet are insignificant in the overall scheme of things; the universe does not care. On this view neither life nor death has any cosmic significance; life has no external purpose. Many workers, especially exobiologists, believe life will prove to be common in the observable universe. The discovery of life elsewhere in our universe would surely rank as the most outstanding event in human intellectual history.

Nevertheless, the possibility exists that life on this planet is unique in our universe (see above quote from Brady and Weil). The transition from randomness to the great intrinsic order of life anywhere seems astonishing. Darwin thought that life originated in a small warm pond. But the complexity of life suggests (to me) that it may have evolved in several quite different successive environments, perhaps first in hot hydrothermal vents in the ocean floor, then at cooler clay interfaces, and perhaps lastly in warm ponds or lakes. A sequential transfer of prototypical life-forms essentially intact from one environment to other quite different ones could have been exceedingly improbable (relative to, say, the probability of finding liquid water on other planets). A version of this same argument, termed *panspermia*, posits that life originated elsewhere in the universe and was transported to earth by stellar winds as spores or microorganisms. On the other hand, if life on our planet eventually proves to be unique, insofar as that can be determined, that should provide a compelling reason beyond self-interest to preserve and possibly enhance it.

PLANET EARTH AND THE BIOSPHERE

The masses and relative sizes (in percent) of the atmosphere, biosphere, hydrosphere, and pedosphere in relation to the *mass of the earth* are summarized in Table 25.4. Note that the four "spheres" are each much less than 1% of our planet's mass (column three). Relative to the earth as a whole, the biosphere is a precious but highly perishable veneer.

THE ATMOSPHERE

Our planet's atmosphere consists chiefly of nitrogen (75.5%), oxygen (23.1%), water (0.2%), and carbon dioxide (0.06%), each expressed relative to the *reference* mass of the dry atmosphere (Table 25.5, columns b and c). This composition is very different

Table 25.4 The mass of the earth and those components most relevant to life

Entity	Mass (g)	Percent	Source
Earth	6×10^{27}	100	[10]
Atmosphere (g)	5×10^{21}	8×10^{-5}	[10]
Biosphere	1.8×10^{18} [a]	3×10^{-8}	[11]
Hydrosphere	1.4×10^{24}	2×10^{-2}	[10]
Pedosphere [b]	$10^{19} - 10^{20}$	$1.7 \times (10^{-7} - 10^{-6})$	

[a] Mass derived from estimated quantity of global dried plant material.
[b] Assumed mean soil thickness of 30–300 cm; assumed mean soil density of 1.25 g/cm^3; for area of cultivable land see Table 25.12, row two.

Table 25.5 The mass of the atmosphere and some of its components

Column/Row	Entity	Mass (g)	Percent	Source
	a	b	c	d
1	Dry atmosphere	5.1×10^{21}	100	[10]
2	Carbon dioxide [a]	3×10^{18}	0.06	[10]
3	Nitrogen	3.9×10^{21}	75.5	[10]
4	Oxygen	1.2×10^{21}	23.1	[10]
5	Water	1.2×10^{19}	0.2	[10]

[a] Adjusted to 2013 levels.

from those of neighboring Mars and Venus, which are mostly carbon dioxide [12]. It is believed that the current concentration of oxygen in the atmosphere was reached about one billion years ago [13]. The high oxygen level in our atmosphere (relative to our fellow planets) reflects the photosynthetic activity of plants. Whereas (hyperbaric) oxygen is generally toxic for living organisms (Chapter 18), it is nevertheless essential for aerobic metabolism, permitting much higher levels of energy transduction than anaerobic (oxygen-free) metabolism.

Carbon dioxide

Present-day atmospheric burdens of these two important greenhouse gases [carbon dioxide, methane] *seem to have been unprecedented during the last 420,000 years.*

Petit et al. 1999

The carbon dioxide *concentration* in the atmosphere has been stable (at about 280 parts per million (ppm)) for hundreds of thousands of years; it started to increase rapidly around 1800 AD, at the beginning of the industrial revolution (Figure 25.1). See above quote from Petit *et al.* Recall that the atmosphere is essentially transparent to short wavelengths of sunlight. Sunlight absorbed at the earth's surface is (in part) radiated back into the atmosphere at longer wavelengths (infra-red), where it is trapped by greenhouse gases such as water vapor, carbon dioxide, and methane. Methane (Figure 25.1, panel b) is a more deleterious greenhouse gas than is carbon dioxide; however, a molecule of methane is only resident in the atmosphere on average for 12 years, whereas the mean residence time of carbon dioxide in the atmosphere is measured in centuries. Accordingly, the steadily increasing level of carbon dioxide in the atmosphere is of great concern now and will remain so well into the future (Figure 25.1, panel a).

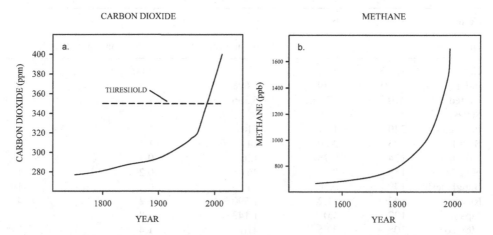

Figure 25.1 Atmospheric levels of carbon dioxide and methane over time [15, 19]. ppb = parts per billion (10^9); ppm = parts per million (10^6).

The bygone level of carbon dioxide in the atmosphere (280 ppm) long maintained our planet at a mean surface temperature of 15 °C [15]. Without this level of atmospheric carbon dioxide, the earth's surface would freeze over. Variations in the earth's *surface temperature*, based on an analysis of ice cores, are found to have closely tracked carbon dioxide levels in the atmosphere for at least the last 400,000 years [16].

The present rise in carbon dioxide levels is due mainly, although not exclusively, to the burning of *fossil fuels*, some 95% of which occurs in the northern hemisphere [15]. About 70% of the global warming to date is attributable to carbon dioxide [15]. The threshold shown in Figure 25.1a indicates a level (already passed) at which an increase (of say 1 °C) in the earth's mean surface temperature may be effectively irreversible for centuries to come [17]. Over a sufficiently long period of time a new *stable* level of carbon dioxide in the atmosphere will appear, along with much larger stores in the increasingly acidified oceans and seas. Many forms of marine life are already slated for extinction.

The largest current annual emitters of carbon dioxide on our planet are identified in Table 25.6 [18]. Note that the entries in column c are computed as the product of the entries in columns a and b. In column d, the values in column c are expressed as a percent of annual global carbon dioxide emissions (i.e., relative to $30,303 \times 10^{12}$ g) (see Table 25.6, column c, bottom row).

As shown in the first row (columns a, c, d), China has the world's largest population, the largest absolute carbon dioxide emissions, and the largest percent of total carbon dioxide emissions. On a per capita basis, the United States is the world's largest emitter

Table 25.6 Annual (2010) carbon dioxide (CO_2) emissions for those 11 countries with populations greater than 100 million

Country	Population	CO_2 emissions per capita	Total CO_2 emissions	CO_2 emissions	Rank order
	(millions)	(10^6 g)	(10^{12} g)	(% of total)	(by column d)
Column	a	b	c	d	e
Operation	-	-	(column a×b)	(100×column c/30,303)	-
China	1,338	5.43	7,265	24.0	1
India	1,171	1.39	1,628	5.4	3
United States	310	17.31	5,366	17.7	2
Indonesia	240	1.71	410	1.4	8
Brazil	195	2.29	447	1.5	6
Pakistan	174	0.78	136	0.4	9
Nigeria	158	0.29	46	0.2	11
Bangladesh	149	0.36	54	0.2	10
Russia	142	11.2	1,590	5.2	4
Japan	127	9.0	1,143	3.8	5
Mexico	108	3.85	416	1.4	7
World	6,825	4.44	30,303	61.1	-

(column b, row three), this figure being three times larger than the emissions of China. These 11 countries shown in Table 25.6 account for 61% of total global emissions (column d, bottom row).

Taken together by *rank order* (column e) we find that China (1), the US (2), India (3), and Russia (4) account for 52% of total current global emissions (columns d). This figure, however, tends to disguise what is actually happening. By subcontracting the manufacture of goods to China (and other developing countries), often with lower environmental standards, the US and other developed countries in effect export their carbon emissions [16].

Keep in mind that contemporary emissions by themselves do not account for the current level of atmospheric carbon dioxide shown in Figure 25.1a. That curve is the *cumulative* result of carbon dioxide emissions, chiefly by Europe, North America, and Japan, over the past two centuries or so. It will be decades before the accumulated atmospheric burden of carbon dioxide due to burning fossil fuels in China and India rivals that due to the developed nations [16]. It is for this reason that the proposal by the West for China and India to substantially reduce greenhouse gas emissions is perceived as intrinsically unfair and hypocritical.

Methane

A doubling of northern methane emissions could be realized fairly easily. Much larger increases cannot be discounted.

Clark and Weaver 2008

A similar plot to that in Figure 25.1a, for another important greenhouse gas, namely *methane* (CH_4), is shown in panel b [19]. See above quote from Clark and Weaver. Methane has a greater immediate impact on global warming than carbon dioxide; this impact, however, declines much more rapidly with time than does the impact of carbon dioxide (see above). It is estimated that the frozen bogs of Siberia contain 70 billion tonnes of methane [16]. Release of methane into the atmosphere from melting bogs is another example of a *scale effect* (see above). Twenty percent of all anthropogenic (human-induced) methane in the atmosphere is said to come from rotting vegetation in *man-made* reservoirs (usually created to control flooding, generate electricity, and provide water for irrigation) [20]. A possible lesson is that human interventions in the environment often have the potential for serious unintended and unwanted side-effects (see above quote from White).

Air pollution

Beyond injecting greenhouse gases into the atmosphere, we are polluting the atmosphere with an estimated 760,000 tonnes of metals (and non-metals), such as arsenic (2.5%), cadmium (1%), and zinc (17%) per year [21].

Table 25.7 The mass of the hydrosphere and its major components

	Entity	Mass (g)	Percent	Source
Column/Row	a	b	c	d
1	Hydrosphere	1.39×10^{24}	100	[22]
2	Oceans & seas	1.35×10^{24} [a]	97.5	[22]
3	Glaciers, ice fields, snow	2.2×10^{22} [b]	1.6	[23]
4	Total fresh water	3.5×10^{22}	2.5	[22]

[a] Density of sea-water taken as 1.035 g/cm^3; [b] density of ice taken as 0.917 g/cm^3; row four, columns b and c, includes row three.

THE HYDROSPHERE
Distribution of water on earth

The total amount of water (fresh and salt, liquid and solid) on earth now is doubtless much the same as it has been for hundreds of millions of years. Comparing row one of Table 25.7 with row one of Table 25.4, we find that the global mass of water accounts for about 0.02% of the mass of the earth (1.39/60 = 0.02). In Table 25.7, row one, we take the total mass of water at or near the earth's surface as the *reference parameter* (i.e., as 100%). The oceans and seas comprise 97.5% of all the water on earth (row two). Fresh water (row four) makes up only 2.5% of total water; roughly two-thirds (0.64) of that fresh water is "locked up" in glaciers and ice fields (see above).

Distribution of fresh water on earth

Humanity now appropriates for its own use more than half of Earth's accessible renewable fresh water...

Postel 1999

The three water crises – dwindling freshwater supplies, inequitable access to water and the corporate control of water – pose the greatest threat of our time to the planet and our survival.

Barlow 2007

In Table 25.8 we take the global stocks of fresh water as the reference parameter (row one). Of this quantity, perhaps 30% exists in the form of groundwater (row two) although this estimate is uncertain [22]. The principal fresh water lakes of the world account for only about 0.26% of all fresh water (Table 25.8, row three); the Great Lakes (Erie, Huron, Michigan, Ontario, Superior) in turn contain about 1/4 of that amount (row four) (0.06/0.26 = 0.23). It is perhaps not widely appreciated that the *recharge rate* for the Great Lakes is only about 1% per year [24]. A large but uncertain fraction of the earth's total fresh water is located in aquifers. See arresting quotes from Postel and Barlow above.

Table 25.8 Partial account of the global distribution of fresh water

Row	Entity	Mass (g)	Percent	Source
1	Total fresh water	3.5×10^{22}	100	[22]
2	Total fresh groundwater	1.05×10^{22}	30	[25]
3	Principal lakes of world	9.1×10^{19}	0.26	[25]
4	The Great Lakes	2.2×10^{19}	0.06	[25]

Table 25.9 Partial analysis of global annual flows of fresh water on land

	Entity	Mass (g)	Percent	Source
Column/Row	a	b	c	d
1	Precipitation on land	1.19×10^{20}	100	[22]
2	Run-off (streams, rivers, lakes)	1.25×10^{19}	10.5	[22]
3	Annual human withdrawal [a]	4×10^{18}	3.4	[22]

[a] For the year 2000.

Global annual flows of fresh water

In Table 25.9, row one, we take as the *reference* parameter the annual amount of global precipitation on land. The mean annual precipitation, averaged over the earth's entire surface area, corresponds to a cumulative water depth of 1.1 m. An equivalent amount of water evaporates from lakes, rivers, oceans, and seas each year. About 80% of the precipitation on land returns to the atmosphere through ground evaporation and plant respiration [26].

The bulk of water flowing to the sea via streams and rivers is inaccessible, as is the case for most of the water in the Amazon River (having the earth's largest catchment area), and also for those many rivers flowing, for example, into the Arctic Ocean (which covers an area of about 14 million km^2). Only about one-third of *global annual run-off* is potentially accessible for human use. Of the given volume of accessible fresh water, humans are now using some 32% ($100 \cdot 3.4/10.5 = 32$) (see Table 25.9, column c, rows two and three) [27]. Other estimates put the figure at around 40% (see below).

Figure 25.2 is based on the data of Abramovitz [27]. There we see that global fresh water withdrawals (lower curve, left-hand ordinate) in the year 2000 amounted to about 5,000 km^3 per year (equivalent to 5×10^{18} g per year). This is some 32 to 40% of all accessible fresh water on earth. We also see that per capita fresh water withdrawals (Figure 25.2, upper curve, right-hand ordinate) are rising rapidly. Thus, water usage is increasing because of both increasing human population (see Figure 25.4a) and increasing mean per capita withdrawals (Figure 25.2, upper curve).

In 1680 AD, global per capita water usage is estimated at 153 m^3/yr [27]. By 1900, the world's per capita annual water use was around 400 m^3/yr [27]. In 1980 it reached

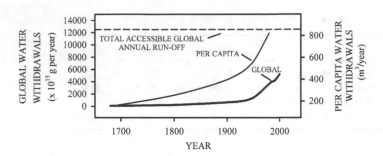

Figure 25.2 Global and per capita annual withdrawals of fresh water over time.

824 m³/yr, corresponding to a 5.4-fold increase in about 320 years (824/153 = 5.4) (Figure 25.2, upper curve, right-hand ordinate).

For comparison, recall that during the growing season, a crop of maize (corn) may require 3.3 million liters (3.3×10^9 g) of water per hectare [28]. In the same vein, it may take some 400,000 liters (4×10^8 g) of water to produce a single car [29]. When we purchase something, rarely are we told how much water was required to produce it. This should be made mandatory for many products.

Hydrologic cycle

Only twenty percent of the precipitation falling on land feeds rivers, streams and underground aquifers...

Postel 1999

Nothing, perhaps not even climate change, will matter more to humanity's future on this planet over the next century than the fate of our rivers.

Pearce 2006

Evaporation from the world's oceans due to solar radiation amounts to about 5×10^{20} g per year [22]. Of this amount 91% falls back into the oceans as rain; the remaining 9% (about 4.5×10^{19} g per year) contributes to precipitation on land. Some 95% of the portion on land makes its way back to the oceans as surface water and another 5% as groundwater.

In addition to the above transfer of fresh water (via evaporation) from the oceans to land, there is also evaporation from soil and vegetation into the atmosphere amounting to about 7.4×10^{19} g per year [22]. Most of this fresh water returns to land as rain. Thus total precipitation on land amounts to about 10^{20} g per year (($4.5 + 7.4) \times 10^{19} = 1.2 \times 10^{20}$ g per year) (Table 25.9, row one) [22]. This account ignores factors such as evaporation from lakes and rivers and the fact that some evaporation from land contributes to precipitation over the seas and oceans. As required for a steady state, the total run-off (ground and surface fresh water) is equal, on average, to the total transfer to land of fresh water due to evaporation from the oceans. The important point is that without the hydrological cycle sketched above there would be little or no life on land. See above quotes from Postel and Pearce.

Irrigation

...only about one-third of the water diverted for irrigation is used effectively by agricultural crops...

Hillel 1991

In most regions of the world, insufficient water is the prime limitation on agricultural productivity.

Brady and Weil 2002

In most countries, about 70% of water usage is associated with agriculture and irrigation [23]. In the US the figure is 41% [23]. The other five nations making the most use of irrigation as of 1995 (in millions of hectares) are India (50), China (50), Pakistan (17), and Iran (7) [26]. China uses irrigation to produce 70% of its food, whereas the US [24] uses 17%.

A substantial part of water for irrigation is taken from aquifers. (For a brief discussion of the land area under irrigation see "Pedosphere," below.) The Nubian aquifer (eastern Sahara) is the world's largest known aquifer; its mass of fresh to slightly brackish water is put at 1.5×10^{20} g or about 0.4% of earth's total fresh water (1.5/3.5 = 0.4%) (Table 25.8, row one). The Ogallala aquifer, stretching underground from South Dakota to northern Texas, contains about 0.01% of the earth's total fresh water. Water withdrawals from the Ogallala aquifer are currently said to exceed the recharge rate by a factor of 14 [29].

Bear in mind that the amount of water we actually use domestically, agriculturally, and industrially is less than the amount withdrawn from all sources by 1.8 times, on average [30]. See above quotes from Hillel as well as Brady and Weil.

Water pollution

North American facilities reported releases of 257 pollutants to water in 2006 [amounting to 228,000 tonnes].

Commission for Environmental Cooperation: *Taking Stock* 2011

Waste water (Table 25.10, bottom row) tends to be dumped without treatment into public waterways (rivers, lakes, and aquifers). It is known that a given volume of contaminated water can degrade a volume of clean fresh water that is larger by a factor of 8 to 10 [25].

Annual human deaths from drinking contaminated water are estimated at 2 – 5 million [30]. See above quote from *Taking Stock*.

Water shortages

Water scarcity is now the single biggest threat to global food production.

Postel 1999

Table 25.10 Global distribution of fresh and waste water

Row	Entity	Mass (g)	Percent	Source
1	Annual accessible run-off	1.25×10^{19}	100	[22]
2	Annual human withdrawal [a]	4×10^{18}	32	[22]
3	Annual waste water production	1.4×10^{18}	11	[25] [b]

[a] For the year 2000; [b] Data shown are for the year 1995 in Europe, North America, Asia, and Africa. Waste water includes agricultural, industrial, and domestic sources.

The water crisis is the most pervasive, most severe and most invisible dimension of the ecological devastation of the earth.

Shiva 2002

...water is deeply embedded in the global economy. We are all part of a virtual worldwide trade in water, through the global flow of goods that require water to produce.

Turner 2010

When fresh water is diverted to agriculture or industry or municipalities, it is not literally "used up"; it is, however, often significantly degraded. This loss of water through pollution, together with increasing population and increasing per capita consumption, points to imminent shortages of clean fresh water. Serious water shortages are already apparent in China and India and these will likely become more serious in the near future. China [26] has around 20% of the world's population (Table 25.6) but only 7% of its renewable fresh water. Water tables in the North China breadbasket are dropping at the rate of 1.5 m per year [29]. In several northern states of India the water tables are dropping by roughly a meter every three years. Some 500 million people currently face water shortages [22]. Recall that in recent years the waters of the Colorado River often fail to reach the Gulf of Mexico, mainly because of excessive withdrawals for agriculture. With passing time the signs of serious global water shortages become more and more apparent. See above quotes from Postel, Shiva, and Turner.

THE PEDOSPHERE

Pedology is the branch of science devoted to the study of soil. The term pedosphere (G. "pedon" = soil) refers to the portion of the earth's surface associated with soil; the natural processes by which soil forms are termed pedogenesis. The first primitive soils are thought to have appeared about 400 mya. The soils that made possible the rise of agriculture (about 10,000 years ago) were likely formed during the retreat of the last glaciation [32, 33]. Estimates for the total mass of soil on our planet are given in Table 25.4 (row five). The composition of soil varies greatly from one region of the earth's surface to another. The average gross composition of the earth's soil by volume is given in Table 25.11 [34]. Observe that air and water account for about half of soil volume on average.

Table 25.11 Composition of soil

Component	Volume composition (%)
Air	25
Minerals	45
Organic matter	5
Water	25

Table 25.12 The area of the global land surface and each of three major components

Entity	Area (km^2)	Percent	Source
Global land surface	149,450,000	100	[37]
Global cultivable land	30,000,000	20	[28]
Global cropland	13,640,000	9	[28]
Global irrigated cropland	2,720,000	2	[28]

Distribution of land

Salinization, erosion, denudation of watersheds, silting of valleys and estuaries, degradation of arid lands, depletion and pollution of water resources, abuse of wetlands and excessive population pressure – all are now occurring more intensively [than for any prior civilization] *and on an ever-larger scale.*

Hillel 1991

Under tropical and temperate agricultural conditions, at least 500 years are required for the formation of 2.5 cm of topsoil...

Pimentel 1993

Our global food supply depends largely (97%) on terrestrial agriculture, which in turn depends directly on soil. As Charles Darwin (1809–1882) knew, the natural processes by which 2 cm or so of soil form require centuries. Soil should not be regarded, on a time scale of "three score and ten" years, as a renewable resource [35]. The global distribution of land and cropland is summarized in Table 25.12. See also Figure 25.3. Cultivable land is estimated at some 20% of the earth's total land area (row two). Of that 20%, something less than half is actually used to grow crops (row three). The global *irrigated* cropland, responsible for a disproportionate fraction of our food supply, has an area only about one-fifth that of actual cropland (2.7/13.6 = 0.20) (Table 25.12, rows three and four). As shown in Figure 25.3a, the levels of global cropland are now some 38% of the estimated limit to useful cropland (100·(14/37 = 38) [30]. See above quotes from Hillel and Pimentel.

In some places, for example the Near East and North Africa, virtually all of the land useful for agriculture is already under cultivation [36]. The amount of land subject to irrigation has increased vastly since 1800 AD (Figure 25.3b). Barring greater efficiency in water use, shortages of water will likely impede further expansion of irrigation.

Table 25.13 Annual global loss of cropland attributable to desertification and salinization

Entity	Area (km^2)	Percent	Source
Global cultivable land	30,000,000	100	[28]
Global desertification	60,000	0.2	[15]
Global salinization	10,000	0.03	[29]

Table 25.14 Global fertilizer use as a function of time

Year	Fertilizer (10^6 tonnes)	Year	Fertilizer (10^6 tonnes)
1950	13	1990	131
1960	24	1994	109
1970	60	2000	123
1980	101	2010	145 ^{a}

^{a} Estimated; other data obtained by graphical interpolation.

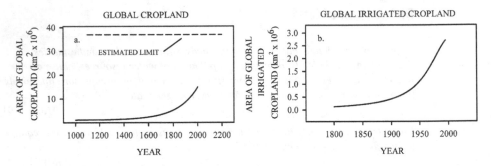

Figure 25.3 Area of global cropland and global irrigated cropland over time.

Desertification

The current rate of desertification (i.e., the effectively irreversible loss of soil in semi-arid regions owing to erosion, usually following loss of vegetation and/or climate change) has been put at some 60,000 km^2/yr (see Table 25.13, row two) [15]. Much of this desertification is attributed to "overgrazing and over-plowing."

Fertilizers

Humankind has already played the fertilizer card in the earth's best agricultural lands, and further nitrogen and phosphate loadings no longer increase yields.

McNeill 2000

The global use of fertilizer between 1950 and 2010 is summarized in Table 25.14 [38]. Note that between 1950 and 2010, global fertilizer use increased more than ten-fold.

There is a tendency among farmers, perhaps on the dubious principle that more is always better, to use fertilizer in excess of that needed to maximize plant productivity. The excess fertilizer (chiefly phosphates and nitrates) in the run-off appears in lakes, rivers, and streams. This in turn may result in eutrophication, whereby (possibly toxic) surface phytoplankton expand, decreasing oxygen and light concentrations at lower levels, with consequent anoxia and a decrease in subsurface life-forms (including fish). Note the above quote from McNeill.

Green Revolution

The ecological and social costs of the Green Revolution were largely ignored. Through its emphasis on high-yield seeds, this agricultural model displaced drought-resistant local crop varieties and replaced them with water-guzzling crops.

Shiva 2002

The success of the Green Revolution made it possible to feed an additional two billion or so people in the postwar period. This success was achieved in large part by use of high-yield seeds, increased irrigation (see above quote from Shiva), and increased application of chemical fertilizers. For example, in the time period from 1961 to 1984, the global human population expanded by 55%, arable land area by 9%, fertilizer by 320%, and irrigation by 60% [39]. Only later did it come to be widely appreciated that the Green Revolution entailed massive and unsustainable increases in water withdrawals [26].

Land under irrigation

The rapid recent increase in cropland under irrigation is shown in Figure 25.3b [40]. The global amount of land under irrigation went from about 0.08 million km^2 in 1800 to 2.25 million km^2 in 1995 (an increase of about 28 times (2.25/0.08 = 28)).

Pesticides

...5 million tons of pesticide are applied annually to crops worldwide. As a result, pesticide resistance has become a ubiquitous problem...

Matson *et al.* 1997

The well-known Precautionary Principle (prevent harm insofar as possible to people and/or the environment) implies that pesticides, for example, should not be deployed until shown to be safe by well-accepted criteria. This self-evident precept is widely ignored (see above quote from Matson *et al.*). A current example (2014) relates to neonicotinoids (neuro-active insecticides used globally). It is widely believed by bee-keepers that these compounds are responsible for the marked and alarming decline in bee populations worldwide. Bees are responsible for pollinating many common foods such as grapes, lettuce, peaches, sweet potatoes, and the like. Insectivorous birds are also likely to be adversely affected.

Salinization

Salinization is the process by which soil becomes saline, often owing to the use of brackish water for irrigation. (See Table 25.13, bottom row.) Postel [41] estimated that up to 20% of irrigated land globally has been affected by salinization. It is further estimated that globally about 1% of *irrigated* land is lost to agriculture annually because of water-logging *and* salinization [42]. Salinization is a problem for which there appears to be no simple solution [26].

Soil pollution

We are annually injecting an estimated 6.3 million tonnes of toxic metals and non-metals (e.g., arsenic, cadmium, and zinc) into the world's soil [21]. A corollary is that some 12 million tonnes of grain are also contaminated annually by heavy metals [30]. The health of large numbers of people stands to be adversely affected by this ongoing pollution of our food.

Soil erosion and land degradation

Soil erosion is one of the most serious environmental problems in the world today. . . .More than 97% of the world's food comes from the land rather the oceans and other aquatic systems. . .

Pimentel 1993

Estimates of the fraction of land transformed or degraded by humanity. . . fall in the range of 39 to 50%.

Vitousek *et al.* 1997

It is true that a degree of soil erosion is a necessary part of the normal turnover of soil. However, global loss of topsoil is estimated at *24 billion tons* per year. Some 60 to 80% of this loss is attributable to humans [40]. From Table 25.4, bottom row, we find that this global loss of soil amounts to 0.04% of global soil per year (to see this go to Table 25.4, row five; take mean mass of soil as 5×10^{19} g. Convert 24 billion tons to 2.2×10^{16} g. Then compute $100 \cdot 2.2 \cdot 10^{16}/5 \cdot 10^{19} = 0.04$). Soil erosion, due to water and wind, tends to remove the most nutrient-rich portion of the soil; other components of soil are also depleted [38, 39]. In the course of a human lifespan, the erosion of cropland need not be visible to the individual farmer; and what is not seen often tends to be ignored. Unfortunately, this pressing and largely invisible problem is seldom discussed in the mass media. See above quotes from Pimentel and Vitousek *et al.*

Given the steady increase in the human population (Figure 25.4a), there is no longer any doubt that our indifference to land-use policies, if continued much longer, will threaten the global food supply [40]. By progressively destroying the planet's soil we run the risk of committing ecological suicide. Indeed, many past civilizations have collapsed on a time scale of hundreds to a thousand or so years, roughly consistent with the time it takes to erode away soil deposits initially 30 to say 100 cm deep (assuming average erosion rates of about 1 mm per year) [44]. The outstanding exception to this rule, until recently, has been the Egyptian civilization, where the soil forming the Nile river basin has been recharged annually for thousands of years by heavy seasonal rains [35].

HUMAN POPULATION AND PER CAPITA GDP

One of the most important and least heralded events in human history occurred between 1965 and 1970. The world's population growth rate reached an all-time peak and began to drop thereafter.

Cohen 1995

The current annual *growth rate* of the total human population is about 100 million [30]. This corresponds, very nearly, to adding a new city of 275,000 people to our planet each and every day of the year. These 275,000 people require accommodation, clean water, education, food, health care, and so on. It is clear from Figure 25.4, panels a and b, that in terms of the global rates of population and economic growth (as measured by gross domestic product (GDP)), we are living in a period without parallel in human history. Consider also the above quote from Cohen.

Although the increase in global population (Figure 25.4a) by itself puts the biosphere under greater stress, it is the increase in economic activity (panel b), especially in the developed countries, that exacts the greatest demands. Currently food requirements are increasing by about 2% per year, while food production is only increasing by 1% per year [30]. Human population and the associated economic growth [45] are driving global warming, water pollution, and soil erosion, and many of the other challenges we face (e.g., see Table 25.1).

OUR TROUBLED TIMES

...it does not make sense to ask about the value of replacing a life-support system... In making decisions society has to take into account all aspects of interactions with the natural world. Economists do not claim that economic aspects are unique or even primary. ... Markets will never be a panacea.

Heal 2000

Anyone who believes that exponential growth can go on forever in a finite world is either a madman or an economist.

Attributed to Kenneth Boulding 1973

By relentlessly disturbing the delicate balance linking the atmosphere, hydrosphere, and pedosphere – the three major physical components of the biosphere – we invite a global

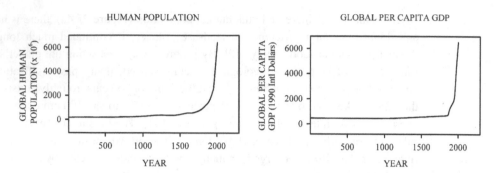

Figure 25.4 Global population and per capita gross domestic product (GDP) over time.

catastrophe. As we inject more and more carbon dioxide into the atmosphere we raise the mean temperature of the earth's surface. This increase in temperature will result in increased evaporation and thus increased precipitation. An increase in precipitation will produce more run-off and hence potentially more soil erosion. Significant changes in any one of the "spheres" is likely to affect the others. Hence the problems we face do not lend themselves to facile one-dimensional solutions. See opening quote from White.

In the short run (one to two decades), the most severe crisis facing our global civilization will likely be *water shortages* (see Figure 25.2 and Table 25.9). In the longer run (25 to 50 years) *climate change* will no doubt come to the fore, perhaps with dire consequences. We have already crossed a modest threshold for effectively *irreversible* change in the level of carbon dioxide in the atmosphere (see Figure 25.1 and Table 25.6). In the intermediate period, over the next 5 to 15 years, *soil erosion* may begin to seriously impact global food production (see above).

Rising temperatures due to burning fossil fuels will exacerbate the problems of inadequate supplies of clean water, owing to greater evaporation. Increasing surface temperatures, persistent drought, and ongoing soil erosion will likely be the new normal. Animals (including our own species) and plants unable to adapt to this relatively abrupt change in climate and habitat will become extinct. (It is estimated that 59 species of mammals have become extinct since 1600 AD. The natural background rate of extinction for mammals is one species per million years [46].) Currently (2014) about one mammalian species in four is considered to be endangered. Population numbers for many species of common animals and plants are now in rapid decline.

A tentative scale for measuring the severity of the calamities we face is suggested in Table 25.15. Both the number and severity of these calamities will probably increase with time for the foreseeable future. Simultaneous catastrophes might soon, if not already, exceed the capacity of relief organizations to respond in a timely fashion.

It is clear that our human civilization faces problems on a scale never before imagined. Scientists especially, and other scholars, have gone to considerable lengths to clarify the nature and extent of these problems. The ground now appears to be shifting, but with the exception of a few countries, chiefly in Northern Europe, little progress has yet been made in addressing these problems systematically. There is little doubt that these problems can be largely resolved using techniques that are already

Table 25.15 Suggested classification of calamities in terms of number of human deaths per event

Class	Number of deaths	Possible cause	Class	Number of deaths	Possible cause
I	10^3	Storms	IV	10^6	Starvation
II	10^4	Floods	V	10^7	Disease
III	10^5	Tsunamis	VI	10^8	Large asteroid impact

available, or that will be forthcoming with the appropriate scientific and technological effort. Many of our problems can be traced back to corporate capitalism and its blind commitment to never-ending growth in material consumption. See above quote from Heal and one attributed to Boulding – both economists.

It may be that more attention needs to be given to *why* the public response to the looming crises has generally been so passive. Part of the explanation may lie with sizeable programs of *disinformation* and *propaganda* relating to climate-change denial arising from wealthy interests who stand to be adversely affected by remediation programs. Chief amongst these climate-change deniers are the fossil fuel industries. Another important factor accounting for the largely submissive reaction to climate change is the pervasive human tendency to repress bad news.

Yet it seems likely that other factors, possibly just as important, are at work. Two specific factors relate to *invisibility* and *temporal scaling*. Many of the influences prone to destabilize our global civilization are effectively invisible. These include CO_2 production by vehicles and power plants, global loss of biodiversity, increasing evaporation of water, ocean warming, many pollutants, salinization, and soil erosion. Similar problems exist with respect to *time horizons*.

Some scientists, for example archaeologists, cosmologists, geologists, and paleontologists, are accustomed to thinking of time periods extending from thousands to millions and even billions of years. Outside of academia, corporate leaders may think primarily in terms of the next quarterly report and politicians the next election. The man or woman in the street may not see much beyond her or his next pay check. By contrast, the problems we urgently need to solve for the sake of ourselves, our children, and our grandchildren involve temporal considerations extending from decades to centuries and even to millennia.

No one can accurately predict the future. Nonetheless, anyone who has studied our contemporary state of affairs in some detail (see above) can predict the general types of changes that must be made if we are to have a future. The gamut of problems we now confront are mainly problems of scale. Regrettably, we live in an age of rampant specialization, leading to an immense fragmentation of knowledge. It seems possible, perhaps likely, that a scaling perspective, with its inherent concerns for integration, its emphasis on quantitation, and its concern for limits and scale effects, could play a useful role both in clarifying the dimensions of the multiple crises we confront and in designing ecologically benign counter-measures. Thus, it may turn out that the future of our global civilization and our species depends in part on the widespread acquisition of an abiding, critical, and informed *sense of scale*.

References

1. Licklider, H. (1966). *Architectural Scale.* New York: George Braziller.
2. Dahl, R.A. and Tufte, E.R. (1973). *Size and Democracy.* Stanford, CA: Stanford University Press.
3. Gallant, J.A. and Prothero, J.W. (1972). Weight-watching at the university: the consequences of growth. *Science*, **175**:381–388.
4. Ferrer i Cancho, R. and Sole, R.V. (2003). Least effort and the origins of scaling in human language. *Proceedings of the National Academy of Sciences*, **100**:788–791.
5. Meentemeyer, V. and Box, E.O. (1987). Scale effects in landscape studies. In: Turner, M.G. (ed.) *Landscape Heterogeneity and Disturbance.* New York: Springer-Verlag, pp. 15–34.
6. Sorge, G.A. (1978). *The Secretly Kept Art of the Scaling of Organ Pipes.* Buren: Frits Knuf.
7. Ball, R.C. (1985). Scaling in polymer physics. In: Pynn, R. and Skjeltorp, A. (eds.) *Scaling Phenomena in Disordered Systems.* New York: Plenum Press.
8. Lang, K.R. and Gingerich, O. (eds.) (1979). *A Source Book in Astronomy and Astrophysics, 1900–1975.* Cambridge, MA: Harvard University Press.
9. Taylor, F.W. (2006). *Elementary Climate Physics.* Oxford: Oxford University Press.
10. Mason, B. (1966). *Principles of Geochemistry.* New York: John Wiley & Sons.
11. Whittaker, R.H. (1975). *Communities and Ecosystems.* 2nd edn. London: Macmillan.
12. Zeilik, M., Gregory, S.A. and Smith, E.v.P. (1992). *Introductory Astronomy and Astrophysics.* Fort Worth, TX: Saunders College Publishing,
13. Alberts, B., Bray, D., Lewis, J. *et al.* (1994). *Molecular Biology of the Cell.* New York: Garland Publishing.
14. Wedepohl, K.H. (1971). *Geochemistry.* New York: Holt, Rinehart and Winston.
15. Houghton, J. (1997). *Global Warming: The Complete Briefing.* Cambridge: Cambridge University Press.
16. Romm, J.J. (2007). *Hell and High Water.* New York: HarperCollins.
17. Hansen, J. (2009). *Storms of my Grandchildren.* New York: Bloomsbury.
18. IEA (2012). *Key World Energy Statistics.* Paris: International Energy Agency,
19. Firor, J. (1990). *The Changing Atmosphere: A Global Challenge.* New Haven, CT: Yale University Press.
20. Pearce, F. (2006). *When the Rivers Run Dry.* Boston, MA: Beacon Press.
21. Nriagu, J.O. (1990). Global metal pollution: poisoning the biosphere. *Environment*, **32**:7–33.
22. Black, M. and King, J. (2009). *The Atlas of Water.* Berkeley, CA: University of California Press.
23. Gleick, P.H. (2009). *The World's Water 2008–2009: The Biennial Report on Freshwater Resources.* Washington, DC: Island Press.
24. de Villiers, M. (2000). *Water.* New York: Charles Scribner.
25. Shiklomanov, I.A. and Rodda, J.C. (eds.) (2003). *World Water Resources at the Beginning of the Twenty-First Century.* Cambridge: Cambridge University Press.
26. Postel, S. (1999). *Pillar of Sand.* New York: W.W. Norton.
27. Abramovitz, J.N. (1996). *Imperiled Waters, Impoverished Future.* Washington, DC: World Watch Institute, Paper 128.
28. Pritchard, S.G. and Amthor, J.S. (2005). *Crops and Environmental Change.* Binghamton, NY: Haworth Press.
29. Barlow, M. and Clarke, T. (2002). *Blue Gold: The Battle Against Corporate Theft of the World's Water.* Toronto: Stoddart.

30. Cribb, J. (2010). *The Coming Famine: The Global Food Crisis and What We Can Do To Avoid It.* Berkeley, CA: University of California Press.

31. Shiklomanov, I.A. (1993). World fresh water resources. In: Gleick, P.H. (ed.) *Water in Crisis. A Guide to the World's Fresh Water Resources.* New York: Oxford University Press, pp. 13–24.

32. Muller, R.A. and Oberlander, T.A. (1978). *Physical Geography Today: A Portrait of a Planet.* New York: Random House.

33. Murphey, R. (1961). *An Introduction to Geography.* Chicago, IL: Rand McNally & Co.

34. Tarbuck, E.J. and Lutgens, F.K. (1984). *The Earth: An Introduction to Physical Geology.* Columbus, OH: Charles E. Merrill Publishing.

35. Hillel, D.J. (1991). *Out of the Earth: Civilization and the Life of the Soil.* New York: Macmillan Inc.

36. Smith, B.D. (1995). *The Emergence of Agriculture.* New York: Scientific American Library.

37. *Desk Reference Atlas* (2001). 4th edn. New York: Oxford University Press.

38. Brown, L.R. (2004). *Outgrowing the Earth.* New York: W.W. Norton.

39. Pimentel, D. (ed.) (1993). *World Soil Erosion and Conservation.* Cambridge: Cambridge University Press.

40. McNeill, J.R. (2000). *Something New Under the Sun: An Environmental History of the Twentieth Century.* New York: W.W. Norton.

41. Postel, S.L. (1998). Water for food production: will there be enough in 2025? *BioScience,* **48**:629–637.

42. Postel, S. (1997). *Last Oasis.* New York: W.W. Norton & Co.

43. Anonymous. (2010). *World Watch,* **23**:32.

44. Montgomery, D.R. (2007). *Dirt: The Erosion of Civilizations.* Berkeley, CA: University of California Press.

45. Maddison, A. (2007). *Contours of the World Economy, 1–2030 AD: Essays in Macro-Economic History.* Oxford: Oxford University Press.

46. Lawton, J.H. and May, R.M. (eds.) (1995). *Extinction Rates.* Oxford: Oxford University Press.

Appendix A Recommended reading

Selected books

Ashcroft, F. (2000). *Life at the Extremes: The Science of Survival*. Berkeley, CA: University of California Press.

Barrow, J.D. (1995). *The Artful Universe*. Oxford: Clarendon Press,.

Bonner, J.T. (2006). *Why Size Matters: From Bacteria to Blue Whales*. Princeton, NJ: Princeton University Press.

Brody, S. (1945). *Bioenergetics and Growth: With Special Reference to the Efficiency Complex in Domestic Animals*. New York: Reinhold.

Brown, J.H. and West, G.B. (eds.) (2000). *Scaling in Biology*. Oxford: Oxford University Press.

Calder, W.A. (1984). *Size, Function, and Life History*. Cambridge, MA: Harvard University Press.

Dusenbery, D.B. (2009). *Living at Micro Scale: The Unexpected Physics of Being Small*. Cambridge, MA: Harvard University Press.

Eisenberg, J.F. (1981). *The Mammalian Radiations*. Chicago, IL: University of Chicago Press.

Huxley, J.S. (1972). *Problems of Relative Growth*. New York: Dover Publications.

Jungers, W.L. (ed.) (1985). *Size and Scaling in Primate Biology*. New York: Plenum Press.

Jürgens, K.D. (1989). *Allometrie als Konzept des Interspeziesvergleiches von physiologischen Grössen*. Berlin: Paul Parey.

Kleiber, M. (1961). *The Fire of Life: An Introduction to Animal Energetics*. New York: John Wiley and Sons.

McGowan, C. (1994). *Diatoms to Dinosaurs: The Size and Scale of Living Things*. Washington, DC: Shearwater Books.

McMahon, T.A. and Bonner, J.T. (1983). *On Size and Life*. New York: Scientific American Books.

McNab, B.K. (2002). *The Physiological Ecology of Vertebrates: A View from Energetics*. Ithaca, NY: Cornell University Press.

Niklas, K.J. (1994). *Plant Allometry: The Scaling of Form and Process*. Chicago, IL: University of Chicago Press.

Pedley, T.J. (ed.) (1977). *Scale Effects in Animal Locomotion*. London: Academic Press.

Peters, R.H. (1983). *The Ecological Implications of Body Size*. Cambridge: Cambridge University Press.

Reiss, M.J. (1989). *The Allometry of Growth and Reproduction*. Cambridge: Cambridge University Press.

Schmidt-Nielsen, K. (1979). *Animal Physiology: Adaptation and Environment*. Cambridge: Cambridge University Press.

Schmidt-Nielsen, K. (1984). *Scaling: Why is Animal Size so Important?* Cambridge: Cambridge University Press.

Thompson, D.W. (1943). *On Growth and Form.* New York: Cambridge University Press.

Weibel, E.R. (1984). *The Pathway for Oxygen: Structure and Function in the Mammalian Respiratory System.* Cambridge, MA: Harvard University Press.

Selected articles

Adolph, E.F. (1949). Quantitative relations in the physiological constitutions of mammals. *Science*, **109**:579–585.

Anderson, P.W. (1972). More is different. *Science*, **177**:393–396.

Bartels, H. (1964). Comparative physiology of oxygen transport in mammals. *Lancet*, **I**:599–604.

Beuchat, C.A. (1993). The scaling of concentrating ability in mammals. In: Brown, J.A., Balment, R.J. and Rankin, J.C. (eds.) *New Insights in Vertebrate Kidney Function.* Cambridge: Cambridge University Press, pp. 259–279.

Biewener, A.A. (2005). Biomechanical consequences of scaling. *Journal of Experimental Biology*, **208**:1665–1676.

Calder, W.A. and Braun, E.J. (1983). Scaling of osmotic regulation in mammals and birds. *American Journal of Physiology*, **244**:R601-R606.

da Silva, J.K.L., Garcia, G.J.M. and Barbosa, L.A. (2006). Allometric scaling laws of metabolism. *Physics of Life Reviews*, **3**:229–261.

Gallant, J.A. and Prothero, J.W. (1972). Weight-watching at the university: the consequences of growth. *Science*, **175**:381–388.

Günther, B. and Guerra, E. (1955). Biological similarities. *Acta Physiologica LatinoAmericana*, **5**:169–186.

Haldane, J.B.S. (1940). On being the right size. In: *Possible Worlds.* London: Evergreen Books, pp. 27–34.

Hill, A.V. (1950). The dimensions of animals and their muscular dynamics. *Science Progress*, **38**:209–230.

Hochachka, P.W. (1989). Upper and lower limits to energy demand in homoiotherms. *Lung Biology in Health and Disease*, **39**:13–26.

Hokkanen, J.E.I. (1986). The size of the largest land animal. *Journal of Theoretical Biology*, **118**:491–499.

Hoppeler, H. and Weibel, E.R. (2005). Scaling functions to body size: theories and facts. *Journal of Experimental Biology*, **208**:1573–1574.

Jürgens, K.D., Fons, R., Peters, T. and Sender, S. (1996). Heart and respiratory rates and their significance for convective oxygen transport rates in the smallest mammal, the Etruscan shrew *Suncus etruscus*. *Journal of Experimental Biology*, **199**:2579–2584.

LaBarbera, M. (1983). Why the wheels won't go. *American Naturalist*, **121**:395–408.

Lin, H. (1982). Fundamentals of zoological scaling. *American Journal of Physics*, **50**:72–81.

McNab, B.K. (1988). Complications inherent in scaling the basal rate of metabolism in mammals. *Quarterly Review Biology*, **63**:25–54.

Morowitz, H.J. (1966). The minimum size of cells. *Ciba Foundation Symposium*, 446–459.

Pirie, N.W. (1973). On being the right size. *Annual Review Microbiology*, **27**:119–132.

Schmidt-Nielsen, K. (1975). Scaling in biology: the consequences of size. *Journal of Experimental Biology*, **194**:287–307.

Stevens, S.S. (1946). On the theory of scales of measurement. *Science*, **103**:677–680.

Taylor, C.R. (1982). Scaling limits of metabolism to body size: implications for animal design. In: Taylor, C.R., Johansen, K. and Bolis, L. (eds.) *A Companion to Animal Physiology*. Cambridge: Cambridge University Press, pp. 161–170.

Weibel, E.R. (1987). Scaling of structural and functional variables in the respiratory system. *Annual Review Physiology*, **49**:147–159.

Western, D. (1979). Size, life history and ecology in mammals. *African Journal of Ecology*, **17**:185–204.

White, C.R. and Seymour, R.S. 2005. Allometric scaling of mammalian metabolism. *Journal of Experimental Biology*, **208**:1611–1619.

Appendix B Data screening guidelines

The following guidelines were developed to facilitate treating widely varying datasets from many different authors in a consistent manner.

1. *Body weight.* Reference data from the standardized body weight table (SBT) (see Chapter 4) were used to exclude data taken from the literature when body weights or masses deviated too far from the reference values. Often the SBT provides minimum and maximum estimates of the normal body weight for a given mammalian species. (Bear in mind that in most cases these minimum and maximum values are themselves based on small samples.) However, it was found that strict use of these minimum and maximum estimates led to the exclusion of an unreasonable amount of data. That led to the adoption of the following pragmatic rule of thumb: if body weights fell into the range 0.8 times the minimum to 1.2 times the maximum then the data were tentatively accepted. Sometimes the SBT only gives the estimated mean weight for a given species. In that case the data were cautiously admitted if they fell into the range 0.64 to 1.44 times the mean.

2. *Gender.* I favored data for males over females in order to reduce possible bias due to pregnancy or lactation.

3. *Data sources.* I preferred primary to secondary sources.

4. *Means.* Given a range of values of some y-variable for varying body weights, I picked those values nearer the mean.

5. *Seasons.* For field data I generally preferred summer values over other seasons.

6. *Wild vs domestic.* I favored data from wild animals as opposed to domesticated ones.

7. *Consistency.* I gave preference to datasets where y-variables and body weight or mass were determined in the same animals.

8. *Single laboratories.* I favored datasets obtained in one laboratory for multiple y-variables as opposed to equivalent datasets assembled from several laboratories.

9. *Species diversity.* Preference was given to datasets showing greater species diversity and greater size range.

10. *Known laboratories.* In some cases I favored data coming from laboratories where I knew the people involved personally.

11. *Anesthesia.* I gave preference to data from un-anesthetized as opposed to anesthetized animals.

12.   *Interpolated data*. I gave preference to "raw" data as opposed to interpolated values.
13.   *Species identification*. If I could not identify a given species in the taxonomic literature available to me I generally omitted the data.
14.   *Confidence limits*. If after carrying out a LSQ analysis of a given dataset I found that the 99% confidence limits on the slope exceeded ± 0.1, I have (with a very few exceptions) excluded the dataset from this study.

Summary

Data in the scaling literature for mammals are presented in different ways: text, tables, plots. A number of considerations (see above) were factored into the decision as to whether numerical data for a given species from a particular paper should be included in a given dataset. The decision is most challenging for small datasets (say, number of species less than 30). Here one wishes to enlarge the sample size, diversity, and size range as much as possible. But it is for small datasets that one or a few data points may exert the greatest bias. In these cases some judgement is called for, which may involve relaxing one or more of the above guidelines somewhat, in the interest of increasing sample size, diversity, or size range. On the whole, however, I tried throughout this study to follow the above rules as closely as practicable. The result is that the datasets employed for calculations are likely to be more coherent and self-consistent than would have been the case otherwise.

Appendix C Summary of the findings of Samuel Brody

I have focused here on the work of Samuel Brody because, as far as I know, he published scaling results in mammals in a uniform manner for more y-variables than anyone else up to that point in time (1945).

Table C1 Adult interspecific comparisons between the present results and the slopes reported by Brody [1]

Column/Row	Slope (b) $\pm$ s.e.		Entity
	Brody [1] (1945)	Present study	
	a	b	c
1	0.987 $\pm$ 0.017	0.996 $\pm$ 0.016	Blood mass/volume
2	0.986 $\pm$ 0.012	1.000 $\pm$ 0.010	Lung mass
3	0.984 $\pm$ 0.009	0.953 $\pm$ 0.008	Heart mass
4	0.941 $\pm$ 0.044	0.962 $\pm$ 0.027	GI tract mass
5	0.928 $\pm$ 0.017	0.945 $\pm$ 0.024	Thyroid mass
6	0.867 $\pm$ 0.009	0.895 $\pm$ 0.007	Liver mass
7	0.846 $\pm$ 0.010	0.880 $\pm$ 0.008	Kidney mass
8	0.798 $\pm$ 0.014	0.801 $\pm$ 0.020	Adrenal mass
9	0.762 $\pm$ 0.018	0.727 $\pm$ 0.012	Pituitary mass
10	0.734 $\pm$ 0.005	0.667 $\pm$ 0.011	SMR [a]
11	0.707 $\pm$ 0.009 [b]	0.777 $\pm$ 0.025 [c]	DMP
12	0.697 $\pm$ 0.009	0.759 $\pm$ 0.005	Brain mass

[a] Nominal slope (see Chapter 17); SMR = standard metabolic rate; [b] computed from data given in Brody [1], p. 857; [c] slope for present study is for peak daily milk production (DMP).

Except for milk production (p. 857), all of the slopes in column a are taken from Brody [1], p. 591. The slope given in column a for milk production is based on three species (rat, goat, cow).

The MPD between the slopes obtained here and those reported by Brody [1] is 4.1$\pm$3.1 (STD), showing broad agreement between the two sets of results. Using 99% CL on the present slopes, one finds that six of Brody's slopes (column a) agree with the slopes given in column b (i.e., lie within the 99% CL). Another three slopes of Brody (rows three, six, seven) are in fair agreement with the present results. Three of

Brody's slopes (rows 10, 11, and 12) do not agree with the present results. Given that the two datasets are separated in time by nearly 70 years, and that Brody's results are mainly for domesticated animals, the overall results are encouraging. The slope of 0.667 given in column b, row ten, reflects corrections of SMR for body temperature (see Chapter 17).

1. Brody, S. (1945). *Bioenergetics and Growth*. New York: Reinhold.

References for quotations

Books

Ackerknecht, E.H. (1968). *A Short History of Medicine.* New York: Ronald Press, p. 114.

Andrew, W. and Hickman, C.P. (1974). *Histology of the Vertebrates: A Comparative Text.* Saint Louis, MO: C.V. Mosby, pp. 249, 337.

Ashcroft, F. (2000). *Life at the Extremes.* Berkeley, CA: University of California Press, p. 304.

Ball, P. (1999). *A Biography of Water: Life's Matrix.* New York: Farrar, Straus and Giroux, p. 213.

Barlow, M. (2007). *Blue Covenant.* New York: New Press, p. 142.

Bernard, C. (1957). *An Introduction to the Study of Experimental Medicine.* New York: Dover Publications, p. 76. (Reprint of text from 1865.)

Boitani, L. and Bartoli, S. (1983). *Simon & Schuster's Guide to Mammals.* Simon & Schuster: New York, p. 10.

Brady, N.C. and Weil, R.R. (2002). *The Nature and Properties of Soils,* 13th edn. Upper Saddle River, NJ: Prentice Hall, p. 2.

Brody, S. (1945). *Bioenergetics and Growth.* New York: Reinhold, pp. 580, 640, 641.

Calder, W.A. (1984). *Size, Function, and Life History.* Cambridge, MA: Harvard University Press, p. 38.

Carey, N. (2012). *The Epigenetics Revolution: How Modern Biology is Rewriting our Understanding of Genetics, Disease, and Inheritance.* New York: Columbia University Press, p. 180.

Carroll, R.L. (1988). *Vertebrate Paleontology and Evolution.* New York: W.H. Freeman, p. 495.

Chorley, R.J. and Kennedy, B.A. (1971). *Physical Geography: A Systems Approach.* London: Prentice-Hall International, p. 1.

Cohen, J.E. (1995). *How Many People Can the Earth Support?* New York: W.W. Norton, p. 54.

Colinvaux, P. (1978). *Why Big Fierce Animals are Rare: An Ecologist's Perspective.* Princeton, NJ: Princeton University Press, p. 18.

Commission for Environmental Cooperation (2011). *Taking Stock. North American Pollutant Releases and Transfers.* http://www.cec.org/

Cox, B. and Forshaw, J. (2011). *The Quantum Universe.* Boston, MA: DaCapo Press, p. 2.

Crick, F. (1966). *Of Molecules and Men.* Seattle, WA: University of Washington Press, pp. 10, 20.

Darwin, C. (undated). *The Origin of Species and The Descent of Man.* New York: Modern Library, p. 436. (*The Origin of Species* first published 1859; *The Descent of Man* first published 1871.)

Dawkins, R. (1978). *The Selfish Gene.* New York: Oxford University Press, p. 23.

Dejours, P. (1981). *Principles of Respiratory Physiology.* Amsterdam: Elsevier/North-Holland Biomedical Press, p. 83.

Eliade, M. (1958). *Patterns in Comparative Religion.* New York: New American Library. (See Foreword.)

Ellis, B. (1966). *Basic Concepts of Measurement.* Cambridge: Cambridge University Press, p. 148.

Eisenberg, J.F. (1981). *The Mammalian Radiations.* Chicago, IL: University of Chicago Press, p. 284.

Firor, J. (1990). *The Changing Atmosphere: A Global Challenge.* New Haven, CT: Yale University Press, p. 9.

Galilei, G. (1954). *Dialogues Concerning Two New Sciences.* Evanston, IL: NorthWestern University, p. 130.

Ganong, W.F. (1965). *Review of Medical Physiology.* Los Altos, CA: Lange Medical Publications, p. 5.

Garven, H.S.D. (1965). *A Student's Histology.* Edinburgh: E. & S. Livingstone, p. 118.

Gorbman, A. and Bern, H.A. (1962). *A Textbook of Comparative Endocrinology.* New York: John Wiley & Sons, p. 7.

Greene, B. (2004). *The Fabric of the Cosmos.* New York: Vintage Books, p. 334.

Haldane, J.B.S. (1940). *Possible Worlds.* London: Evergreen Books, p. 30.

Hardy, R.N. (1972). *Temperature and Animal Life.* London: Edward Arnold, p. 56.

Hays, H.L. (1973). *Statistics for the Social Sciences,* 2nd edn. New York: Holt, Rinehart and Winston, p. 385.

Heal, G. (2000). *Nature and the Marketplace: Capturing the Value of Ecosystem Services.* Washington, DC: Island Press, p. 124.

Heisenberg, W. (1962). *Physics and Philosophy: The Revolution in Modern Science.* New York: Harper & Row, p. 58.

Helmholtz, H. (1873). *Popular Lectures on Scientific Subjects.* London: Longmans Green & Co., p. 227.

Hillel, D.J. (1991). *Out of the Earth: Civilization and the Life of the Soil.* New York: Macmillan, p. 135.

Hiss, T. (1991). *The Experience of Place.* New York: Random House, p. xvi.

Jerison, H.J. (1973). *Evolution of the Brain and Intelligence.* New York: Academic Press, p. 141.

Kemp, T.S. (1982). *Mammal-like Reptiles and the Origin of Mammals.* London: Academic Press, p. 308.

Kleiber, M. (1961). *The Fire of Life: An Introduction to Animal Energetics.* New York: John Wiley & Sons, p. 191.

Lowenstam, H.A. and Weiner, S. (1989). *On Biomineralization.* New York: Oxford University Press, p. 229.

Massey, B.S. (1971). *Units, Dimensional Analysis and Physical Similarity.* London: Van Nostrand Reinhold, p. 87.

McNeill, J.R. (2000). *Something New Under the Sun: An Environmental History of the Twentieth Century.* New York: W.W. Norton, p. 49.

Minium, E.W. and Clarke, R.B. (1982). *Elements of Statistical Reasoning.* New York: John Wiley & Sons, p. 78.

Monod, J. (1972). *Chance and Necessity: An Essay on the Natural Philosophy of Modern Biology.* New York: Vintage Books, p. 100.

Montgomery, D.R. (2007). *Dirt: The Erosion of Civilizations.* Berkeley, CA: University of California Press, p. 239.

Moroney, M.J. (1982). *Facts from Figures.* Baltimore, MD: Penguin Books, p. 3.

Nalbandov, A.V. (1976). *Reproductive Physiology of Mammals and Birds: The Comparative Physiology of Domestic and Laboratory Animals and Man.* San Francisco, CA: W.H. Freeman, p. 1.

Pauling, L. (1970). *General Chemistry.* New York: Dover Publications, p. 420.

Pearce, F. (2006). *When the Rivers Run Dry.* Boston, MA: Beacon Press, p. xi.

Pimentel, D. (1993). *World Soil Erosion and Conservation.* Cambridge: Cambridge University Press, pp. 1, 2.

Postel, S. (1999). *Pillar of Sand.* New York: W.W. Norton, pp. 2, 262.

Riedl, R. (1978). *Order in Living Organisms.* Chichester: John Wiley & Sons, p. 40.

Roosen-Runge, E.C. (1977). *The Process of Spermatogenesis in Animals.* Cambridge: Cambridge University Press, p. 163.

Rowell, L.B. (1986). *Human Circulation: Regulation During Physical Stress.* New York: Oxford University Press, p. 9.

Schepartz, B. (1980). *Dimensional Analysis in the Biomedical Sciences.* Springfield, IL: Charles C. Thomas, p. 148.

Schmidt-Nielsen, K. (1984). *Scaling: Why is Animal Size so Important?* Cambridge: Cambridge University Press, pp. 17, 22, 140, 157, 165.

Shiva, V. (2002). *Water Wars: Privatization, Pollution and Profit.* Toronto: Between the Lines, pp. 1, 9.

Smith, H.W. (1951). *The Kidney: Structure and Function in Health and Disease.* New York: Oxford University Press, p. 3.

Tanford, C. and Reynolds, J. (2001). *Nature's Robots: A History of Proteins.* Oxford: Oxford University Press, p. 5.

Thompson, D.W. (1943). *On Growth and Form.* New York: Cambridge University Press, pp. 2, 15, 53.

Turner, C.D. (1966). *General Endocrinology.* Philadelphia, PA: W.B. Saunders, p. 63.

Welsch, U. and Storch, V. (1976). *Comparative Animal Cytology and Histology.* Seattle, WA: University of Washington Press, p. 224.

Weibel, E.R. (1984). *The Pathway for Oxygen. Structure and Function in the Mammalian Respiratory System.* Cambridge, MA: Harvard University Press, pp. 12, 347.

Wright, R. (2004). *A Short History of Progress.* New York: Carroll & Graf, p. 30.

Book chapters and journal articles

Adamson, J.W. and Finch, C.A. (1975). Hemoglobin function, oxygen affinity, and erythropoietin. *Annual Review Physiology*, **37**:351–369.

Aiello, L.C. and Wheeler, P. (1995). The expensive-tissue hypothesis: the brain and the digestive system in human and primate evolution. *Current Anthropology*, **36**:199–221.

Anderson, P.W. (1972). More is different. *Science*, **177**:393–396.

Bartels, H. (1982). Metabolic rate of mammals equals the 0.75 power of their body weight? *Experimental Biology Medicine*, **7**:1–11. (Question mark was omitted during printing.)

Bartholomew, G.A. (1986). The role of natural history in contemporary biology. *BioScience*, **36**:324–329.

Calder, W.A. (1987). Scaling energetics of homeothermic vertebrates: an operational allometry. *Annual Review of Physiology*, **49**:107–120.

Campbell, J.H. (1987). The new gene and its evolution. In: Campbell, K.S.W. and Day, M.F. (eds.) *Rates of Evolution*. London: Allen & Unwin, pp. 283–309.

Carroll, S.B. (2001). Chance and necessity: the evolution of morphological complexity and diversity. *Nature*, **409**:1102.

Clark, P.U. and Weaver, A. (2008). *Report on Abrupt Climate Change: Summary and Findings*. Reston, VA: US Geological Survey.

Courtois, P.-J. (1985). On time and space decomposition of complex structures. *Communications of the ACM*, **28**:590–603.

da Silva, J.K.L., Garcia, G.J.M. and Barbosa, L.A. (2006). Allometric scaling laws of metabolism. *Physics Life Reviews*, **3**:229–261.

Garcia, G.J.M. and da Silva, J.K.L. (2006). Interspecific allometry of bone dimensions: a review of theoretical models. *Physics of Life Reviews*, **3**:188–209.

Gingerich, P.D., Smith, B.H. and Rosenberg, K. (1982). Allometric scaling in the dentition of primates and prediction of body weight from tooth size in fossils. *American Journal of Physical Anthropology*, **58**:81–100.

Gould, S.J. (1975). On the scaling of tooth size in mammals. *American Journal of Zoology*, **15**:351–362.

Gould, S.J. (1992). Ontogeny and phylogeny – revisited and reunited. *Bioessays,* **14**:275–279.

Hardy, J.D., Stolwijk, J.A.J. and Gagge, A.P. (1971). Man. In: Whittow,G.C. (ed.) *Comparative Physiology of Thermoregulation*. New York: Academic Press, pp. 327–380.

Hill, A.V. (1950). The dimensions of animals and their muscular dynamics. *Science Progress*, **38**:209–230.

Hoppeler, H. and Weibel, E.R. (2005). Scaling functions to body size: theories and facts. *Journal of Experimental Biology*, **208**:1573–1574.

Irving. L. (1934). On the ability of warm-blooded animals to survive without breathing. *Scientific Monthly*, **38**:422–428.

Jürgens, K.D., Fons, R., Peters, T. *et al.* (1996). Heart and respiratory rates and their significance for convective oxygen transport rates in the smallest mammal, the Etruscan shrew *Suncus etruscus*. *Journal of Experimental Biology*, **199**:2579–2584.

Ladell, W.S.S. (1965). Water and salt (sodium chloride) intakes. In: Edholm, G.O. and Bacharach, A.L. (eds) *The Physiology of Human Survival*. London: Academic Press, pp. 235–299.

Leise, E.M. (1990). Modular construction of nervous systems: a basic principle of design for invertebrates and vertebrates. *Brain Research Reviews,* **15**:1–23.

Linzbach, A.J. (1960). Heart failure from the point of view of quantitative anatomy. *American Journal of Cardiology,* **5**:370–382.

Loudon, A.S.I. (1987). The reproductive strategies of lactation in a seasonal macropodid marsupial: comparison of marsupial and eutherian herbivores. *Symposium Zoological Society London*, **57**:127–147.

Martin, R.D. (1996). Scaling of the mammalian brain: the maternal energy hypothesis. *News Physiological Science*, **11**:149–156.

Matson, P.A., Parton, W.J., Power, A.G. and Swift, M.J. (1997). Agricultural intensification and ecosystem properties. *Science*, **277**:504–509.

McNab, B.K. (1973). Energetics and the distribution of vampires. *Journal of Mammalogy*, **54**:131–144.

Pantin, C.F.A. (1956). Comparative physiology of muscle. *British Medical Bulletin*, **12**:199–202.

Pattee, H.H. (1973). The physical basis and origin of hierarchical control. In: Pattee, H.H. (ed.) *Hierarchy Theory: The Challenge of Complex Systems*. New York: George Braziller, pp. 71–108.

Pearson, O.P. (1953). The metabolism of hummingbirds. *Scientific American*, **188**:69–72.

Petit, J.R., Jouzel, J., Raynaud, D. *et al.* (1999). Climate and atmospheric history of the past 420,000 years from the Vostok ice core, Antarctica. *Nature*, **399**:429–436.

Poczopko, P. (1980). Relations of metabolic rate and body temperature. In: Schmidt-Nielsen, K., *et al.* (eds.) *Comparative Physiology: Primitive Mammals.* Cambridge: Cambridge University Press, pp. 155–162.

Pond, C.M. (1977). The significance of lactation in the evolution of mammals. *Evolution*, **31**:177–199.

Pond, C.M. (1984). Physiological and ecological importance of energy storage in the evolution of lactation: evidence for a common pattern of anatomical organization of adipose tissue in mammals. *Symposium Zoological Society London*, **31**:1–32.

Promislow, D.E.L. (1991). The evolution of mammalian blood parameters: patterns and their interpretation. *Physiological Zoology,* **64**:393–431.

Ricklefs, R.E., Konarzewski, M. and Daan, S. (1996). The relationship between basal metabolic rate and daily energy expenditure in birds and mammals. *American Naturalist*, **147**:1047–1071.

Ruben, J.A. and Bennett, A.A. (1987). The evolution of bone. *Evolution*, **4**:1187–1197.

Rushmer, R.F., Buettner, K.J.K., Short, J.M. and Odland, G.F. (1966). The skin. *Science*, **154**:343–348.

Schmidt-Nielsen, K. (1954). Heat regulation in small and large desert mammals. In: Cloudsley-Thompson, J.L. (ed.) *Biology of Deserts.* London: Institute of Biology, pp. 182–187.

Snapper, J.R., Tenney, S.M. and McCann, F.V. (1974). Observations on the amphibian "diaphragm". *Comparative Biochemistry and Physiology*, **49A**:223–230.

Stephan, H. and Andy, O.J. (1964). Quantitative comparisons of brain structures from insectivores to primates. *American Zoologist*, **4**:59–74.

Tenney, S.M. and Bartlett, D. (1967). Comparative quantitative morphology of the mammalian lung: trachea. *Respiratory Physiology*, **3**:130–135.

Thach, W.T., Goodkin, H.P. and Keating, J.G. (1992). The cerebellum and the adaptive coordination of movement. *Annual Review Neuroscience*, **15**:403–442.

Trimble, V. (1997). Origin of the biologically important elements. *Origins of Life and Evolution of the Biosphere*, **27**:3–21.

Turner, J. (2010). Editorial (April 30), *Toronto Star.*

Tuttle, M.D. and Stevenson, D. (1982). Growth and survival of bats. In: Kunz, T.H. (ed.) *Ecology of Bats.* New York: Plenum Press, pp. 105–149.

Vitousek, P.M., Mooney, H.A., Lubchenco, J. and Melillo, J.M. (1997). Human domination of earth's ecosystems. *Science,* **277**:494–499.

White, L.J. (1967). The historical roots of our ecological crisis. *Science*, **155**:1203–1207.

Widdowson, E.M. (1968). Minerals in the animal body. *Proceedings Nutrition Society*, **27**:138–143.

Widdowson, E.M. and Dickerson, J.W.T. (1964). Chemical composition of the body. In: Comar, C.L. and Bronner, F. (eds.) *Mineral Metabolism: An Advanced Treatise.* New York: Academic Press, pp. 1–217.

Wigner, E.P. (1964). Events, laws of nature, and invariance principles. *Science,* **145**:995–998.

Wright, G.L. (1976). Possible mechanisms involved in death in laboratory animals due to thermal stresses (heat and cold). In: Johnson, H.D. (ed.) *Progress in Biometeorology.* Amsterdam: Swets & Zeitlinger, pp. 167–173.

Index

Printed in the United States
By Bookmasters